# 2nd International Conference on Mechanical, Manufacturing, Modeling and Mechatronics (IC4M 2017) and 2nd International Conference on Design, Engineering and Science (ICDES 2017)



MATEC Web of Conferences Volume 104 (2017)

Kortrijk, Belgium
24 - 26 February 2017

**Editors:**

**Jan Detand**
**Du Ruxu**
**James Andrew Self**
**Jos Gunsing**

ISBN: 978-1-5108-3938-0

**Printed from e-media with permission by:**

Curran Associates, Inc.
57 Morehouse Lane
Red Hook, NY 12571

**Some format issues inherent in the e-media version may also appear in this print version.**



Printed by Curran Associates, Inc. (2017)

For additional information, please contact EDP Sciences – Web of Conferences
at the address below.

EDP Sciences – Web of Conferences
17, Avenue du Hoggar
Parc d'Activité de Courtabœuf
BP 112
F-91944 Les Ulis Cedex A
France

Phone: +33 (0) 1 69 18 75 75
Fax: +33 (0) 1 69 28 84 91

contact-edps@webofconferences.org

**Additional copies of this publication are available from:**

Curran Associates, Inc.
57 Morehouse Lane
Red Hook, NY 12571 USA
Phone: 845-758-0400
Fax: 845-758-2633
Email: curran@proceedings.com
Web: www.proceedings.com

# TABLE OF CONTENTS

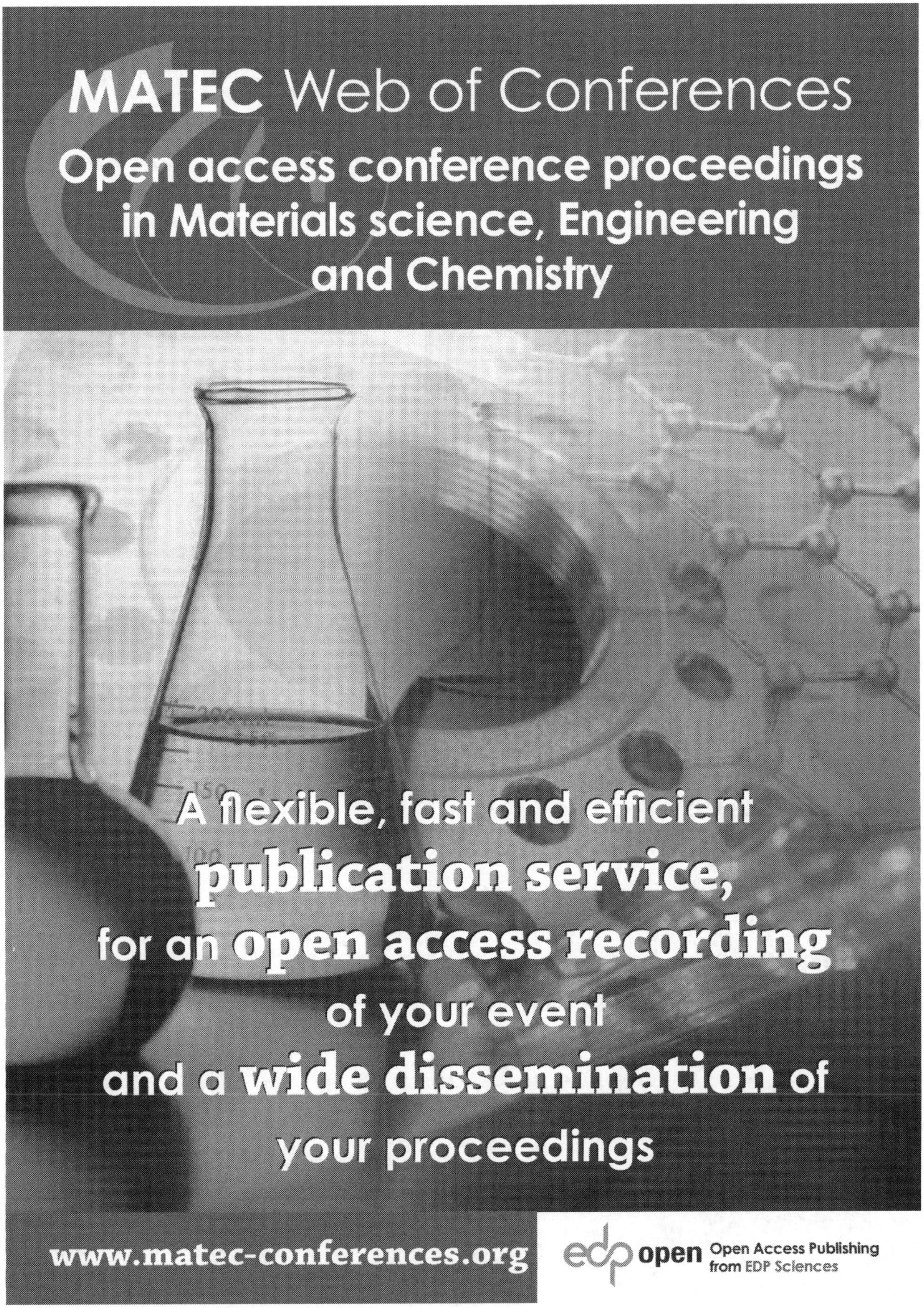
MATEC Web of Conferences
Open access conference proceedings in Materials science, Engineering and Chemistry
A flexible, fast and efficient publication service, for an open access recording of your event and a wide dissemination of your proceedings
www.matec-conferences.org
edp open Open Access Publishing from EDP Sciences

ISSN (electronic edition): 2261-236X

# PREFACE

2017 2nd International Conference on Mechanical, Manufacturing, Modeling and Mechatronics (IC4M 2017) & 2017 2nd International Conference on Design, Engineering and Science (ICDES2017) were held in University Gent, Campus Kortrijk, Belgium during February 24-26, 2017. The proceeding is a compilation of the accepted papers and represent an interesting outcome of the conference.

IC4M 2017&ICDES 2017 were co-organized by Hong Kong Society of Mechanical Engineers (HKSME) and Gent University, Kortrijk Campus. The conferences provide a useful platform both for showing the latest research directions and for exchange of research results and ideas in broad Mechanical, Manufacturing, Engineering and Design areas.

We would like to acknowledge all of those who supported IC4M 2017&ICDES 2017. Each individual and institutional help were very important for the success of this conference. Especially we would like to thank the organizing committee for their valuable advices in the organization and helpful peer review of the papers.

We hope that IC4M 2017&ICDES 2017 will be a forum for fruitful discussions that will shed light on new ideas and foster collaborative researches. We are sure that the proceedings will serve as an important research source of references and the knowledge, which will lead to not only scientific and engineering progress but also new products and processes.

Prof. Jan Detand
Ghent University, Belgium
Prof. Du Ruxu
The Chinese University of Hong Kong, Hong Kong
Prof. James Andrew Self
Ulsan National Institute of Science and Technology (UNIST), South Korea
Prof. Jos Gunsing
Avans University of Applied Science, Netherlands

# IC4M 2017&ICDES 2017 COMMITTEE

**General Chair**

Prof. Jan Detand, Ghent University, Belgium

**Conference Committee**

Prof. Mohd Hamdi Bin Abd Shukor, University of Malaya, Malaysia

Prof. Du Ruxu, The Chinese University of Hong Kong, Hong Kong

Prof. Cees de Bont , Hong Kong Polytechnic University, Hong Kong

**Program Committee**

Prof. Yong Zeng, Concordia University, Canada

Prof. Mohd Nasir Tamin, Universiti Teknologi Malaysia, Malaysia

Prof. Lucienne Blessing, University of Luxembourg, Germany

Prof. Yan Wang, Georgia Institute of Technology, USA

Prof. James Andrew Self, Ulsan National Institute of Science and Technology (UNIST), South Korea

**Local Chairs**

Prof. Johannes Cottyn, Ghent University, Belgium

Prof. Kurt Stockman, Ghent University, Belgium

Prof. Jelle Saldien, Ghent University, Belgium

**Technical Committee**

Prof. Yuliang Chen, National Defense University, Taiwan

Prof. Sim Kok Swee, Multimedia University, Malaysia

Prof. Amaresh Chakrabarti, Indian Institute of Science, India

Prof. Katz Zvi, University of Pretoria, South Africa

Prof. Zengtao Chen, University of Alberta, Canada

Prof. Kai Cheng, Brunel University, UK

Prof. Ismail Fidan, Tennessee Tech University, USA

Prof. John Gero, George Mason University, USA

Prof. Zaidi bin Mohd Ripin, Universiti Sains Malaysia, Malaysia

Prof. Tetsuo Tomiyama , Cranfield University, UK

Prof. Gopinath Chintala, Alliance University, India

# Statement of Peer review

In submitting conference proceedings to MATEC Web of Conferences, I certify to the Publisher that I adhere to the **Policy on Publishing Integrity** of the journal in order to safeguard good scientific practice in publishing.

1. All articles have been subjected to peer review administered by the proceedings editors.
2. Reviews have been conducted by expert referees, who have been requested to provide unbiased and constructive comments aimed, whenever possible, at improving the work.
3. Proceedings editors have taken all reasonable steps to ensure the quality of the materials they publish and their decision to accept or reject a paper for publication has been based only on the merits of the work and the relevance to the journal.

Title, date and place of the conference

2017 2nd International Conference on Mechanical, Manufacturing, Modeling and Mechatronics (IC4M 2017)
2017 2nd International Conference on Design, Engineering and Science (ICDES2017)
February 24-26, 2017 University Gent, Campus Kortrijk, Belgium

Proceedings editor(s):

Prof. Jan Detand, Prof. Du Ruxu, Prof. James Andrew Self, Prof. Jos Gunsing

Date and editor's signature 13.03.17

17, avenue du Hoggar - PA de Courtabœuf – BP 112 - 91944 Les Ulis Cedex A (France)
Tél. : 33 (0)1 69 18 75 75 - Fax : 33(0)1 69 07 45 17 – www.edpsciences.org

MATEC Web of Conferences **104**, 01001 (2017) DOI: 10.1051/matecconf/201710401001
*IC4M & ICDES 2017*

# Finite Element Analysis as a response to frequently asked questions of machine tool mechanical design-engineers

Gerhard Kehl[1,a], David Blank[1] and Christiane Höger-Riedel[1]

[1]*University of Applied Sciences, Faculty of Engineering Management, Flandernstr. 101, Germany*

**Abstract.** The finite element analysis (FEA) nowadays is indispensable in the product development of machining centres and production machinery for metal cutting processes. It enables extensive static, dynamic and thermal simulation of digital prototypes of machine tools before production start-up. But until now less reflection has been made about what are the most pressing questions to be answered in this application field, with the intention to align the modelling and simulation methods with substantial requirements. Based on 3D CAD geometry data for a modern machining centre (Deckel-Maho-Gildemeister DMG 635 V eco) merely the basic steps of a static analysis are reconstructed by FEA. Particularly the two most frequently asked questions by the design departments of machine tool manufacturers are discussed and highlighted. For this authentic simulation results are used, at which their selection is a consequence of long lasting experience in the industrial application of FEA in the design process chain. Noticing that such machine tools are mechatronic systems applying a considerable number of actuators, sensors and controllers in addition to mechanical structures, the answers to those core questions are required for design enhancement, to save costs and to improve the productivity and the quality of machined workpieces.

## 1 Introduction

The implied vertical machining centre (VMC) is a classical CNC machine tool for metal cutting. The shown VMC (Fig. 1) offers three Cartesian feed axes (X-, Y- and Z-axis) with a work space of 635 x 510 x 460 $mm^3$ and is equipped with one milling spindle, cooling lubricant equipment and automatic chip conveyor.

Fig. 1 shows the supporting base structure of the examined machine with the arrows of movable feed axes. The machine bed rests on the ground via three elastic, damping and height adjustable levellers. The column is bolted down on the machine bed, composing the static machine frame, serving as a base for the automatic tool changer and the tool magazine. The workpiece table (X-axis) is located on the machine bed.

On the cutting tool side the linear movement in Y-direction is realized by the horizontal slide over the column. The vertical slide (Z-axis) is bounded on the front end of the horizontal slide and contains the spindle. To enable the linear motions accurately, each slide is guided by roller linear guides, driven by servo motor with belt-drive and attached ball screw.

[a] Corresponding author : gerhard.kehl@hs-esslingen.de

 DOI: 10.1051/matecconf/201710401001

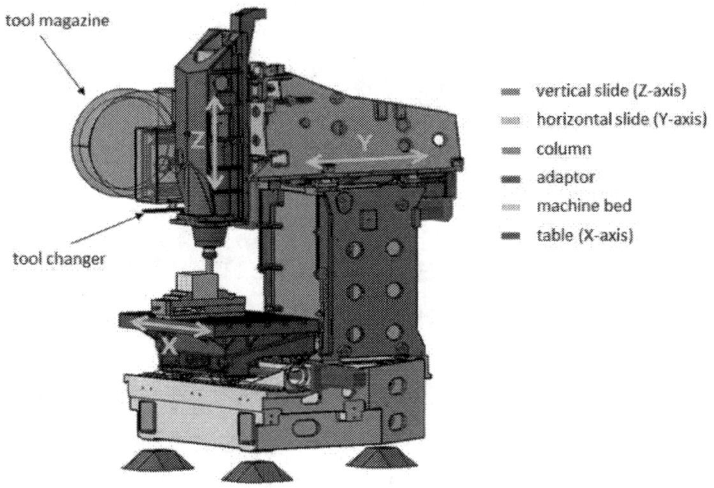


**Figure 1.** Axes configuration of VMC DMG 635 V eco.

## 2 Import and preparation of 3D CAD Geometry Data

U Machining process forces (from milling, boring, threading …) at the tool centre point between tool (T) and workpiece (W) lead to displacements, which are caused by every participating machine component. The relevance of each machine component can be seen from the flux of force, which has to be determined at first.

Machine components with no influence on the deformation at the tool or the workpiece are suspended from the consideration and are not part of the supporting structure. Protective cover, chip conveyor, cooling fluid tank, control panel and service unit are famous examples for such peripheral components.

In terms of short FE modelling time as part of the design process chain it is recommended to take an up-to-date copy of the original CAD geometry data as basis and import a defeatured version of the supporting structure (as in Fig. 1) into the FEA software. For this purpose peripheral components have to be suppressed or deleted. Once the geometry is transferred to the FEA software material properties for each structural component are allocated dependent on the employed material (steel, grey cast iron, polymer concrete, brass …). Typical analyses of machine tools are restricted to compliance behaviour and because of this the necessary geometry dimensioning of structural parts generates no strength of materials problems [1]. As a consequence only a few material properties are enough for such linear FEA models of machine tools, that is to say Young's modulus, Poisson ratio and material density.

## 3 FE modelling of the machine tool

The modelling process is divided into the volume meshing of the structural components (usually with parabolic hexahedral or tetrahedral finite elements) and the modelling of force transmitting machine elements by equivalent mechanical systems (linear spring, mass, viscous damper). This is relevant if modelling as volume is not reasonable or requires complex contact conditions. For the correct parameterization of equivalent mechanical systems (e.g. for bearings, ball screws, linear guides, flexible couplings, gears, machine levellers) measurement data usually is consulted, which can be delivered by notable suppliers or are common for experienced FEA engineers in the machine tool sector, gained by extensive calibration of experiment and simulation data [2].

For a general, non-customized FEA model it is suitable to represent cutting tool, workpiece and fixture as virtual rigid parts, e.g. the bar elements in Fig. 2. With this the points of load application at tool and workpiece can be linked to the machine structure and can be utilized as representative coordinates in the machines workspace.

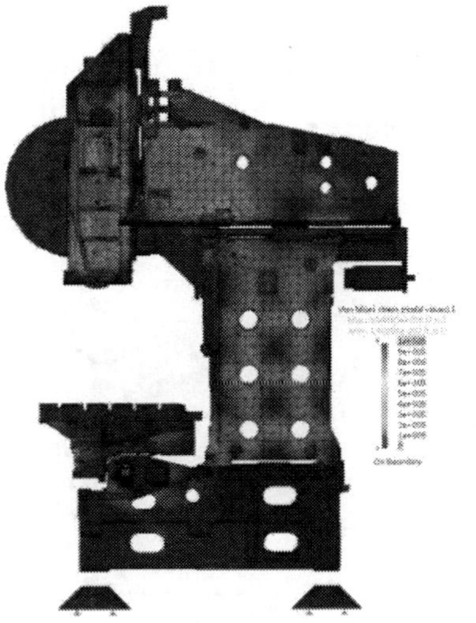

**Figure 2.** FE mesh of VMC with mechanical stress representation.

The FE model built up in the mentioned approach is the basis for answering typical questions of machine tool designers, which have to be solved in every machine tool design process.

## 4 Frequently asked questions of machine tool design-engineers

"WHAT IS THE MOST SUITABLE GEOMETRY FOR MECHANICAL STRUCTURES LIKE MACHINE BEDS, COLUMNS … ?"

Obviously casted or welded metal components for VMCs should be configured at its best like space frame structures. This results from fundamental considerations about frame structures which should preferably possess a high degree of stiffness combined with low mass just like it is the case with constructions, bridges, antenna masts as well as with aerospace structures.

In this connection, bar triangles in contrast to bar rectangles are statically and kinematically determined inside and therefore possess a high level of stiffness as a result of the assumption that with flexible connection of the bars only tensile- and pressure loads are existent but no bending exposure [3]. Concerning the force P on point B, Fig. 3 demonstrates the comparably stiff characteristics of the framework consisting of triangular geometric basic element, here shown at the triangle ABC and ACD. In contrast to that, the rectangular bar composition ABCD (without the diagonal brace AC) needs to be considered as flexible, so called four bar linkage. The fundamental difference in stiffness between both variants is obvious for flexible connections as well as for fixed connections of the bars.

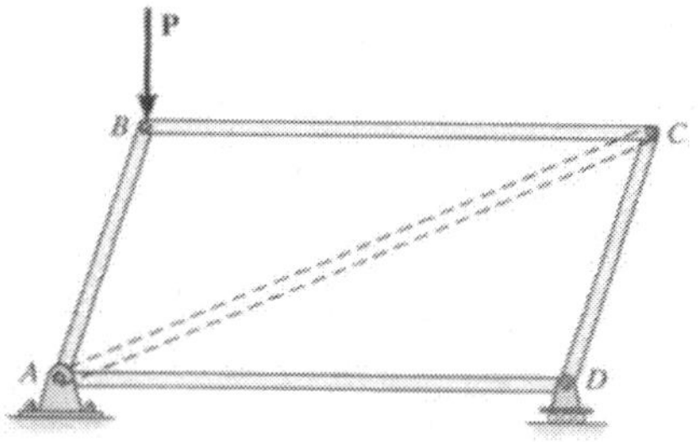


**Figure 3.** Geometric basic elements: triangle vs. rectangle [4].

 DOI: 10.1051/matecconf/201710401001

Even though structure components from machine tools are usually casted, welded together, glued or screwed structures, the ideal geometric basic element always remains triangular.

Table 1 shows by means of one piece casted hollow blocks of similar mass and material, how stiffness can be increased through spatial ideal application of triangular basic elements (B) in comparison to the application of rectangular basic elements (A).

In all cases, the box shaped construction parts are unilateral rigidly mounted and the stiffness concerning bending and torsion are determined through simulative methods.

**Table 1.** Comparison of mechanical structures with rectangular and triangular base elements.

| | **improvement of stiffness**<br><br>stiffness B / stiffness A | **A**<br>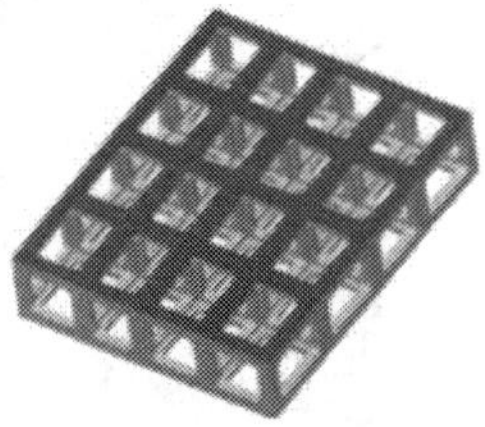 | **B**<br>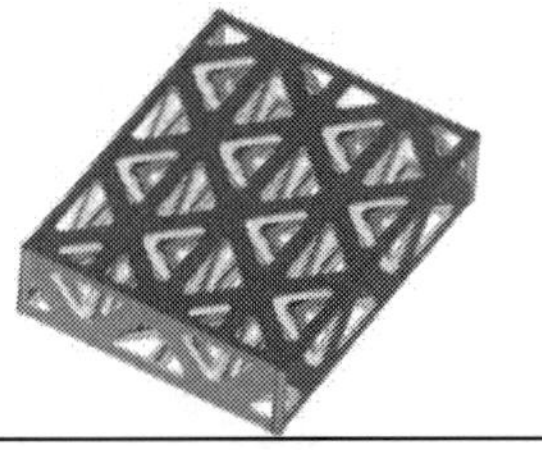 |
|---|---|---|---|
| bending upwards | 1,53 | 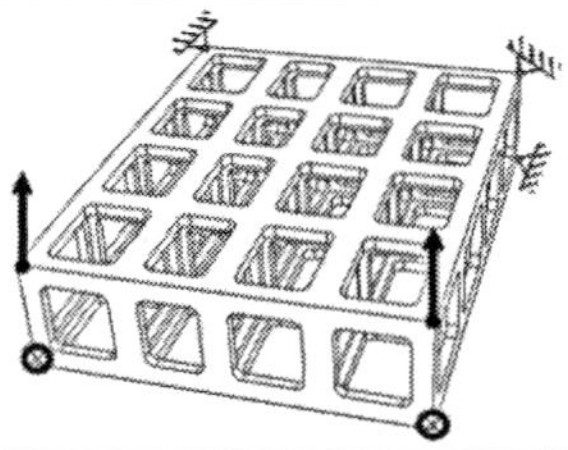 | 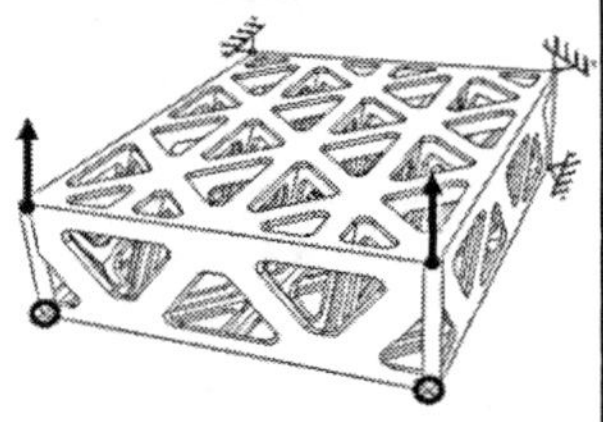 |
| bending sidewards | 2,24 | 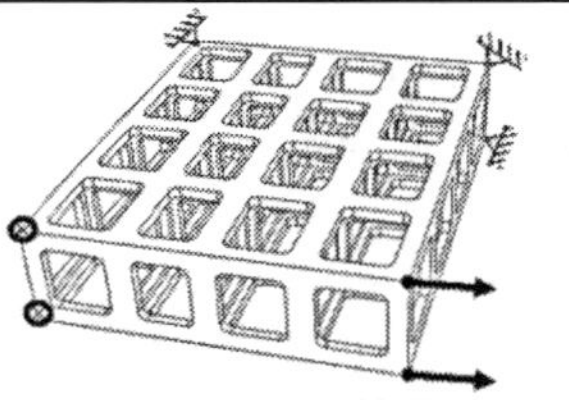 | 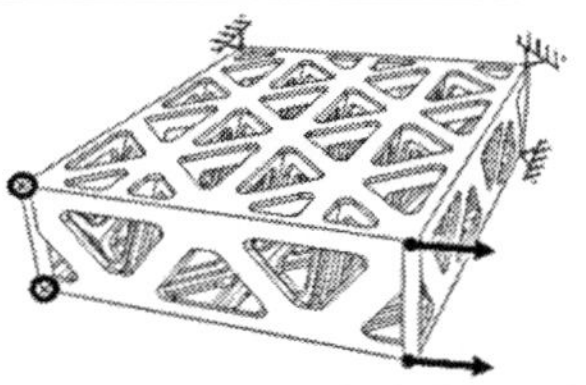 |
| torsion | 3,95 | 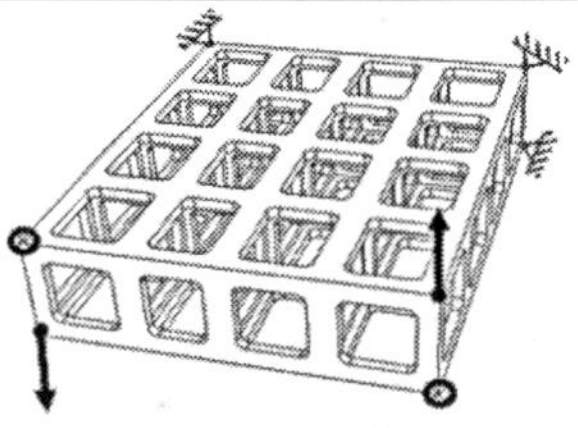 | 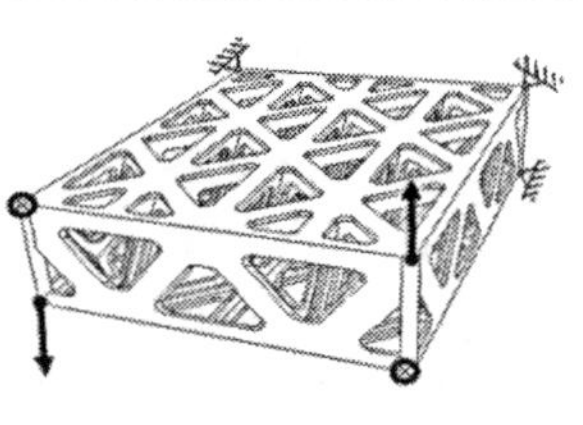 |

The above presented improvements are admittedly seldom achieved in practice, because structure B usually cannot be implemented in real machines parts. The reasons are strong design restrictions (e.g. functional surfaces or contact areas to neighbouring parts have to be provided) or manufacturing restrictions (e.g. minimum and maximum wall thicknesses are given by the casting process). However, also improvements in stiffness by 10…40 % improve the machine tool and it's hereby produced workpieces significantly.

To illustrate that this supposed trivial design guideline even nowadays is not followed consequently, Fig. 4 shows two design states for a new machine bed. The revised design on the right side by applying triangular base elements has a 30 % higher stiffness/mass ratio over the first design on the left side, which is based on rectangular geometry elements. As a consequence the machine bed with higher stiffness/mass ratio tends to result in better stability in cutting processes and higher quality of machined workpieces [5].

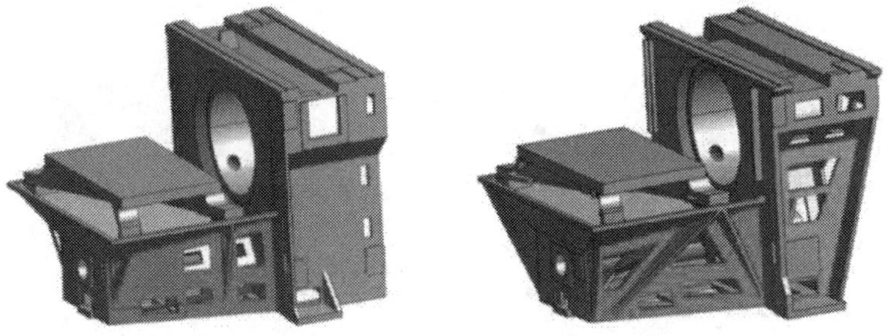

**Figure 4**. Two design states of a machine bed (mainly triangular/rectangular based).

Apart from generally applying triangular base elements the finding of load-capable geometry for machine structures can be done automatically by topology optimization, taking multiple load cases into account [6]. Topology optimization always utilizes the finite element analysis and leads to more detailed designs, but is not basically necessary, if the machine tool design-engineer in general tries to assemble mechanical structures by triangular base elements.

## "DOES THE GRAVITY OF THE MOVING SLIDES ON THE STATIC MACHINE FRAME AFFECT THE STRAIGHTNESSES OF THE LINEAR FEED AXES ?"

In favour of analysing the feed axes straightness of the VMC a virtual precision test by FEA is suitable. For this purpose exclusively all horizontally moving slides are considered and represented by their masses. According to Fig. 2 those are the table with workpiece and fixture moving together horizontally in X-direction, whereas on the tool side the horizontal slide, the vertical slide with spindle and tool move together horizontally in Y-direction.

For comparison Fig. 5a shows the equivalent measurement setup for a precision test of a VMC in terms of the feed axes straightness. At this junction a dial indicator in the spindle is moved alongside a metal ruler mounted on the table surface tracing orthogonal deviations to the linear movement in the Y-direction.

Fig. 5b illustrates the emulated situation by FE simulation, in which only the movable masses in Y-direction and the opposing table are visible. The straightness error between tool and workpiece is calculated in defined intervals of the Y-axis stroke (e.g. every 50 mm). For this static load cases under gravity of the movable masses are solved and the results are systematically interpreted.

The exemplary treated VMC shows straightness errors induced by self-loaded deformation primarily in the Y-axis. Fig. 6 outlines the orthogonal deviations in Z-direction over the Y-stroke for tool and workpiece and relatively between them ($\Delta Z$). The latter reside with $\Delta Z \approx 48$ µm in an inadmissible range, possibly leading to unreasonable straightness errors on machined workpiece surfaces. The dominant reason for this is the deformation of the overhanging Y-slide.

Upon detailed analysis of causes for this behaviour applicable design measures (e.g. stiffening the static machine frame and slides by adequate ribs, alternative linear guides) can be derived, improving the straightness of the Y-axis. Remaining straightness errors possibly may be compensated subsequently in the first place by providing uneven machining of the linear guide mounting surfaces during machine tool manufacturing or in the second place by using control algorithms during operation.

The machine foundation can also have remarkable influences on the straightness of the feed-axes, especially if the machine bed has poor inherent rigidity. However in this case the machine bed shows sufficient inherent rigidity leading to max. 10 % straightness improvement, if the number of foundation points is increased significantly or if rigid levellers are used.

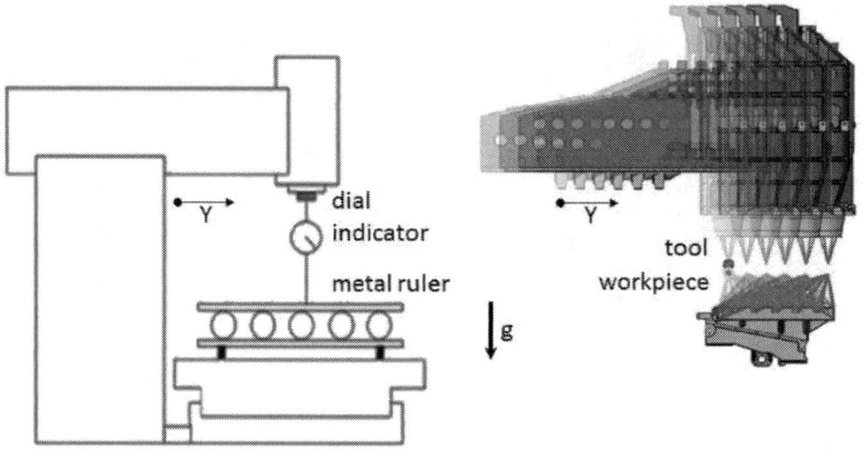


**Figure 5. a)** Measurement setup for precision test. **b)** Simulation set up for precision test.

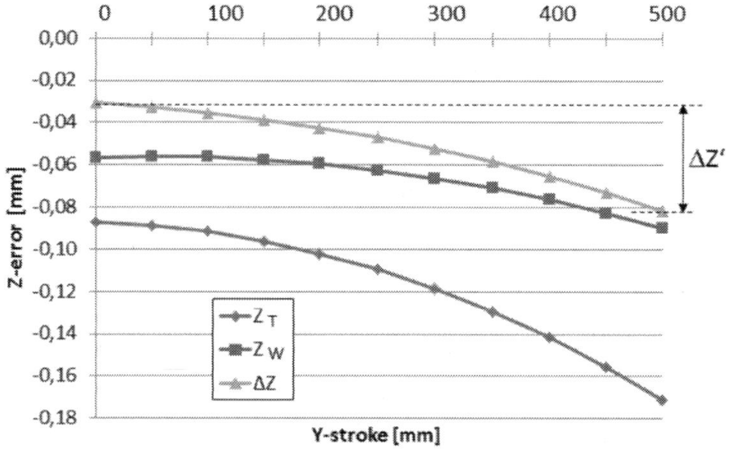


**Figure 6.** Straightness errors in Z-direction over Y-stroke.

## 5 Conclusions

The FEA is indispensable in the product development of machining centres and production machinery for metal cutting processes. Thus important questions, which occur during the design process, can be answered entirely targeting cost savings, improvement of productivity and high quality of machined workpieces.

## References

1. M. Weck and C. Brecher, *Fertigungssysteme* 2, **7** (2002)
2. A. Rao, *Factory and Plant Equipment.* ANSYS Solutions – The Premier Magazine for Design Innovation 7, **4**, pp.6-10 (2006)
3. K. Kabus, *Mechanik und Festigkeitslehre*, p. 52, **7** (2013)
4. R.C. Hibbeler, *Technische Mechanik 1*, **12**, p. 303 (2012)
5. P. Wagner and G. Kehl, *"Will it Cut or Won't it?" – Crucial Question in Simulation-Aided Machine Tool Development,* pp. 1-4 (CIM, Biograd, 2007)
6. G. Kehl, P. Jickeli, M. Schietinger, D. Blank, *Topology Optimization of Rotating Automation Components for Machine Tools*, pp. 208-217 (SIMULTECH, Colmar, 2015)

### 3.2 The optimization model of kernel parameter is constructed

The main idea of KPCA transformation is to map training sample data into a high dimensional feature space, then to establish an optimal separating hyperplane for the distance between two kinds of data points and the hyperplane being the largest. However, in traditional pattern recognition, the idea of Fisher discriminate function is just opposite. Its basic thought is to find an optimal projection direction, along which original data is projected into a straight line in order to distinguish better data after projection; the distance between different data modes is as big as possible whereas the data discrete degree between the same patterns is as small as possible. Therefore an optimization method of kernel parameters can be achieved by thought of Fisher discriminate functions [6].Kernel parametric model optimized by PSO based on fisher norm is established as follows.

Assumption that parameters $X_1\left(x_{11},x_{12},...,x_{1i}\right)$ and $X_2\left(x_{21},x_{22},...,x_{2j}\right)$ are two kinds of characteristic samples in a feature space, among which $i=1,2,...,n_1, j=1,2,...,n_2$ .The mean vectors for the two kinds of samples in feature space are respectively as equation (4)(5).

$$t_1=\frac{1}{n_1}\sum_{i=1}^{n_1}\Phi(x_{1i}) \tag{5}$$

$$t_2=\frac{1}{n_2}\sum_{j=1}^{n_2}\Phi(x_{2j}) \tag{6}$$

The square of the distances between classes is as equation (6).

$$\begin{aligned} d_b &= \|t_1-t_2\| = (t_1-t_2)^T(t_1-t_2) \\ &= \frac{1}{n_1^2}\sum_{i=1}^{n_1}\sum_{j=1}^{n_2}k(x_{1i},x_{1j})-\frac{2}{n_1n_2}\sum_{i=1}^{n_1}\sum_{j=1}^{n_2}k(x_{1i},x_{2j})+\frac{1}{n_2^2}\sum_{i=1}^{n_1}\sum_{j=1}^{n_2}k(x_{2i},x_{2j}) \end{aligned} \tag{7}$$

The square of inner-class discrete degree is as equation (7)(8).

$$d_{w1}=\frac{1}{n_1}\sum_{i=1}^{n_1}\|\Phi(x_{1i})-t_1\|^2=\sum_{i=1}^{n_1}k(x_{1i},x_{1i})-\frac{1}{n_1}\sum_{i=1}^{n_1}\sum_{j=1}^{n_2}k(x_{1i},x_{1j}) \tag{8}$$

$$d_{w2}=\frac{1}{n_1}\sum_{j=1}^{n_2}\|\Phi(x_{2i})-t_2\|^2=\sum_{j=1}^{n_2}k(x_{2j},x_{2j})-\frac{1}{n_2}\sum_{i=1}^{n_1}\sum_{j=1}^{n_2}k(x_{2i},x_{2j}) \tag{9}$$

According to fisher discriminate criterion, fitness function of PSO is set up as equation (9).

$$fitness=f(\sigma)=d_{w1}+d_{w12}/d_b \tag{10}$$

Seeking for the minimum value of fitness function $f(\sigma)$ can ensure that the between-class scatter is maximal; at the time the parameter $\sigma$ can be $\sigma^*$ on the condition that $f(\sigma)$ changes to be steady at the outset.

### 3.3 Steps of kernel parameters optimized by PSO

DOI: 10.1051/matecconf/201710401002

The step of the kernel function parameter optimized by PSO is as follows:
1) Through equation (7), (8), (9), the square value of distance between classes and that of within-class scatter are calculated, respectively;
2) Fisher discriminate function is constructed by the formula (10), and regarded as a fitness for PSO;
3) Some conditions are given such as value range of the parameters $\omega_{\max}$ , $\omega_{\min}$ ,$m$, $T_{max}$, $c_1$and $c_2$,,$V_{max}$;
4) Initial population $\sigma$ is randomly generated; similarly, the value of individual fitness and the global fitness value of population are calculated;
5) Whether iterative step meets the termination condition is determined. If $t < T_{max}$,, the step (4) and (5) will be repeated; otherwise, the current solution will be output as the optimal solution, and the algorithm is terminated.

# 4 Rolling bearing of bridge cane fault recognition based on PSO-KPCA

## 4.1 Construction of bearing feature parameter set

Faults simulation experiment were done through the rolling bearing of bridge crane researched as object in laboratory, for example normal condition, damage of bear cage, scratch of bear inner race, and flaw of bear outer ring. Figure1shows test system of simulated faults experiment, and Figure 2 shows two typical failure forms of rolling bearing .Based on measuring the acceleration signal from different rotational speeds of six measuring points in the gearbox cover , some operations are done such as signal processing, extraction of characteristic parameters for time domain and frequency domain as the original feature parameters set. 60 groups of different faults samples (30 groups of training samples and 30 groups of test samples) extracted from features sets are applied to KPCA and identification of bearing faults condition [7].

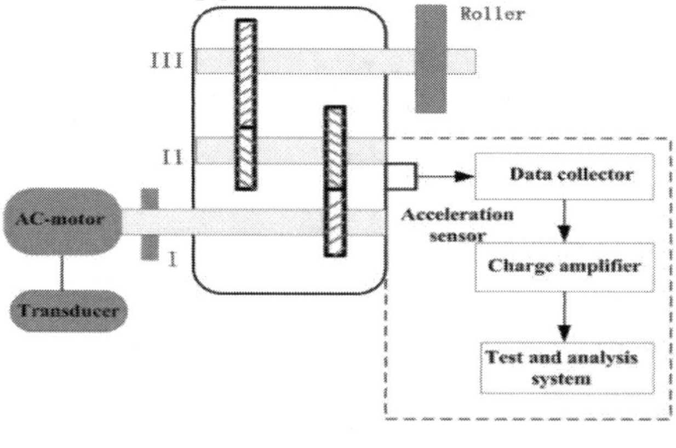


**Figure 1.** Test system diagram of simulated fault diagnosis experiment of rolling bearing

( a) Damage of bear cage ( b) Flaw of bear outer ring

**Figure 2** Typical failure forms of rolling bear in bridge cane

 DOI: 10.1051/matecconf/201710401002

## 4.2 Kernel function parametric optimization implemented by PSO

In KPCA, the width of the radial basis kernel function $\sigma$ can ensure a good performance of kernel function. For training samples, PSO is adopted to optimize kernel parameters, among which parameters are set as $m$=20,$T_{max}$=50,$V_{max}$=1, $\omega_{\max} = 1.2$ , $\omega_{\min} = 0.4$ ,$c_1$=$c_2$=2.The optimization width values $\sigma^*$ of radial basis function between condition model of rolling bearing are obtained, which are shown in table 1. The optimal values based PSO are the basis for parameters setting of bearing fault feature sets by KPCA.

**Table1.** Kernel parameter optimization results of different condition samples of bearing

| condition model of rolling bearing | $\sigma^*$ |
|---|---|
| Normal condition and scratch of bear inner race | 10.24 |
| Damage of bear cage and flaw of bear outer ring | 0.81 |
| Damage of bear cage and scratch of bear inner race | 17.48 |
| Scratch of bear inner race and flaw of bear outer ring | 3.22 |

## 4.3 Fault condition recognition of based on PSO- KPCA

In the research, 30 original feature sets on normal condition are trained, and tests are done through 30 different groups of data features such as damage of bear cage and the failures of bearings inner race and outer ring; Figure 3 shows the projection from the first principal component to the second principal component, extracted by KPCA. Among which, non-linear principal component for KPCA are unified dimension parameters.

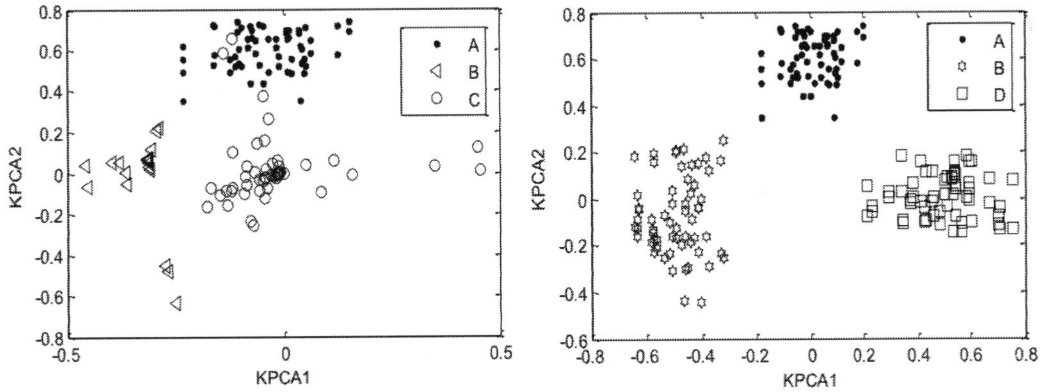


(a) (b)

A: Normal condition, B: Damage of bear cage, C: Scratch of bear inner race, D: Flaw of bear outer ring
**Figure 3.** The first and second principal component projection KPCA of different conditions of bearing

As can be seen from Figure 3 (a)(b), KPCA-PSO analysis can clearly distinguish the bearing fault types. The principal component characteristics of normal condition data are self-assortment, and those of damage of bear cage and scratch of bear inner race characteristics of the are relatively scattered, but away from the normal centre of clustering, clearly belong to different categories. Additionally, the inner-class distance of KPCA is small, whereas the distance between classes of KPCA is larger. So, KPCA optimized by PSO shows a greater advantage in the non-linear analysis, and can accurately identify the fault condition of rolling bearing.

## 5 Conclusions

PSO algorithm is used to optimize the parameters of kernel function for reduction the blindness of KPCA in parameters set and applied into bearing fault conditions recognition to improve the performance of KPCA. It can effectively identify bearing fault modes such as normal condition, damage of bear cage, scratch of bear inner race, and flaw of bear outer ring. The effect of failure identification is obviously superior. Research results show that kernel principal component analysis based on PSO has advantages in the nonlinear feature extraction of mechanical fault and is very suit for light and incipient mechanical fault condition recognition.

## References

1. Sh .Renping, H. Wentao, W. Yayun, Q. Xiankun .The fault feature extraction and classification of gear using principal component analysis and kernel principal component analysis based on the wavelet packet transform. *IMEKO*.**54** (2014) 118 - 132
2. H. Qingbo, K. Fanrong, Y. Ruanqiang. Subspaced-based gearbox monitoring by kernel principal component analysis . *MSSP*, **21** ( 2007) 1755-1772
3. X. Yingqun Xiao,F. Lianggui. A novel neural-network approach of analog fault diagnosis based on kernel discriminate analysis and particle swarm optimization. *WFSC* , **12** (2012) 904 - 920
4. B. Scholkopf, A. Smola, K.R. Muller.Nonlinear principal component analysis as a kernel eigenvalue problem. *NC* , **10** (1998)1299-1319
5. H. Xuewen. Study on theory and method of fault intelligent diagnosis based on support vector machine [D]. South University Of China, Changsha, **12** (2004).
6. Zh.Ying , Z. Hongfu, B.Fang. Classification of fault location and performance degradation of a roller bearing. *IMEKO*, **46** (2013) 1178 - 1189
7. W. Xinfeng, Q. Jing , L. Guangjun. Kernel Function Optimization in Kernel Principle Component Analysis and Its Application to Feature extraction of gear faults. *Journal of Vibration,Measurement &Diagnosis*,**27**,1( 2007) 62-64

# Design and Optimization of Contralateral-type Half-rotating Wing Based on Local Constraint

Li Congmin[1], Wang Xiaoyi[1,a], Li Qian[1], Wang Huixing[1], Qiu Zhizhen[1] and Shan Jianhua[1]
[1]*School of Mechanical Engineering, Anhui University of Technology, Ma'anshan, 243002, P.R.China*

**Abstract.** Half-rotating wing (HRW) is a kind of similar-flapping wing system based on half-rotating mechanism which could perform rotating-type flapping instead of oscillating-type flapping. Contralateral-type half-rotating wing (CHRW) could run more steadily and smoothly than previous HRW by newly layout of local constraint that mainly consists of moving sheet, circular stator and straight stator. The work principle of CHRW were firstly given in this paper. The design method of key parts local constraint in CHRW was proposed. The contact force model and smooth running condition of moving sheet were especially derived from contact model between moving sheet and circular stator and geometric relation model between moving sheet and straight stator. On the other hand, compared result between theoretical calculation and simulation verified the correctness of contact force model. After optimizing the parameters of some parts of local constraint, the value of contact force was greatly reduced to improve operation stability of CHRW. The research results mentioned above could provide theoretical guidance for design structure of CHRW.

## 1 Introduction

Flapping-wing air vehicle has been a topic of considerable interest due to its excellent maneuverability and hovering ability [1]. Gradually, many universities and institutes had many research achievements. For instance, the "Micro Bat" [2,3] was developed by California Institute of Technology, and the "DELFLY" [4], a micro flapping wing aircraft, was manufactured by Delft University of the Netherlands. Besides, the "MFI" flapping wing aircraft [5] was from University of California at Berkeley, and a controllable flight flapping-wing machine has successfully developed by Nanjing University of Aeronautics and Astronautics [6]. Driving mechanism is the core component of air vehicle, while excellent driving mechanism is benefit for the stability of aircraft and the reduction of energy consumption. At present, it is difficult for the flapping-wing driving mechanism to get the component simplicity and driving symmetry, which affect the stability of movement and the development of large scale.

Half-rotating mechanism (HRM) [7], an asymmetric rotating mechanism evolved from the basic crank rocker mechanism, provides a new idea for the development of flapping wing driving mechanism. Previously, the traditional HRM was widely used in ship propeller [8] and bionic walking mechanism [9]. With the reasonable improvement and innovation of the HRM, the work team from Anhui University of Technology designed a new kind of HRM that was suitable for flight. It is

[a] Corresponding author: wangxy@ahut.edu.cn

prominent for the development of large scale due to its excellent symmetry and well simplification. Therefore, a more simplified layout and a new local constraint was proposed in this paper on the basis of the original HRW [10]. The CHRW could meet the design requirement by actual operation. Furthermore, the parameters of component which function on the range of local constraint was optimized to improve the life of moving sheet and the stability of CHRW.

## 2 The layout of CHRW and work principle of local constraint

### 2.1 The work principle of local constraint

The sketch of HRM was shown in Figure 1, where *H* was a stagnation point. The center point of half-rotation rod (point *C*) whose orientation was always held in the direction of the connection between point *C* and point *H* was connected to the crank endpoint. Half-rotating rod would turn half round when the crank turned one round. The constralateral-type half-rotating mechanism (CHRM) was shown in Figure 2, which was composed of moving sheet, straight stator and circular stator. Both one side of crank (point *C*) and the center of positioning sliding shaft (point *H*) were fixed in frame. In addition, the crank and blade was hinged at point *C* and the blade would slide in groove of positioning sliding shaft when crank was rotated around point *O*. However, the gap between wing blade and groove of positioning shaft brought in the angular deviation which would resulted in irregular movement of blade. Essentially, the contact between moving sheet and circular stator could correct the trajectory of blade by eliminating angular deviation. Meanwhile, the straight stator would play a key role to avoid indeterminate movement of blade when point *C* overlapped with point *H*.

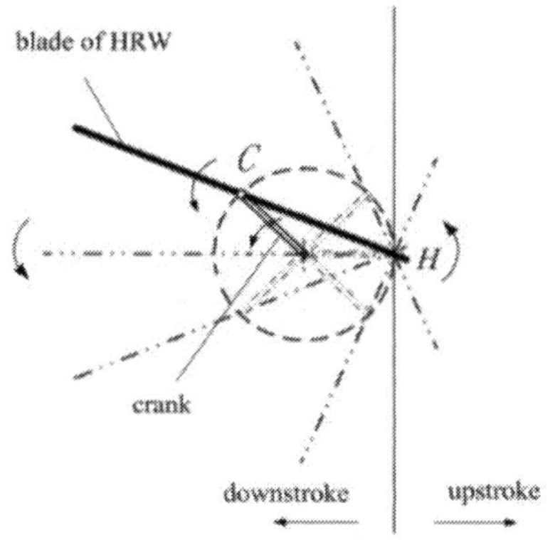


**Figure 1.** Work principle of HRM

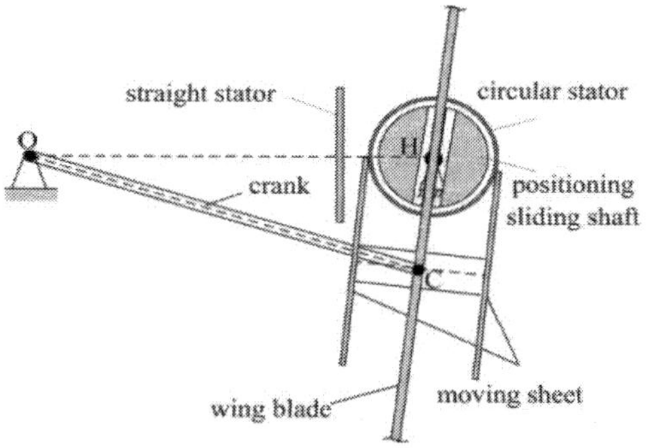


**Figure 2.** The CHRM with local constraint

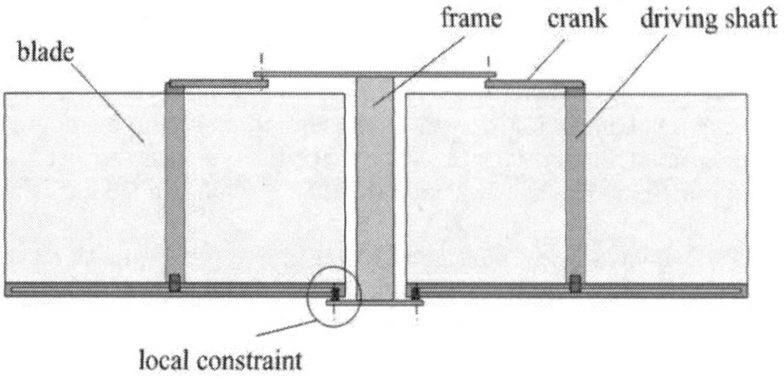


**Figure 3**. The layout of CHRW

### 2.2 The layout and movement of CHRW

The layout of CHRW was shown in Figure 3. The crank, blade and local constraint were distributed in both sides of frame, which improved the stress state of the whole frame and mitigated the vibration of CHRW.

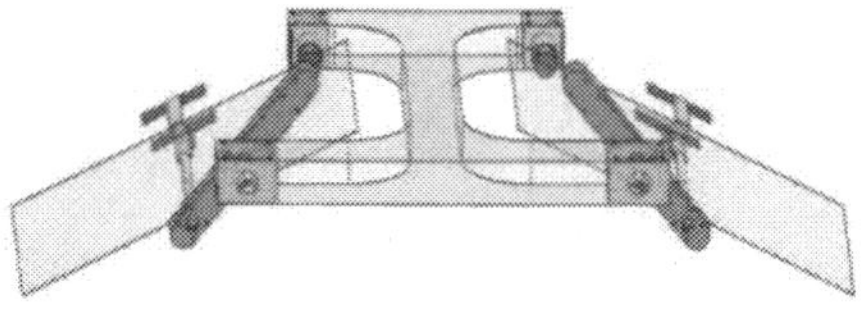

(a) Downstroke

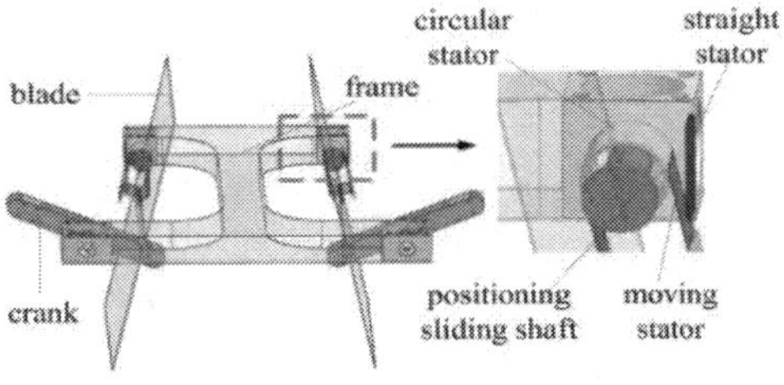


(b) The beginning of local constraint

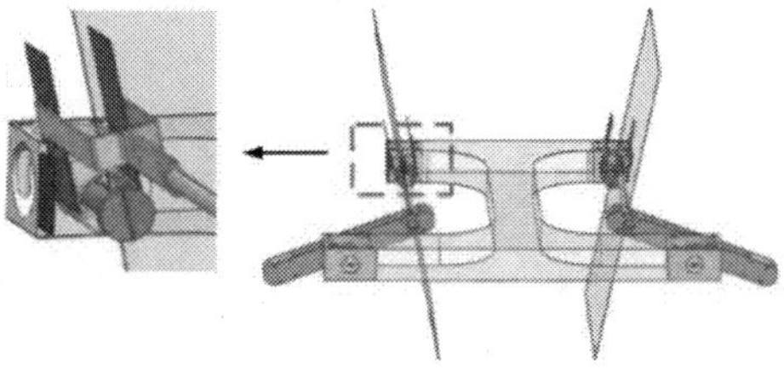

(c) The end of local constraint

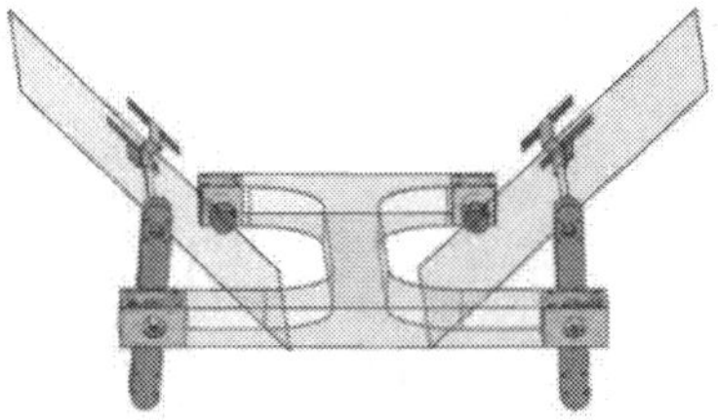

(d) Upstroke

**Figure 4.** The movement of simplified constralateral-type HRW in different time

The movement process was shown in Figure 4, which was based on the state of movement and the principle of HRW.

## 3 Design of local constraint of CHRW

The local constraint was made up of moving sheet, straight stator and circular stator. In this paper, the design of moving sheet was the research emphasis.

### 3.1 The performing range of local constraint

The gap between blade and groove could be insured with the smooth sliding of blade, but simultaneously it brought in angular deviation which would damage the accuracy of movement for blade.

The angular deviation model was shown in Figure 5. $R$ was the length of crank, $\varphi$ was the rotating angle of crank, $\delta$ was the width of groove and $\theta$ represented the angular deviation. The angular deviation could be calculated as follows

$$\theta = \frac{\delta - t}{4R\sin\left|\frac{\varphi}{2}\right|} \tag{1}$$

The angular deviation $\theta$ would be increased with the rise of $\varphi$. Therefore, a maximum angular deviation $\theta_m$ could be set to limit the range of local constraint $\varphi'$ as

$$-2\arcsin(\frac{\delta - t}{4R\theta_m}) \le \varphi' \le 2\arcsin(\frac{\delta - t}{4R\theta_m}) \tag{2}$$

### 3.2 Design of circular stator

The angular deviation increased quickly when point *C* closed to the point *H*. Consequently, it is necessary to correct the trajectory of blade and calculate the envelope line of moving sheet primarily.

The coordinate system was established as shown in Figure 6, whose origin was set in point *H*. Moving sheet was located in the line *AB* whose linear equation could be calculated as

$$-x\cos\frac{\varphi}{2}+y\sin\frac{\varphi}{2}=p \tag{3}$$

Where $p$ was the perpendicular distance between point *H* and line *AB*.

The equation of envelope line could be got as

$$x^2+y^2=p^2 \tag{4}$$

Obviously, the envelope line of moving sheet was a circle whose center and radius were respectively point *H* and *p*. Consequently, a circular stator was designed according to the above calculation.

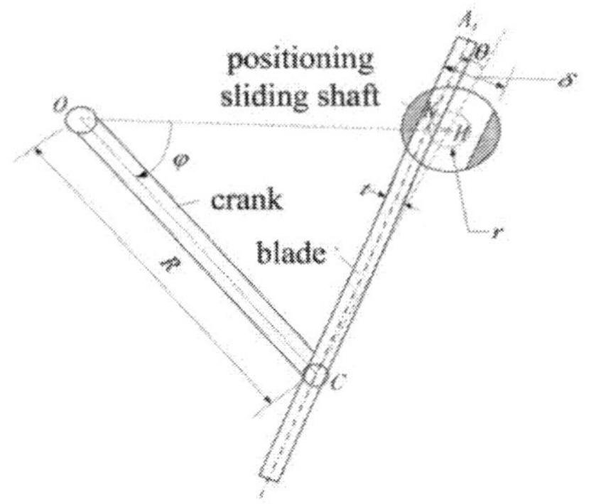


**Figure 5.** Geometric model of angular deviation

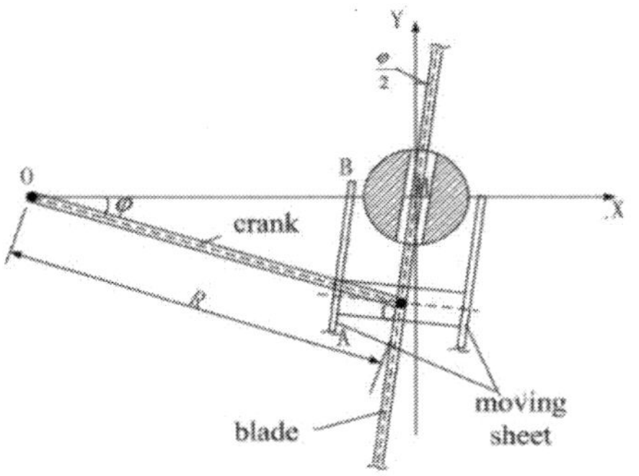


**Figure 6.** Geometric position of moving sheet

### 3.3 Design of moving sheet

#### *3.3.1 The velocity of contact for moving sheet*

The geometric position of blade in groove was shown in Figure 7. The solid line was the actual position of moving sheet while the dashed line was the theoretical position of moving sheet. The velocity of moving sheet on actual position and theoretical position could be compared to simplify calculation because the calculation of actual velocity for moving sheet endpoint was too complex.

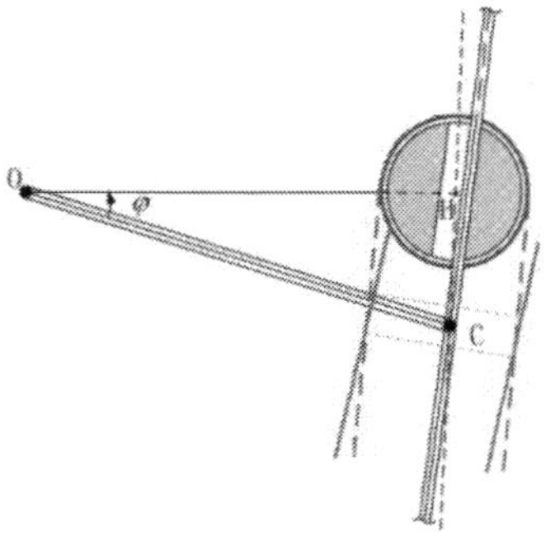

**Figure 7**. Two kinds of positions for moving sheet

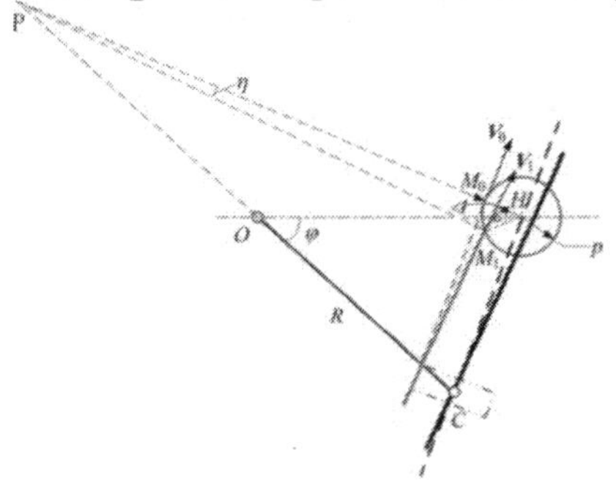

**Figure 8.** Velocity of moving sheet endpoint

DOI: 10.1051/matecconf/201710401003

The sketch of velocity was shown in Figure 8. Then point $P$ was the velocity instantaneous center of blade. $M_0$ was the theoretical position of moving sheet endpoint whose velocity was $\boldsymbol{V}_0$. $M_1$ was the actual position of moving sheet endpoint whose velocity was $\boldsymbol{V}_1$. In addition, $\Delta$ was the deflection of blade. $PM_0$ was vertical with $\boldsymbol{V}_0$ while $PM_1$ was vertical with $\boldsymbol{V}_1$. $\eta$ was the included angle of $PM_0$ and $PM_1$. Thereby $\boldsymbol{V}_0$ and $\boldsymbol{V}_1$ were calculated by velocity instantaneous method as

$$\begin{cases} V_0 = PM_0 \cdot \dfrac{\omega}{2} = \dfrac{\omega}{2}\left(2R\cos\dfrac{\varphi}{2} - p\right) \\ V_1 = PM_1 \cdot \dfrac{\omega}{2} = \dfrac{\omega}{2}\left(2R\cos\dfrac{\varphi}{2} - p + \Delta\right) \end{cases} \tag{5}$$

$\boldsymbol{V}_0$ and $\boldsymbol{V}_1$ were compared according to formula (5) and actual parameters of other components to get the result as follows

$$\begin{cases} \dfrac{V_1 - V_0}{V_0} = \dfrac{\Delta}{2R\cos\dfrac{\varphi}{2} - p} \approx 0 \\ \tan\eta = \dfrac{\sqrt{2p\Delta - \Delta^2}}{2R\cos\dfrac{\varphi}{2} - p} < \dfrac{\sqrt{2p\Delta}}{2R\cos\dfrac{\varphi}{2} - p} = \dfrac{\sqrt{\dfrac{2\Delta}{p}}}{\dfrac{2R}{p}\cos\dfrac{\varphi}{2} - 1} \approx 0 \end{cases} \tag{6}$$

The magnitude and orientation of $\boldsymbol{V}_0$ and $\boldsymbol{V}_1$ were extremely close from the above equation, thus $\boldsymbol{V}_0$ could be used as a substitute for $\boldsymbol{V}_1$ in the following calculations.

### *3.3.2 The contact force between moving sheet and circular stator*

The moving sheet would be bended when contacted circular stator since the moving stator was made from elastic metal. The contact model was established to calculate the contact force as shown in Figure 9, where the moving sheet could be regarded as many particles connected by springs.

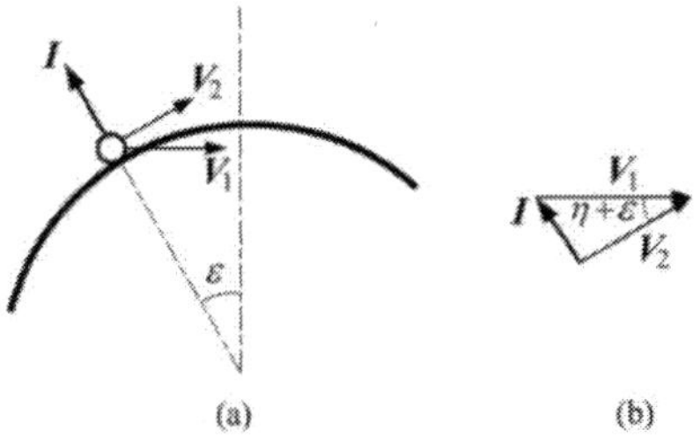


**Figure 9.** The model of contact

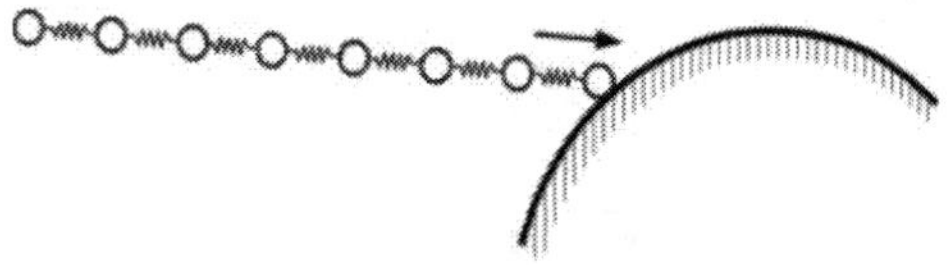

**Figure 10.** The velocity of contact

The contact velocity was shown in Figure 10. $\boldsymbol{V}_1$ was the velocity before contact whose direction was as same as moving sheet. And then, $\boldsymbol{V}_2$ was the velocity after contact whose direction was tangent to the surface of circular stator. However, it would result in more than one contact if the bending angle of $\boldsymbol{V}_1$ could not reach to $\eta+\varepsilon$. Therefore, $\boldsymbol{V}_2$ was the minimum velocity by the analysis on the above. Momentum theorem could be used to calculate $\boldsymbol{V}_2$ as

$$\begin{cases} I = mV_2 - mV_1 \\ I = \rho A V_0^2 \sin\varepsilon \mathrm{d}t \end{cases} \tag{7}$$

Where $\boldsymbol{I}$ was impulse, $m$ was the mass that participated in the contact part of moving sheet.

In addition, $\boldsymbol{I}$ could also be calculated by equation $I = Fdt$. $F$ was instantaneous contact force of the moving sheet and circular stator. Therefore, $F$ could be calculated as follows

$$F = \rho A V_0^2 \sin(\varepsilon+\theta) = \frac{1}{4}\rho A\omega^2\left(2R\cos\frac{\varphi}{2}-p\right)^2 \sin\varepsilon \tag{8}$$

Where $\rho$ was the density of moving sheet, $A$ was the cross sectional area, $\omega$ was the angular velocity of crank.

### 3.3.3 Calculation of thickness of moving sheet

It was concluded in section 2.3.2 that $V_2$ was the reasonable velocity for moving sheet when contacted circular stator. The geometric model of contact force was shown in Figure 11.

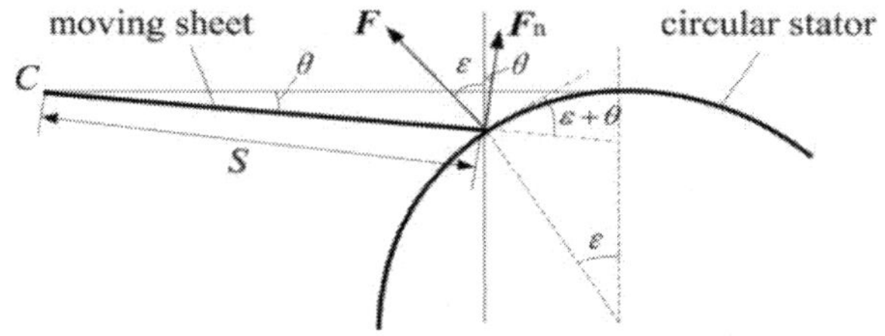


**Figure 11.** The geometric model of contact force

$F_n$ was the force who bent the moving sheet, it could be calculated as follows

$$F_n = \frac{1}{4}\rho A\omega^2\left(2R\cos\frac{\varphi}{2}-p\right)^2 \sin\varepsilon\cos(\varepsilon+\theta) \tag{9}$$

The condition of smooth movement for moving sheet was $\frac{F_n S^2}{2EI} = \varepsilon+\theta$, so it was got as

$$\frac{\rho A\omega^2 S^2\left(2R\cos\frac{\varphi}{2}-p\right)^2 \sin\varepsilon\cos(\varepsilon+\theta)}{8EI} = \varepsilon+\theta \tag{10}$$

Where $I = \frac{1}{12}bh^3$, $b$ was the width of moving sheet, and $h$ was the thickness of moving sheet, so it could be got in this form

$$h = \sqrt{3}\omega S(2R\cos\frac{\varphi}{2}-p)\sqrt{\frac{\rho\sin\varepsilon\cos(\varepsilon+\theta)}{2E(\varepsilon+\theta)}} \tag{11}$$

## 3.4 Designing of the straight stator

The constraint for blade was disappeared when point $C$ was overlapped with point $H$, which was not allowed in the design. The straight stator was used to avoid the free movement of blade by missing constraint.

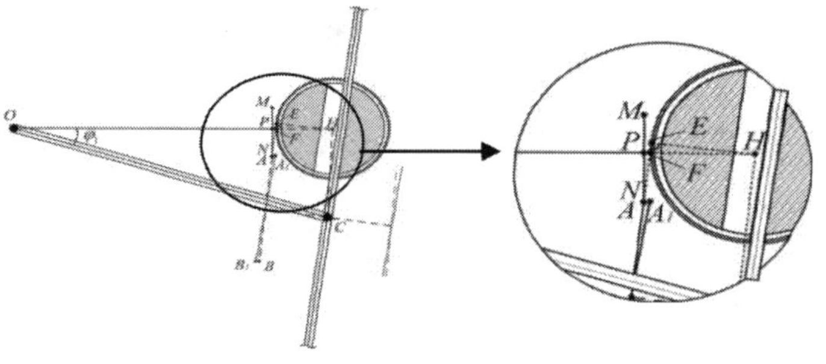


**Figure 12.** The geometric relation of moving sheet and straight

The geometric relation of moving sheet and straight stator was shown in Figure 12. *AB* was the theoretical position of moving sheet and $A_1B_1$ was the actual position of moving sheet. Both positions were almost the same in terms of the calculation by the parameters in above design. Therefore, the actual position could be placed by the theoretical position as

$$l = 2\cos\frac{\varphi_1}{2}(2R\sin\frac{\varphi_1}{2} - S - p\tan\frac{\varphi_1}{2}) \tag{12}$$

Where $l$ was the length of straight stator, $\varphi_1$ was the angle of crank when moving sheet and straight stator contacted.

The distance between straight stator and circular stator could be calculated as

$$d = \frac{p}{\cos\frac{\varphi_1}{2}} + \sin\frac{\varphi_1}{2}(2R\sin\frac{\varphi_1}{2} - S - p\tan\frac{\varphi_1}{2}) - p \tag{13}$$

The parameters of straight stator could be calculated from parameters of moving sheet and circular stator. Besides, $\varphi_1$ could be provided in accordance with technical condition.

## 4 Simulation and optimization of contact force between moving sheet and circular stator

### 4.1 Simulation of contact force

The contact force could be calculated by formula (8). Four sets of parameters were proposed in Table 1, which was calculated by four groups parameters based on theoretical analysis and calculation in section 2. Meanwhile, ADAMS was used in simulating the contact force in order to verify the theoretical result.

**Table 1.** Comparison of theoretical contact force and simulation contact force under different local constraints

| Group | $b$ (*mm*) | $t_1$ (*mm*) | $l_1$ (*mm*) | $r$ (*mm*) | $\delta$ (*mm*) | $t_2$ (*mm*) | Contact force value by calculation (*N*) | Contact force value by simulation (*N*) |
|---|---|---|---|---|---|---|---|---|
| a | 26.5 | 0.18 | 25 | 10 | 0.6 | 0.4 | 0.27 | 0.30 |
| b | 30 | 0.21 | 28 | 12 | 0.7 | 0.4 | 0.34 | 0.40 |
| c | 40 | 0.3 | 26 | 13 | 0.6 | 0.2 | 0.63 | 0.60 |
| d | 30 | 0.22 | 25 | 10 | 0.65 | 0.4 | 0.42 | 0.45 |

In the Table 1, $b$, $t_1$ and $l_1$ were respectively the width, thickness and length of moving sheet. Besides, $r$ was the radius of circular stator, $\delta$ was the width of groove and $t_2$ was the thickness of blade.

After comparing theoretical calculation and simulation results, the conclusion could be got that the contact force model was reasonable.

### 4.2 Optimization of contact force

High-speed movement of blade in operation caused the high-frequency contact between moving sheet and circular stator. It was necessary to optimize the contact force to extend the life of moving sheet and improve the stability of half-rotating blade.

#### *4.2.1 Design variables*

The design variables must be independent as

$$\boldsymbol{X} = (X_1, X_2, X_3, X_4, X_5, X_6) = (b, t_1, l_1, r, \delta, t_2) \tag{14}$$

#### *4.2.2 Objective function*

High-frequency contact was harmful to moving sheet and circular stator so that the minimum contact force was chosen as objective function as

$$\min f(X) = F \tag{15}$$

#### *4.2.3 Constraint condition*

$$g_1(x) = \sqrt{\frac{X_5 - X_6}{X_4}} + \frac{X_5 - X_6}{2X_3} - \frac{6.4\times10^{-10}}{X_2^2}[X_3^2(120\sqrt{1-\frac{X_3^2}{14400}} - X_4)^2\sqrt{\frac{X_5 - X_6}{X_4} - \frac{(X_5 - X_6)^2}{4X_4^2}}] \le 0$$

$$g_2(x) = \sqrt{X_4^2 + X_3^2} - 60 \le 0$$

$$g_2(x) = \sqrt{X_4^2 + X_3^2} - 60 \le 0$$

$$g_4(x) = \frac{X_4}{\sqrt{1-\frac{X_3^2}{14400}}} - X_3 \le 0 \tag{16}$$

$$g_5(x) = X_4 - X_3 \le 0$$

#### *4.2.4 Comparison of the contact force before and after optimization*

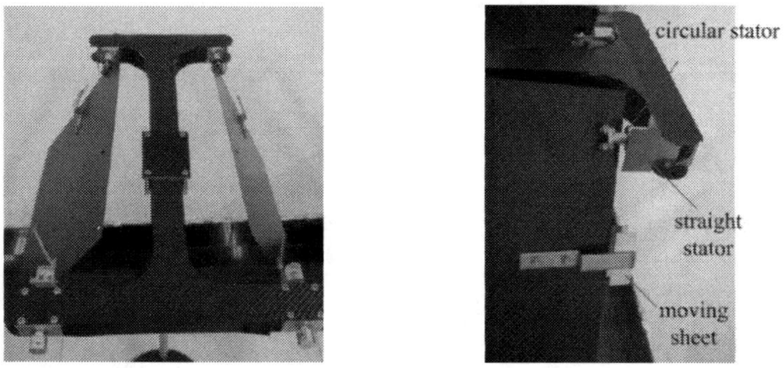


**Figure 13.** The prototype of CHRW

The prototype of CHRW was shown in Figure 13. Then the function of "fmincon" from MATLAB was used to optimize the parameters of the driving mechanism and the parameters before and after optimization were shown in Table 2. It could be indicated that the contact force reduced about 33.9% after optimization.

**Table 2.** The result before and after optimization

| | $b$ (mm) | $t_1$ (mm) | $l_1$ (mm) | $r$ (mm) | $\delta$ (mm) | $t_2$ (mm) | force (N) |
|---|---|---|---|---|---|---|---|
| before | 26.5 | 0.2 | 26 | 10 | 0.8 | 0.5 | 0.321 |
| after | 28 | 0.145 | 30 | 9.3 | 0.63 | 0.42 | 0.212 |

## 5 Conclusion

(1) The contralateral-type layout and local constraint was designed based on the principle of HRM, which included calculation of size and position for moving sheet, circular stator and straight stator.

(2) The contact model of the moving sheet and circular stator was established based on the momentum theorem, which was verified by four contact models in ADAMS.
(3) The dimension parameters of working part who toke part in the local constraint was optimized in MATLAB. The contact force reduced about 33.9%, which obviously extended the lives of moving sheet and circular stator, at the same time improved the stability of CHRW.

## Acknowledgments

This work is supported by the National Natural Science Foundation of China (Grant No. 51375014 and No. 51405001)

## References

1. Wood RJ. The First Take off a Biologically Inspired At-scale Robotic Insect [J]. IEEE Transactions on Robotics, 24(2): 341-347(2008)
2. Mueller J. Fixed and Flapping Wing Aerodynamics for Micro Air Applications [M]. Virginia: American Institute of Aeronautics and Astronautics(2001)
3. Pornsin-Sirirak TN, Lee SW. MEMS Wing Technology for a Battery-powered Arnithopter [C]. The 13th IEEE Annual International Conference on MEMS, 799-804(2000)
4. De Croon G C H E, De Wagter C. Design, Aerodynamics and Vision-Based Control of the DflFly [J]. Internatioanl Journal on Micro Air Vehicles. 1(2) : 71-97(2009)
5. Schenato L, Deng X, Wu W C, et al. Virtual Insect Flight Simulator (VIFS): A software tested for insect flight [C]. Seoul(2001)
6. Zhu B. Research on flapping-wing flight mechanism and bionic flapping-wing mechanism [D]. Nanjing: Nanjing University of Aaeronautics and Astronautics(2010) (in Chinese)
7. Qiu Z Z, Half-Rotating Mechanism: Structure, Characteristics and Application, Hefei: University of Science and Technology of China(2011) (in Chinese)
8. Y H Zhang, Z Z Qiu. Propeller with a Blade and its Motion Characteristics [J]. Chinese Journal of Mechannical Engineering, 42 (3): 193~196(2006) (in Chinese)
9. X Y Wang, L L Mei. Analysis and Application of Motion Geometry Feature for a New Biped-imitating Walking Mechanism [J]. Mechanical Science and Technology for Aerospace Engineering, 30 (6): 1020~1024(2011) (in Chinese)
10. Wang X Y, Zhang Y Q, Wang H X, et al. Research of Local Constraint for Simplified Half-Rotating Mechanism[J]. Applied Mechanics & Materials, 826:35-39(2016) (in Chinese)

# Analysis on the Workspace of Six-degrees-of-freedom Industrial Robot Based on AutoCAD

Jin-quan Li[1], Rui Zhang[1,a], Fang Cui[1], Qi Guan[1] and Yang Zhang[1]

[1]*School of Automation, Beijing University of Posts and Telecommunications, 100876 Beijing, China*

**Abstract.** This research discusses the workspace of the industrial robot with six degrees of freedom(6-DOF) based on AutoCAD platform. Based on the analysis of the overall configuration of the robot, this research establishes the kinematic mathematical model of the industrial robot by using DH parameters, and then solves the workspace of the robot consequently. In the AutoCAD, Auto Lisp language program is adopted to simulate the two-dimensional(2D) and three-dimensional(3D) space of the robot. Software user interface is written by using the dialog box control language of Visual LISP. At last, the research analyzes the trend of the shape and direction of the workspace when the length and angle range of the robot are changed. This research lays the foundation for the design, control and planning of industrial robots.

## 1 Introduction

At present, researches from China and abroad on the workspace of industrial robots have made some achievements [1-4]. The study of workspace of robot is inclined to use Monte Carlo method [5-7]. Hou Yulei [8] proposes a method for the workspace boundary extraction called 'limit fixed interval angle', based on the principles of bar limit combined with fixed joint variable. Tian Haibo [9] et al analyze manipulator's workspace by using Monte Carlo method and brings adaptive-divided mesh method to calculate the volume of workspace. Yi Jun [10]et al develop dynamic simulation system of robot based on AutoCAD. In this research, robot kinematics model by DH parameter method is improved, using Auto Lisp programming language to achieve six-degrees-of-freedom industrial robot workspace simulation. The influence of structure parameters on the shape and position of workspace and position is analyzed. This method also provides a powerful help for the design and research of industrial robot.

## 2 Description of 6-DOF industrial robot workspace

According to the modified DH parameter, the six-degree-of-freedom robot link coordinate system is established. See Figure 1.

[a] Corresponding author : zhangrui0908@bupt.edu.cn

DOI: 10.1051/matecconf/201710401004

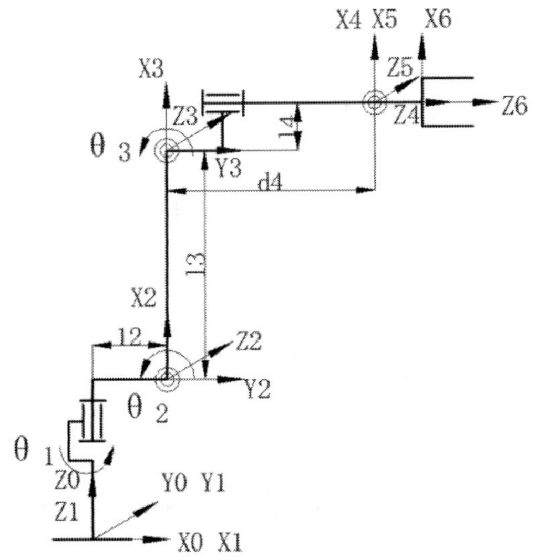


**Figure 1.** Schematic diagram of the robot.

The parameters of the robot D-H can be obtained as table 1:

**Table 1.** Robot DH parameter table.

| $i$ | $a_{i-1}$ [m] | $\alpha_{i-1}$ [°] | $d_i$ [mm] | $\theta_i$ [°] |
|---|---|---|---|---|
| 1 | 0 | 0 | 0 | $\theta_1$ |
| 2 | $l_2$ | –90° | 0 | $\theta_2$ |
| 3 | $l_3$ | 0 | 0 | $\theta_3$ |
| 4 | $l_4$ | –90° | $d_4$ | $\theta_4$ |
| 5 | 0 | 90° | 0 | $\theta_5$ |
| 6 | 0 | –90° | $d_6$ | $\theta_6$ |

The position and attitude of the end of manipulator can be obtained through the robot coordinate system:

$$\boldsymbol{T}=\begin{bmatrix} n_x & o_x & a_x & p_x \\ n_y & o_y & a_y & p_y \\ n_z & o_z & a_z & p_z \\ 0 & 0 & 0 & 1 \end{bmatrix} \tag{1}$$

Where: $\begin{bmatrix} n_x & o_x & a_x \\ n_y & o_y & a_y \\ n_z & o_z & a_z \end{bmatrix}$ is the rotation matrix (attitude) of the robot end coordinate system with respect to base coordinate system; Formula (2) is the center of end position for robot. $c_i=\cos\theta_i$, $s_i=\sin\theta_i$, $c_{ij}=\cos(\theta_i+\theta_j)$, $s_{ij}=\sin(\theta_i+\theta_j)$.

$$\begin{bmatrix} p_x \\ p_y \\ p_z \end{bmatrix}=\begin{bmatrix} l_2c_1-d_4c_1s_{23}+l_4c_1c_{23}+d_4s_1+l_3c_1c_2 \\ l_2s_1-d_4s_1s_{23}-d_4c_1+l_4s_1c_{23}+l_3c_2s_1 \\ -l_4s_{23}-d_4c_{23}-l_3s_2 \end{bmatrix} \tag{2}$$

 DOI: 10.1051/matecconf/201710401004

## 3 Implementation of AutoCAD plug-in in the workspace of robot

Because of the simplicity of Visual LISP language syntax, the user can directly call the AutoCAD to draw through commends, and debug and check easily. Other languages will increase the complexity of development. Therefore, Visual LISP is used to calculate the software development.

In AutoCAD development environment with Visual Lisp, using Auto Lisp language file "*.lsp" is written to realize simulation of the robot workspace program. The user dialog interface file "*.dcl" format is generated, which is achieving the loading and display. Add a custom menu "Industrial Robot" to AutoCAD menu bar. Then load the "*.lsp" source code files and "*.dcl" dialog box file through the custom command macro. After the user clicks "OK" button, it gets the edit box and checks the input data. And then, it completes the follow-up program on the drawing commands of the workspace part at the time of closing the dialog box. Ultimately, two development of the software interface in AutoCAD is shown in Figure 2.

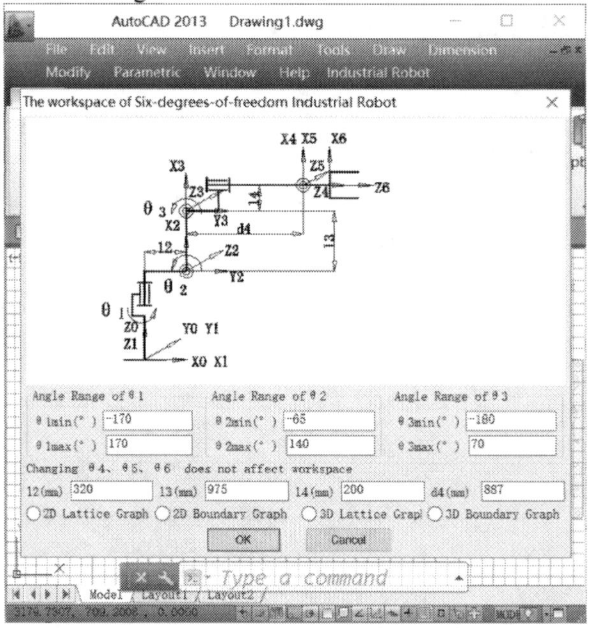


**Figure 2.** Robot workspace dialog box.

By the Formula (2), we can calculate the 2D workspace of the robot in which end reference point is formed on the base coordinate system. Combined with the scope of the angle and the initial value of length parameters, the section of workspace in the YOZ direction can be gotten when $\theta_1$=90° . See the use of Auto LISP programming in Figure 3. On the basis of 2D workspace, changing $\theta_1$ from the minimum value gradually increased to the maximum value, the 3D workspace of the robot with six degrees of freedom can be obtained.

DOI: 10.1051/matecconf/201710401004

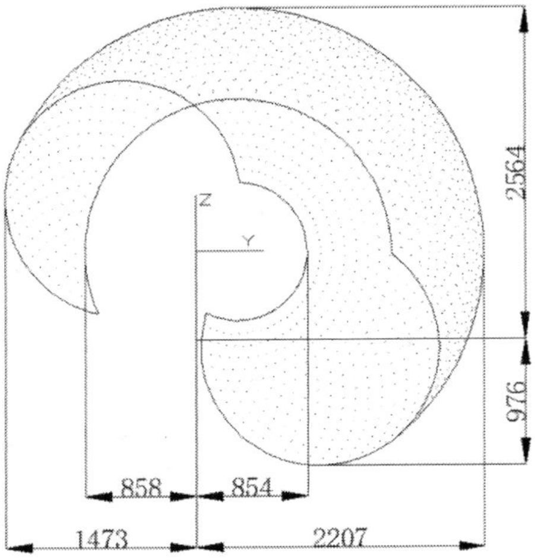


**Figure 3.** 2D space diagram with the solid boundary range.

## 4 The influence of structural parameters on the workspace of the robot

The workspace of the robot is drawn by the software written by LISP Visual to understand the influence of the change of the structure parameters on the workspace. The main variables are the length variable $l_2$、$l_3$、$d_4$ , and angle variable $\theta_2$ 、 $\theta_3$. Because of the small impact of $l_4$, it is not considered. The original image and the picture after changing the mechanism parameters will be put together for the convenience of comparison

### 4.1 The influence of the changing rod length on workspace

The boundary map of the workspace is obtained by setting value of $l_2$ in turn as shown in Figure 4. Analysis shows that, when other parameters are kept constant, only increasing or decreasing the $l_2$ , the size and shape of the YOZ working plane remain unchanged, that is to say only changing the position of the workspace. When the increase $l_2$ , the YOZ working plane has a tendency to move upward; otherwise, it is downward.

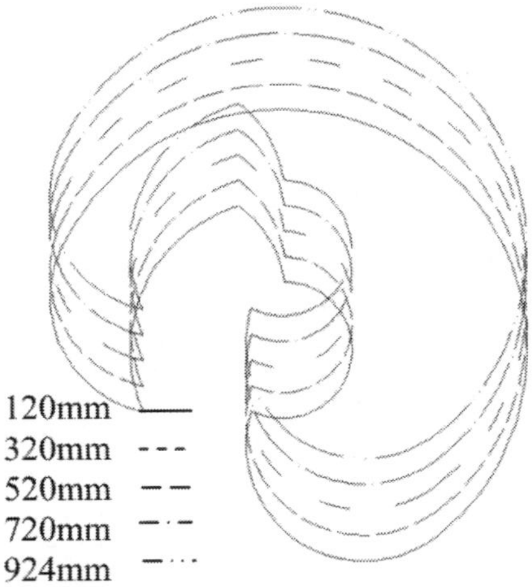


**Figure 4.** YOZ plane change trend chart when changing $l_2$ .

Figure 5 is the boundary map of the workspace when $l_3$ is changed. The analysis shows that when other parameters remain unchanged and increase $l_3$, the YOZ working plane has a tendency to move upward, meanwhile, the area increases; otherwise the opposite. Figure 6 is the boundary map when $d_4$ is changed. We can know that when other parameters remain unchanged, only increase the $d_4$, the robot YOZ working plane area increased, otherwise decreased.

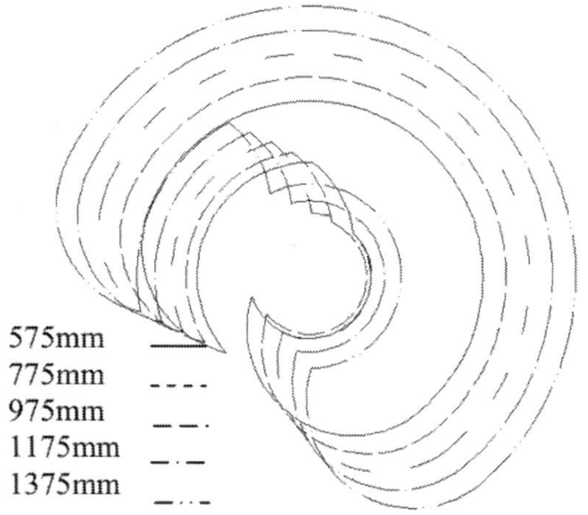


**Figure 5.** YOZ plane change trend chart when changing $l_3$

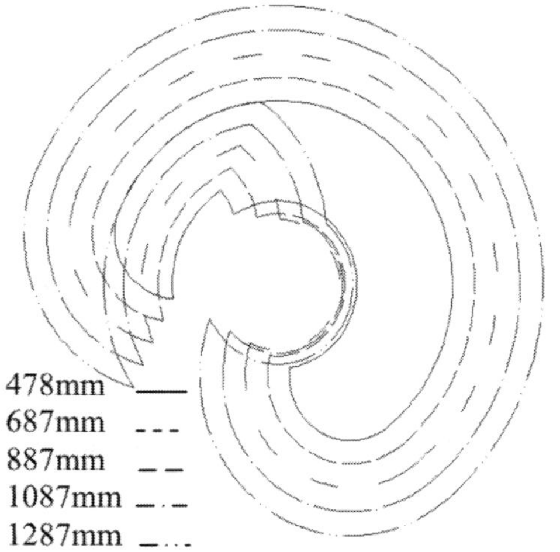


**Figure 6.** YOZ plane change trend chart when changing $d_4$

## 4.2 The Influence of the Changing Angle Range on the Workspace

As shown in Figure 7, the boundary map of workspace is obtained after the change of scope of $\theta_2$. Analysis shows that when other parameters are kept constant and only the angel range of $\theta_2$ is increased, the YOZ working plane of the robot is expanded gradually and the area increases, otherwise it will decrease.

Figure 8 is obtained from changing of $\theta_3$. Analysis shows that, when other parameters remain unchanged, the YOZ working plane area of the robot is increased with the increase of the angle; otherwise it will decrease. And when the angle range increases to a certain degree, the YOZ working plane interior will change from the depression into the bulge. This situation should be paid attention to when designing.

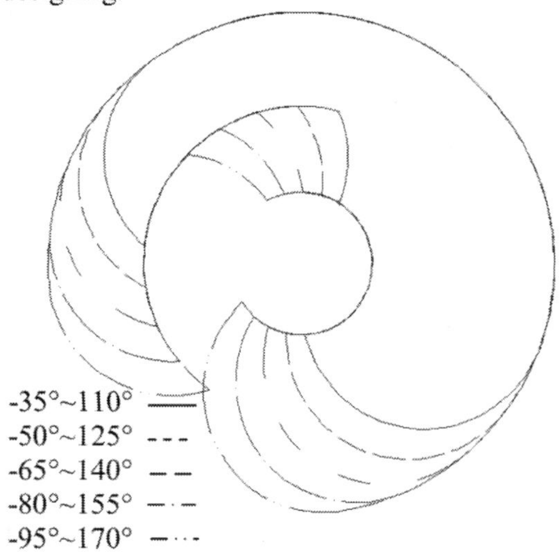


**Figure 7.** YOZ plane change trend chart when changing $\theta_2$.

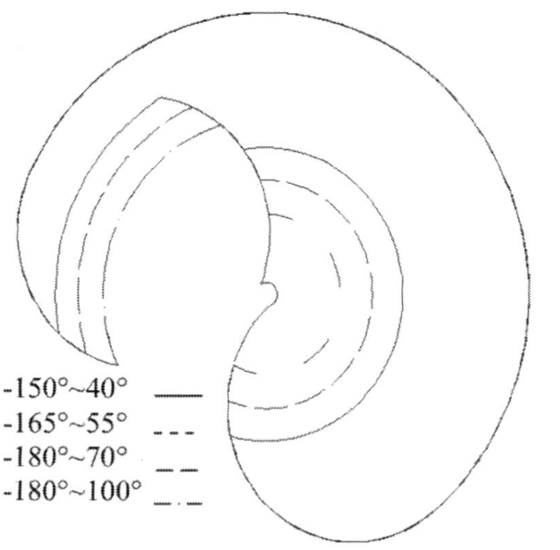


**Figure 8.** YOZ plane change trend chart when changing $\theta_3$.

## 5 Conclusion

Through secondary development on AutoCAD platform, the workspace of six degrees of freedom industrial robot is studied. According to the modified DH parameter, the modeling of the robot is carried out. Using Visual LISP the dialog interface program is written to simulate the workspace of the robot. By changing the input parameters, the workspace can be quickly and accurately drawn and the range of motion of the robot can be calculated. So the influence of structural parameters on the workspace can be analyzed directly. This is a clear, simple and effective method to determine the workspace of industrial robot. The method can quickly determine the structure size and the range of motion of the robot according to the design requirements of the work, so as to improve the working efficiency. At the same time, it also creates a good basic condition for the robot dynamics analysis, trajectory planning and motion control.

## References

1. Z. Dong, C. Yin, X. Li. Research on characteristics and movement space of palletizing robot mechanical arm[J]. Journal of Chinese Agricultural Mechanization, **35(6)**, 53-56, 64 (2014)
2. J. Behnoush Rezaeian, O. Kambiz Ghaemi, M. Ali. Optimization of kinematic redundancy and workspace analysis of a dual-arm cam-lock robot[J]. Robotica, **34**, 23-42 (2016)
3. Li Jinquan, Duan binglei, Li Zhongming, et.al. Analysis on Workspace and Influence Coefficient of a Proposed Palletizing Robot[J]. Journal of Beijing University of Posts and Telecommunications, **34(6)**, 78-81 (2011)
4. W. F. Xu, L. T. Li, B. Liang, et al. Workspace analysis of space 3R robot[J]. Journal of Astronautics, **28(5)**, 1389-1394 (2007)
5. Z. Xie, C. Li, W. Ni, et.al. Workspace analysis of a 6-DOF installing-calibrating robot[J]. Advanced Materials Research, **422**, 75-78 (2012)
6. Z. Liu, H. Liu, Z. Luo, et.al. Improvement on Monte Carlo method for robot workspace determination[J]. Transactions of the Chinese Society for Agricultural Machinery, **44**, 230-235 (2013)
7. P. Adrián, G.Arturo, M.José María. Monte-Carlo workspace calculation of a Serial-Parallel biped robot[J]. Advances in Intelligent Systems and Computing, **418**, 157-169 (2016)
8. Y. Hou, C. Wang, X. Hu, et.al. A New Method for Workspace Boundary Extraction for Jointed Serial Robot[J]. China Mechanical Engineering, **26(3)**, 308-312, 318 (2015)
9. Tian Haibo, Ma Hongwei, WeiJuan, et.al. Workspace and Structural Parameters Analysis for Manipulator of Serial Robot[J]. Transactions of The Chinese Society of Agricultural Machinery, **44(4)**, 196-200 (2013)
10. J. Yi, Y. Ding. System of Robot Dynamics Simulation Based on AutoLISP[J]. Coal Technology, **32(5)**, 158-160 (2013)

# Key Technologies in Large-scale Rescue Robot Wrists

Zhidong Tang[1] and Chao Yun[2,a]

[1]*School of Mechanical Engineering and Automation, Beihang University, 100191 Beijing, China*
[2]*Robot Institute, Beihang University, 100191 Beijing, China*

**Abstract.** The full-Automatic Quick Hitch Coupling Device (full-AQHCD for short) is used as the starting point, key technologies in a large-scale rescue robot wrist, which is constituted by integrating a quick hitch coupling device, a turning device, and a swaying device together, are reviewed respectively. Firstly, the semi-AQHCD made domestically for the main-Arm Claw Wrist (main-ACW for short) is introduced, and the full-AQHCD imported from Oil Quick company in Sweden for the vice-Arm Cutter Wrist (vice-ACW for short) is presented. Secondly, aiming at three key technologies in the full-AQHCD including rotary joint technology, automatic docking technology and precise docking technology for quick action coupling, are concisely expressed. Thirdly, the hydraulic motor driving gear type slewing bearing technology of the turning device made domestically for the main-ACW is introduced, and the hydraulic motor driving worm type slewing bearing technology of the turning device imported from HKS company in Germany for the vice-ACW is presented, especially, the existing gap in the similar domestic technology is discussed. Subsequently, the hydraulic cylinder driving 4-bar linkage technology of the swaying device made domestically for the main-ACW is introduced, and the hydraulic double spiral swing cylinder technology of the swaying device imported from HKS company in Germany for the vice-ACW is presented, especially, the existing gap in the similar domestic technology is discussed. Finally, it is emphasized that these technological gaps have seriously restricted the ability of the vice-ACW to successfully work in future actual rescue combats, therefore, it must be highly valued in the follow-up research and development (R&D) through cooperating with professional manufacturers in China, thereby making technological advances.

## 1 Introduction

By comparison with the no-automatic quick hitch coupling device, the semi-Automatic Quick Hitch Coupling Device (semi-AQHCD for short) has been popularized and applied widely so far. The concept of full-Automatic Quick Hitch Coupling Device (full-AQHCD for short) is introduced for the first time into "Double Powers, Intelligent, and Double Arms Series of Large-Scale Rescue Robot (DPIDASLSRR for short)", which is the national "12th 5-Year" science and technology support project. Due to the full-AQHCD belongs to the new things without precedent in China, for shortening the development cycle and reaching the international advanced level, among 20 ton, 40 ton and 60 ton (20 t, 40 t and 60 t for short, respectively) three series of prototype for "DPIDASLSRR", the 20 t prototype imported a full-AQHCD product OQ65-3[1] from Oil Quick company in Sweden and 2

[a] Corresponding author : cyun18@vip.sina.com

 DOI: 10.1051/matecconf/201710401005

Degrees of Freedom (DOF/DOFs for short) of turning and swaying product TR-K180[2-3] from HKS company in Germany for the vice-Arm Cutter Wrist (vice-ACW for short) which mainly undertakes a dexterous mission with the help of the main-Arm Claw Wrist (main-ACW for short) which mainly undertakes a strength mission, this sort of three-in-one wrist is constituted by integrating a Quick Hitch Coupling Device (QHCD for short), a turning device, and a swaying device together. The 20 t prototype subsequently participated in an actual rescue combat[4] of 7 magnitude earthquake in Ya'an of Sichuan province in April 20, 2013. The evaluation of the actual rescue combat has verified that the OQ65-3 and the TR-K180 have excellent quality, reliable performance and are suitable for China's natural conditions, therefore, realizing localization of them for 40 t and 60 t prototypes should be most desirable.

## 2 Key technologies in large-scale rescue robot wrists

### 2.1 Wrist assemblies

Figure 1 shows the wrist1 of main arm1 claw and the wrist2 of vice arm2 cutter of the 20 t prototype. Wrists of 40 t and 60 t prototypes are respectively corresponding to and similar to the ones of the 20 t prototype. Claw and cutter are collectively referred to as attachments. Figure 2 shows the 20 t prototype participating in the actual rescue combat of 7 magnitude earthquake in Ya'an of Sichuan province in April 20, 2013.

**Figure 1.** Two wrists of 20 t prototype for large-scale rescue robot

**Figure 2.** 20 t prototype in actual rescue combat of 7 magnitude earthquake in Ya'an in April 20, 2013

Set in the prototype development description book, the technical specifications of the main-ACW undertaking strength mission are shown in Table 1, the technical specifications of the vice-ACW undertaking dexterous mission are shown in Table 2, and *tn* is the tonnage for prototype. The performance of the 20 t prototype in actual rescue combat proved that the swing speed of the swaying device and the rotary speed of the turning device both in equivalent to 5 r/min are appropriate, if more than 5 r/min are too fast and not conducive to the posture adjustment for attachments, if less than 5

**Table 1.** Technical specifications of main-ACW for large-scale rescue robot

| $tn$ /t | $d_{s1}$/ ° | $v_{s1}$/ (°/s) | $m_{s1}$/ (t·m) | $d_{t1}$/ ° | $v_{t1}$/ (°/s) | $m_{t1}$/ (N·m) | $ld_{t}$/ t |
|---|---|---|---|---|---|---|---|
| 20 | ±45 | 20 | Nil | 350 | 45 | Nil | 4 |
| 40 | ±45 | 20~30 | 4.5 | 350 | 45 | 1063 | 10.2 |
| 60 | ±45 | 20~30 | 6 | 350 | 45 | Nil | 14 |

**Table 2.** Technical specifications of vice-ACW for large-scale rescue robot

| $tn$ /t | $d_{s2}$/ ° | $v_{s2}$/ (°/s) | $m_{s2}$/ (t·m) | $d_{t2}$/ ° | $v_{t2}$/ (°/s) | $m_{t2}$/ (N·m) | $ld_2$/ t |
|---|---|---|---|---|---|---|---|
| 20 | ±55 | 18.3 | 1.5 | 360 | 48 | 3500 | 4 |
| 40 | ±55 | 20~30 | 2.5 | 360 | 18~30 | Nil | 5.8 |
| 60 | ±55 | 20~30 | 4 | 360 | 18~30 | Nil | 6 |

r/min are too slow. In the actual rescue combat, that is to say that time is the life, if the operating movement is very slow, then it will mean that fewer people will be saved and more people will dead.

$d_{s1}/d_{s2}$ are the counter-clockwise(+) and clockwise(-) maximum swing angles of the swaying device for the main/vice-ACW, it can be seen that the maximum swing range for the main/vice-ACW is 90°/110°, a difference of 20°. $v_{s1}/v_{s2}$ is the swing speed of the swaying device for the main/vice-ACW, which can be seen that it is equivalent to the rotation speed of 3~5 r/min. $m_{s1}/m_{s2}$ is the swing moment of the swaying device for the main/vice-ACW (Nil indicates that it is not specified in the overall design of the prototype, the following is the same). $d_{t1}/d_{t2}$ is the rotary angle of the turning device for the main/vice-ACW, the main arm's 350° means that it cannot fully rotate, the vice arm's 360° means that it can fully rotate. $m_{t1}/m_{t2}$ is the rotary moment of the turning device for the main/vice-ACW. $v_{t1}/v_{t2}$ is the rotary speed of the turning device for the main/vice-ACW, which can be seen that the main arm's is equivalent to the rotation speed of 7.5 r/min. The vice arm's speed is equivalent to the rotation speed of 3~5 r/min, which means that the operating movement is adjustable between a bit too slow and appropriate, 20 t prototype's speed can be a bit too fast up to 8 r/min. $ld_1/ld_2$ is the maximum static load on the main/vice-ACW when it works alone in operation.

Figure 3 shows the wrist assembly of the main arm claw for the 40 t prototype. Figure 4 shows the wrist assembly of the vice arm cutter for the 40 t prototype. The upper coupling parts is used to link the bucket rod and the connection rod at the end of the main/vice arm, for achieving the longitudinal swaying freedom of the main/vice-ACW. The lower coupling parts is used to link with attachments. However, the main-ACW does not need the lower coupling parts, but directly links with attachments. The main/vice-ACW assembly for the 60 t prototype is correspondingly similar to the ones for the 40 t prototype.

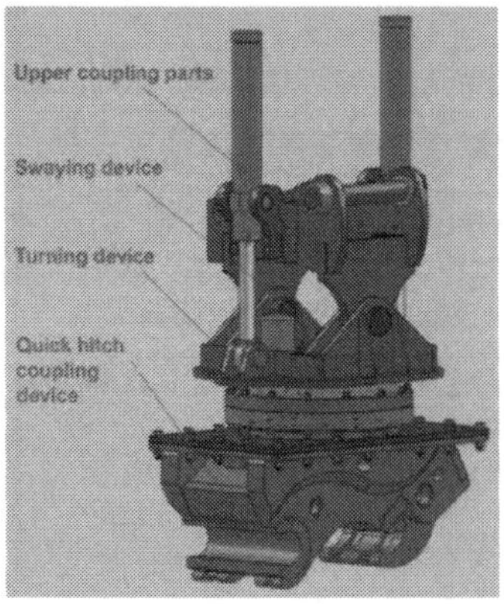


**Figure 3.** Wrist assembly of main arm claw of 40 t prototype for large-scale rescue robot

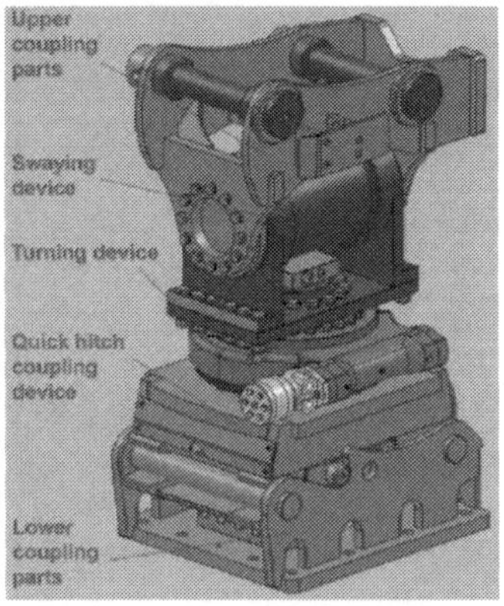


**Figure 4.** Wrist assembly of vice arm cutter of 40 t prototype for large-scale rescue robot

## 2.2 Key technology in QHCD for wrist

### *2.2.1 Semi-AQHCD for main-ACW*

In 1996, the structure design added at the end of the mechanical arm of the construction machinery for conveniently hanging the attachments should be the early embryonic form of the QHCD[5]. Soon afterwards it began to popularize among excavators and loaders. The QHCD for loader at home and abroad is actually a kind of hydraulically controlled quick coupling bracket with four groups of hinge hole structure, through the hook to hang and the pin shaft to lock the attachments[6-7] as shown in Figure 5.a. Now this kind of quick coupling bracket with such structure has been widely used for loader in China. But because the quick coupling bracket is difficult to integrate 2 DOFs for both swaying and turning, so it is not suitable to be used as an integration platform for wrist.

Because the semi-AQHCD popular for excavator (as shown in Figure 5.b) has unsafe factors caused by the artificial participation still required in changing attachments, which has directly resulted some injuries in the UK, then the British introduced guidance documents for safe use of this kind of semi-AQHCD[8]. And, a death accident in Britain directly caused by this kind of semi-AQHCD made the British government pay attention to a non-mandatory ban again, namely new construction machinery shall not continue to be equipped with this kind of semi-AQHCD, those that have been equipped in the past may continue to be used for old construction machinery[9].

According to the trend of development, the semi-AQHCD for the main-ACW, which is shown in Figure 3 and Figure 5.b, will eventually fall into disuse because of occurrence of the full-AQHCD, but there will be a long period of time for its existence in China.

### *2.2.2 Full-AQHCD for vice-ACW*

So far, at home and abroad the definition of full-AQHCD is just those that as long as have the function of automatic locking/unlocking attachments[10-12] as shown in Figure 6. In recent years the foreign improvements regarding this kind of full-AQHCD mainly include structure design for automatic security proof latch off[13-15], structure design for lubrication measures[16-17], design for anti-unlock at hydraulic failure time[18-19], structure design and hydraulic control measures for both ends with movable paws[20-22].

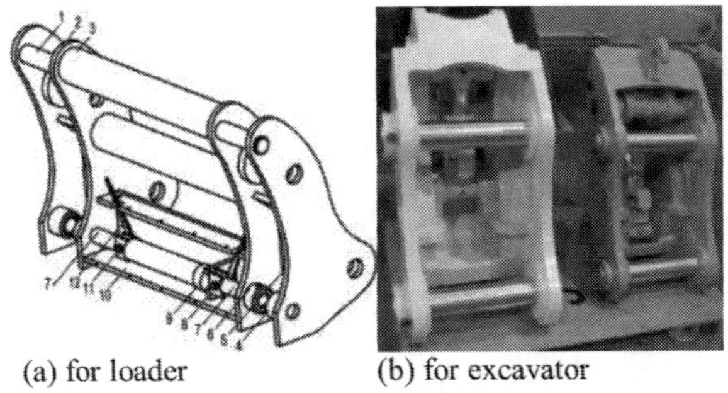

(a) for loader (b) for excavator

**Figure 5.** Quick hitch coupling device

**Figure 6.** Automatic quick hitch coupling device

But the full-AQHCD described in this paper refers to not only the function of automatically locking/unlocking the attachments but also the function of automatically connecting/disconnecting the oil lines of the attachments. The open foreign literature on the automatically connecting/disconnecting the oil lines of the attachments is few, and is limited to this kind of directly automatic connection between the end of the mechanical arm of the construction machinery and the oil lines of the attachments[23-24], and does not involve the QHCD in it. Three key technologies involved in the full-AQHCD for the vice-ACW shown in Figure 4 are studied and listed as follows:

(1) Rotary joint technology as shown in Figure 7. A rotary joint transporting a liquid or gas from a stationary end to a moving side is a very wide range of key components. The study, analysis and comparison of the structural forms of the existing domestic and foreign rotary joints and central turning hydraulic joints are carried out.

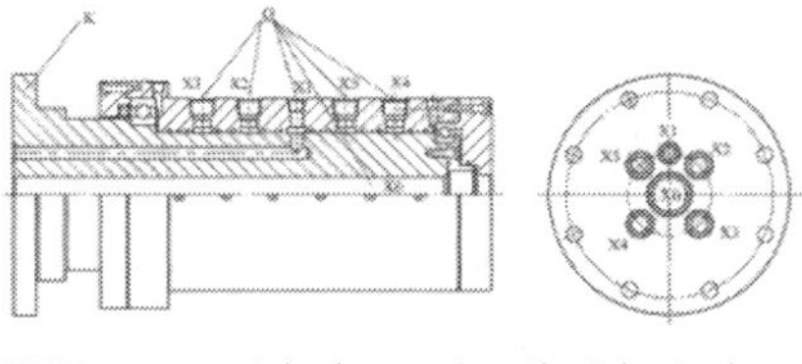

**Figure 7.** National standard Z6F type multi-channel cylindrical elastomer sealing rotary joint

(2) Automatic docking technology of quick action coupling. Aiming at the two quick action couplings, including ISO7241 (as shown in Figure 8) and ISO16028 (as shown in Figure 9), the structural forms and their pros and cons were investigated, analyzed and evaluated in detail, to provide a technical reference for final selection of ISO16028 quick action coupling. The key problem was analyzed and discussed of changing the ISO16028 from manual docking to automatic docking.

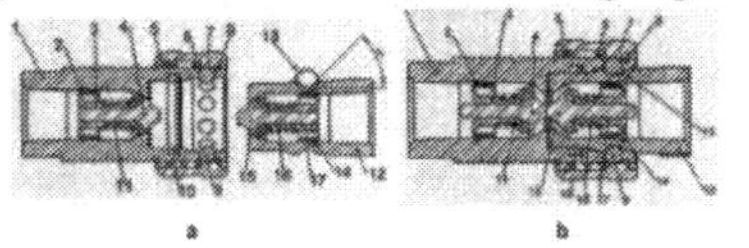

(a) disconnect (b) connect

**Figure 8.** Structure drawing for ISO7241 quick action coupling

(a) disconnect (b) connect

**Figure 9.** Structure drawing for ISO16028 quick action coupling

(3) Precise docking technology of quick action coupling. Inaccurate docking will seriously damage the quick action coupling and lead to failing connection and safety accident.

### 2.3 Key technology in turning device for wrist

The rotational DOF for wrist refers to the QHCD (with or without the attachments), which is driven by the turning device, rotating around the swaying device, in fact, it is realized by the relative rotation between the inner ring and the outer ring of the turning device. Due to the adoption of different inner/outer ring driving technology, the rotation of the QHCD of the main-ACW is driven by the inner ring of the turning device, but the rotation of the QHCD of the vice-ACW is driven by the outer ring of the turning device. The inner ring and the outer ring are supported by rolling elements, therefore, the one that integrates the inner ring, the outer ring and the rolling elements together is called a slewing bearing. A slewing bearing and a hydraulic motor are integrated together to form a turning device.

#### *2.3.1 Gear type turning device for main-ACW*

The supporting types of rolling elements of slewing bearing are mainly the single row 4-point contact ball type, the double row different diameter ball type, the single row cross roller type and the three row roller type, where the single row 4-point contact ball type in recent years developed rapidly to replace the double row different diameter ball type[25].

There are many domestic application researches on the slewing bearing. These application researches have gained extensive experience, especially, the single row 4-point contact ball type is the preferred type for the slewing bearing of small and medium size, but the three row roller type is the preferred type for the rotary support of large size[26], if it is an extra-large size rotary support, then its rolling elements are enlarged to the rolling wheels, such as trolley type conical rolling wheels, multi row rolling wheels and chain rolling wheels, and so on[27].

Because the gear transmission ratio is not large, it will cause too large radial size to the slewing bearing, but the main-ACW does not require compact structure, so the radial size of its slewing bearing is not strictly required, but just meeting the load requirement will be fine for the slewing bearing, therefore, the technology of gear driving slewing bearing with the single row 4-point contact ball type has been used in the turning device for the main-ACW as shown in Figure 10. The outer ring of the slewing bearing is fixedly connected with the swaying device, and the inner ring is actually an inner tooth big gear, which is fixedly connected with the QHCD and is driven by the outer small gear on the hydraulic motor.

### *2.3.2 Worm type turning device for vice-ACW*

The gear pair has high efficiency, long life and high reliability, but when its transmission ratio increases, the dimensions of its space will swell correspondingly, so it is not suitable for the application in large transmission ratio conditions, but the worm pair is just applicable to these conditions. Because under the premise of the vice-ACW meeting the load requirements it should have a structure as compact as possible, a volume as small as possible, and a running as dexterous as possible (i.e., not too fast but also not too slow), so it can't use the gear type turning device of large radial size similar with the one for the main-ACW.

Based on the investigation, it has been found that HKS company in Germany uses the worm type turning device to achieve the goal of compact structure, small volume and dexterous running (it can reach up to a too fast speed of 8 r/min). But the best domestic product actually can only reach an extremely slow speed of 1.5 r/min (corresponding to the flowrate of 30 L/min for the hydraulic motor, but the HKS product can reach 8 r/min at the flowrate of 45 L/min). The inner ring of the slewing bearing of this kind of worm type turning device is fixedly connected with the swaying device, and the outer ring, which is actually a worm gear (with solid model of gear tooth profile, and its appearance is similar to gear) as shown in Figure 11, is fixedly connected with the QHCD, and is meshed with the

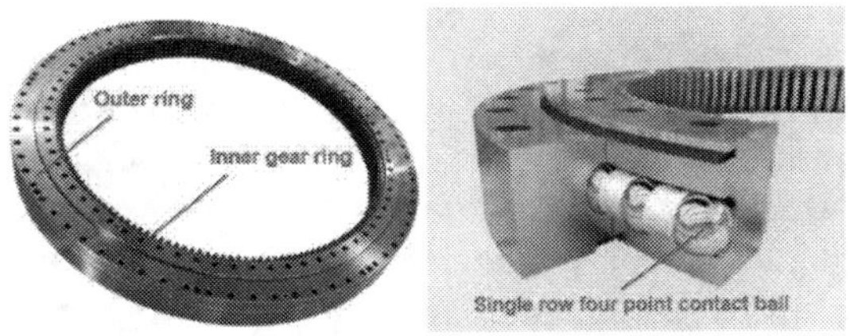


**Figure 10.** Gear driving slewing bearing with single row 4-point contact ball type

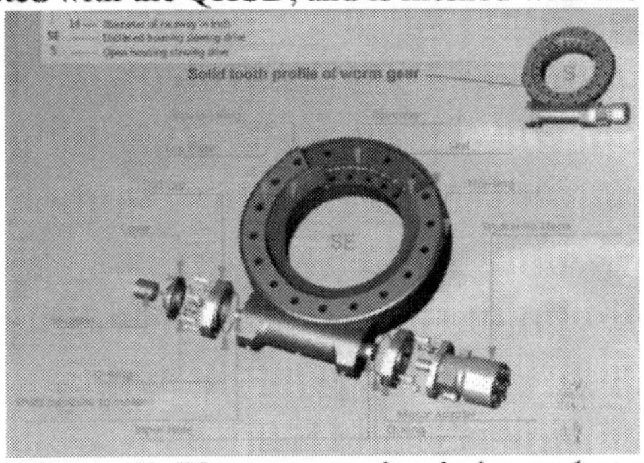


**Figure 11.** Worm type turning device product of domestic company in China

worm driven by the hydraulic motor. It still has some differences in details with the one of HKS company in Germany, which are mainly that, the outer ring is actually a shell and is fixedly connected with the swaying device, the inner ring is a worm gear (with empty model of gear tooth profile, and its appearance is similar to wheel) as shown in Figure 12, which is fixedly connected with the QHCD and is meshed with the worm driven by the hydraulic motor, i.e., the turning device is composed of a disassembled set of 4 including a shell, a worm gear, a worm and a hydraulic motor, and is 1 piece less than the domestic one, which is composed of a disassembled set of 5 including a shell, a worm gear, an inner ring, a worm and a hydraulic motor, therefore, the HKS turning device has a more compact structure. Therefore, both the speed (1.5 r/min or 8 r/min) and the structure (loose or compact) are the main gaps existing between the domestic and foreign products from which the proper turning device is intended to be selected for a large-scale rescue robot wrist.

## 2.4 Key technology in swaying device for wrist

The swing DOF for wrist refers to the swaying device together with the turning device and the QHCD (with or without the attachments), which is driven hydraulically, swinging relative to main/vice arm end, in fact, it is realized by the relative swinging between the swaying device and the upper coupling parts. Due to the adoption of different swaying realizing technology, their structures vary greatly.

### *2.4.1 4-bar linkage cylinder type swaying device for main-ACW*

The swing motion of the 4-bar linkage cylinder type swaying device is similar to the rotating motion of the hoist boom, which is changed from the linear motion of the piston of the hydraulic cylinder by a crane, and which longitudinally swings upward and downward, therefore, the hoist boom, the crane body and the hydraulic cylinder the three actually constitutes a 4-bar linkage mechanism[28] as shown in Figure 13, hereinto only can the length of the straight line segment, which is composed of the

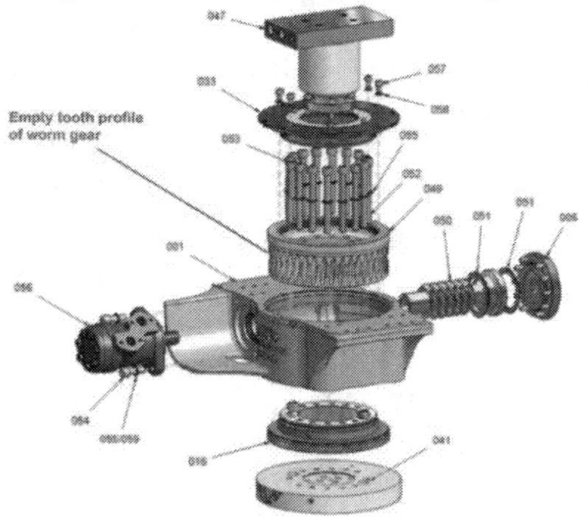


**Figure 12.** Worm type turning device product of HKS company in Germany

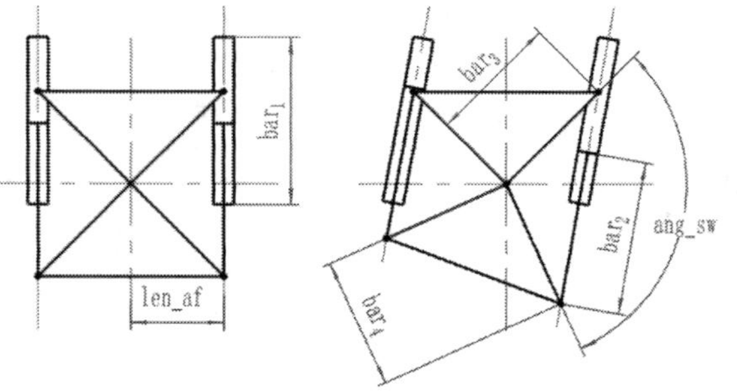


**Figure 13.** 4-bar linkage cylinder type mechanism

corresponding cylinder barrel $bar_1$ and the piston rod $bar_2$ the two bars, be variable, and thus change the corresponding included angle ang_sw of the triangle as shown in Figure 13, i.e., the longitudinal swing angle of the hoist boom, the swing motion of both the big arm and the small arm of the excavator is also realized by the same technology of 4-bar linkage cylinder. The technology of bilateral symmetric 4-bar linkage double cylinders used in the main-ACW of the large-scale rescue robot is shown in Figure 3. Due to the adoption of high pressure double acting single piston rod hydraulic cylinder, so a short swing arm of force (the distance between the hydraulic cylinder axis and the swing axis, i.e., the dimension len_af as shown in Figure 13) can generate a large swing moment. But the shortcoming of this kind of 4-bar linkage cylinder type swaying device is that the maximum swing angle (the dimension ang_sw as shown in Figure 13) is restricted by the limit of piston stroke, if the maximum swing angle is larger, the hydraulic cylinder length will be longer, and the overall dimensions will be bigger, as a result, the overall structure will be more loose rather than compact.

### *2.4.2 Double spiral swing cylinder type swaying device for vice-ACW*

The 4-bar linkage cylinder technology is to indirectly change the linear reciprocating motion of piston into the swing motion of other bars with the aid of 4-bar linkage mechanism, the piston itself has no rotation movement, the existence of the 4-bar linkage mechanism makes this kind of device structure certainly to be loose, so it is not applicable in the vice-ACW which requires that the overall structure should be as compact as possible. The load bearing ability of the vane swing cylinder is not well, it is difficult for the working cavity seal, the internal leakage is large, the volumetric efficiency is low, and it is hard to increase the working pressure, so its usage is very limited, and it is usually only applicable to be used in such low pressure equipment as servo turret and so on. To deliver exactly the same movement and torque, the space occupied by the gear rack swing cylinder is larger than the vane swing cylinder, so the gear rack swing cylinder usage is much more restricted. The spiral swing cylinder can change the linear reciprocating motion of the piston of the hydraulic cylinder into a clockwise and anti-clockwise rotation of the 2 directions and output a swing motion and a swing moment[29]. Through two stage screw pair amplification effect, only need a short piston stroke can output a large swing angle, in addition, a double helix swing cylinder has a more compact structure and a larger output swing angle than a single spiral swing cylinder, thus it becomes the first choice for the vice-ACW as shown in Figure 14.

It began from around 1999, the domestic preliminary study on the spiral swing cylinder technology of this kind of product, manufactured by such foreign companies as Helac company in U.S.A.[30] and HKS company in Germany, has been made[31-34], until 2010 the research gradually increased, up to now it only has less than 10 years of time, and it is mainly theoretical and experimental research, including transmission efficiency, structure design and optimization[35], and a few researches also on robot applications[36-38].

HKS company in Germany can alone complete the 2 DOFs of the vice-ACW for the large-scale rescue robot, and the connecting dimensions between the turning device and the swaying device have good compatibility, therefore, there are a very high integration level and a very compact structure and almost no redundancy volume and weight between the 2 DOFs as shown in Figure 15. But currently there are no such similar domestic manufacturers, thereby needing two different professional manufacturers respectively to complete the 2 DOFs, the poor compatibility of the connecting dimensions between the 2 DOFs is inevitable, when the 2 DOFs are integrated together by the third party, the connecting transition flange has to be required to add additionally between the 2 DOFs, resulting in a lot of redundant volume and weight, together with the gap of the performance and the immature of the technology existing in the 2 DOFs, therefore, they can't be completely replaced by HKS company's products. In addition, because there are no mature products for selection for the actual engineering application of the double spiral swing cylinder on the robot, it often need to be developed separately by such as G.S. Luo et al.[36-38], but the accumulated experience during the period of less than 10 years is limited and there are not enough actual/simulated rescue combats for

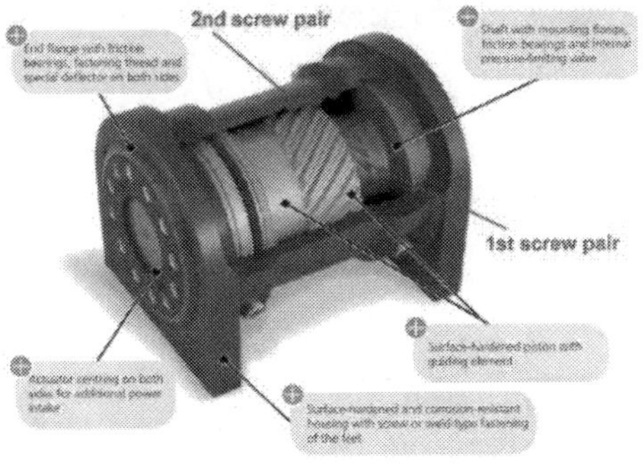


**Figure 14.** Two screw pairs of double spiral

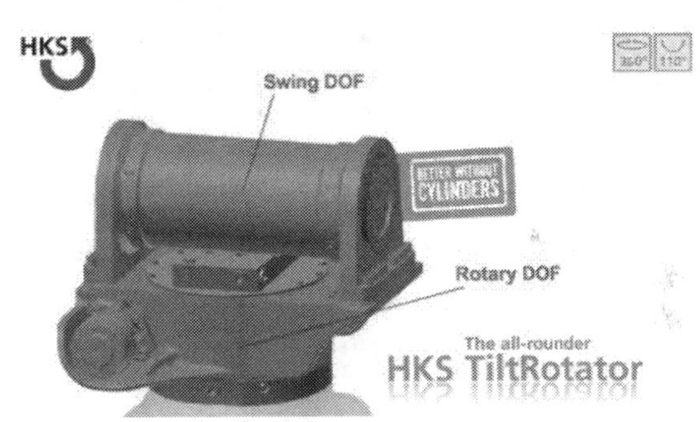


**Figure 15.** Two DOFs product of HKS company in

verifications, which lead to not dare to select for use rashly even if it has been developed out, therefore, this is the main gap existing between the domestic and foreign products from which the proper swaying device is intended to be selected for a large-scale rescue robot wrist.

## 3 Conclusion

(1) The full-AQHCD is the first bright spot in the vice-ACW, three key technologies involved should be tackled for clearing away obstacles of localization.

(2) Key technologies in the main/vice-ACW for the large-scale rescue robot have not the interchangeability, and they have the shortcomings caused by structural reasons themselves which cannot be overcome, at present, no better technologies at home and abroad can be an alternative.

(3) The 2 DOFs product of HKS is the second bright spot in the vice-ACW. If the "DPIDASLSRR" project is going to change prototypes into products in the short term, the suggestion is that the realization of the 2 DOFs for the vice-ACW should be still imported from HKS company in Germany.

(4) Knowing the existing gaps is not shameful unless ignoring them and rejecting to learn the

world, therefore, this is the future work of the vice-ACW for "DPIDASLSRR".

## References

1. OilQuick AB. OQ 65 suitable for machines from 14-22 t [EB/OL]. http://www.oilquick.com/oq-65. (2016-01-01)
2. HKS. Home [EB/OL]. https://www.hks-partner.com/en/home/. (2016-01-01)
3. HKS. TiltRotator [EB/OL]. https://www.hks-partner.com/en/products/construction-technology-series/tiltrotator/. (2016-01-01)
4. L. Chen. A rescue robot should be paid more attention to "Troops are kept a thousand days" than "To be used on one day"[N/OL]. Science and Technology Daily,2013-05-21. http://digitalpaper.stdaily.com/http_www.kjrb.com/kjrb/html/2013-05/21/content_204231.htm?div=-1. (2016-01-01)
5. W.M. Barden. Quick coupling device: US5581917[P]. (1996-12-10)
6. Y.B. Gao. The Optimization design on quick-coupler device of loader [D]. (Daqing: Northeast Petroleum University, 2013)
7. A.L. Vering, K.E. Lindenmuth, R.K. Oswald. Quick coupler: US7882898[P]. (2011-02-08)
8. d&ri. Quick hitch guidance [J]. Demolition & Recycling International. **12**, 2:4-4 (Mar/Apr2010)
9. ce.UK semi-automatic quick hitches ban sought [J]. Construction Europe. **21**, 8:7-7 (Oct2010)
10. X.S. Meng. Study on the standardization of the connection dimensions between the excavator and the quick change device [J]. CONSTRUCTION MACHINERY DIGEST, 03:43-45 (2013)
11. GEITH Inc. Quick coupler provides automatic, double-locking attachment system [J]. Pit & Quarry. **107**, 8:28-28 (Feb2015)
12. GEITH International Ltd. New couplers from Geith [J]. International Cranes & Specialized Transport. **26**, 5:48-48 (Jun2015)
13. P. McCormick, C. McCormick, L. McCormick, et al. A quick hitch coupler: WO2011051923[P]. (2011-05-05)
14. P. McCormick, C. McCormick, L. McCormick, et al. Quick hitch coupler: US2012210614[P]. (2012-08-23)
15. P. McCormick, C. McCormick, L. McCormick, et al. Quick hitch coupler: US8857081[P]. (2014-10-14)
16. T.C. Robl, T.R. Stefek. Quick coupler: US2013322954[P]. (2013-12-05)
17. T.C. Robl, T.R. Stefek. Quick coupler: US8869437[P]. (2014-10-28)
18. J. Parker, H.P. Gaines. Quick coupler: US2013160268[P]. (2013-06-27)
19. J. Parker, H.P. Gaines. Quick coupler: US2013160269[P]. (2013-06-27)
20. T. Friedrich. Quick coupler: US2014030005[P]. (2014-01-30)
21. T. Friedrich. Quick-coupler: US2014356059[P]. (2014-12-04)
22. T. Friedrich. Quick-coupler: US2014373401[P]. (2014-12-25)
23. R.(DE) Mieger, T.(DE) Zitterbart. Quick coupling: US2002157287[P]. (2002-10-31)
24. T.(DE) Zitterbart, R.(DE) Mieger. Quick coupling: US6813851[P]. (2004-11-9)
25. Y.Q. Wang, Q. Fang. Style-choosing calculation and designing of revolving support mechanism of hydraulic excavator [J]. Coal Mine Machinery, **27**, 11:4-6 (2006)
26. J. Dai. Reasonable choice of rotatory bearings [J]. MECHANICAL AND ELECTRICAL EQUIPMENT, 05:85-87 (2006)
27. Y.M. Wang, Y.L. Wang, M.H. Tong. Enabling technologies for new rotary support device of ultra-large-scale rotary floating cranes [J]. CHINESE JOURNAL OF CONSTRUCTION MACHINERY, **7**, 02:180-184 (2009)
28. Y. Xiang, N. Chen, Y.X. Hu, et al. Analysis and study of swing mechanism of boom type roadheader[J]. COLLIERY MECHANICAL & ELECTRICAL TECHNOLOGY, 03:4-6+11 (2001)
29. G. Li. Typical application of swing hydraulic cylinder [J]. Metal Working(Cold), 21:56-57 (2013)
30. Helac Corporation. Home [EB/OL]. http://www.helac.com/. (2016-01-01)

31. Y.P. Shi, L. Yang. Design of large swing angle spiral swing hydraulic cylinder [J]. Chinese Hydraulics & Pneumatics, 06:5-7 (1999)
32. Y.P. Shi, L. Yang. Design of spiral swing cylinder with large swing angle [J]. MACHINE TOOL & HYDRAULICS, 01:62-63 (2000)
33. J.G. Zhang. Design of spiral swing cylinder with large swing angle [J]. Journal of Lianyungang College of Chemical Technology, **13**, 01:16-19 (2000)
34. Z.M. Jin. Dynamic and accurate matters -- HKS oscillating hydraulic cylinder [J]. Modern Components, 08:40-41 (2004)
35. S.G. Xie. Design and optimization of the structure parameters of the double helix rotating cylinder based on virtual prototype technology [D]. (Changsha: Central South University, 2010)
36. G.S. Luo. Research on subsea 7 function maser-slave manipulator and its nonlinear robust control [D]. (Hangzhou: Zhejiang University, 2013)
37. G.S. Luo, J.W. Chen, L.Y. Gu. An elbow of 7-DOF hydraulic manipulator based on double-screw-pair transmission [J]. ROBOT, **36**, 01:36-42 (2014)
38. M.S. Lu. The structural design of underwater manipulator based on hydraulic spiral swing cylinder [D]. (Hangzhou: Hangzhou Dianzi University, 2015)

# Passivation of $HfO_2$/Ge interface with YON fabricated by different approaches

**Z. X. Cheng[1, a], Q. Xu[1], L. Liu[1] and J. P. Xu[1]**

[1]School of Optical and Electronic Information, Huazhong University of Science and Technology, Wuhan 430074, China

**Abstract**. Nitrided $Y_2O_3$ (YON) interfacial passivation layer (IPL) is used to passivate the $HfO_2$/Ge interface for better interfacial and electrical properties of Ge metal-oxide-semiconductor (MOS) capacitor. Two different approaches were used to prepare the YON IPL, one is to deposit YON directly by sputtering $Y_2O_3$ target in Ar+$N_2$ ambient, and the other is to deposit YN first by sputtering Y target in Ar+$N_2$ ambient followed by the annealing in $N_2$+$O_2$ to convert YN to YON. Experimental results indicate that the MOS capacitor fabricated by the latter approach could achieve more excellent interfacial and electrical properties due to more effective suppression of the formation of Ge oxides.

## 1 Introduction

As the Si-based metal-oxide-semiconductor field-effect transistor (MOSFET) is approaching its physical limits, novel substrate materials and devices have been extensively studied [1, 2]. Among the novel substrates with high mobility, Ge is believed to be one of the most promising candidate to replace the traditional Si substrate, due to its much higher and more symmetrical mobilities [3]. However, direct deposition of high-*k* dielectric on Ge substrate usually results in a poor interface with high interface-state density ($D_{it}$). Specifically, when $HfO_2$, which is considered to be the most widely-used dielectric for Si MOSFET, is used for Ge MOSFET, the parasitically grown unstable $GeO_x$ ($x<2$) between $HfO_2$ and Ge substrate severely degrades its interface quality [4]. Moreover, $HfGeO_x$ formed by the intermixing between Ge substrate (or $GeO_x$) and $HfO_2$ dielectric will also result in the increased $D_{it}$ because of the fivefold coordinated Hf in $HfGeO_x$ [5]. Recently, $Y_2O_3$ has attracted much attention to solve this problem as an alternative passivation layer on Ge substrate [6-8], because it can react with Ge substrate to form stable $YGeO_x$ interlayer and suppress the formation and diffusion of $GeO_x$. Moreover, the so-called "valency passivation" could effectively suppress the dangling bonds in the interface and consequently improve the interface quality by considering the trivalent nature of Y [8]. Therefore, $Y_2O_3$, which has high *k* value (12-17) [6, 9] and large bandgap (~5.5 eV) [6, 9] is considered to be one of the most promising IPLs for Ge MOS devices. However, the crystallization of $Y_2O_3$ is still an issue because of its low crystalline temperature (400°C) [10], which generates a large amount of defects in the film and causes the increase of leakage current [11]. N incorporation into $Y_2O_3$ has been suggested to increase the crystalline temperature [11]. Moreover, the barrier effect introduced by N could effectively suppress the diffusion of Ge and O to prevent the formation of $GeO_x$, and further improve the interface quality [12]. In this paper, we further investigate the effect of

[a] Corresponding author : zxcheng@hust.edu.cn

YON, which is used as IPL in the Ge MOS with $HfO_2$ as dielectric. Two different methods were used to prepare YON, one is to deposit YON directly by sputtering $Y_2O_3$ target in Ar+$N_2$, and the other is to deposit YN first by sputtering Y target in Ar+$N_2$ followed by annealing in $N_2$+$O_2$ to convert YN to YON. The interfacial and electrical properties of samples were compared and discussed.

## 2 Experiment

Figure 1 (a) shows the fabrication process of Ge MOS capacitors. N-type (100) Ge wafers with a resistivity of 0.02 ~ 0.1 Ω·cm were used for fabricating the Ge MOS capacitors. The Ge wafers were cleaned in trichloroethylene, acetone and ethanol in sequence, and then the native oxide was removed by a cyclic rinsing in diluted HF (1:50) and deionized water. After drying by $N_2$ gas, the wafers were immediately transferred into a sputter chamber. The stacked gate dielectric of $HfO_2$/YON (denoted as YON sample) was formed by successively depositing 2-nm YON by sputtering $Y_2O_3$ target in Ar + $N_2$ and 6-nm $HfO_2$ by sputtering $HfO_2$ target in Ar ambient. And the stacked gate dielectric of $HfO_2$/YN (denoted as YN sample) was formed by successively depositing 2-nm YN by sputtering Y target in Ar + $N_2$ and 6-nm $HfO_2$ by sputtering $HfO_2$ target in Ar ambient. For comparison, 8-nm $HfO_2$ gate dielectric (denoted as control sample) was also deposited using the same sputtering conditions as above. Then a post-deposition annealing (PDA) was carried out in $N_2$ (500 sccm) + $O_2$ (50 sccm) ambient at 500°C for 5 min to improve the film property and convert YN to YON. Al was evaporated and patterned as gate electrode with an area of $7.85\times10^{-5}$ $cm^{-2}$, followed by forming-gas (5% $H_2$ + 95% $N_2$) annealing at 300°C for 20 min to reduce contact resistance. The schematic of fabricated MOS capacitor is shown in figure 1 (b).

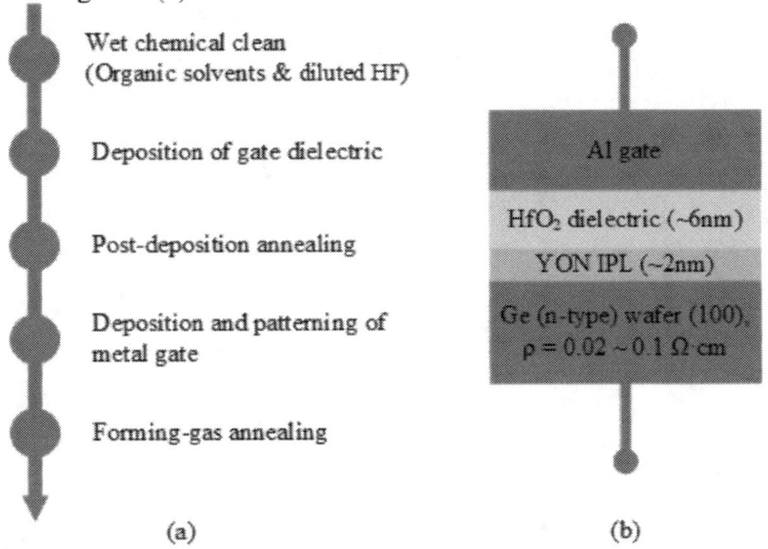


**Figure 1.** (a) Fabrication process flow, (b) Schematic of Al/$HfO_2$/IL/Ge capacitor

Capacitance-voltage (*C-V*) and gate leakage current density versus gate voltage ($J_g$–$V_g$) characteristics were measured under a light-tight and electrically-shielded condition by HP4284A precision LCR meter and HP4156A semiconductor parameter analyzer, respectively. X-ray photoelectron spectroscopy (XPS) was used to analyze the chemical states near/at the high-k/Ge interface, which was performed using a Thermo VG Multilab 2000 photoelectron spectrometer with a monochromatic Al Kα radiation source (hν = 1486.6 eV and width = 0.47 eV) under a vacuum lower than $3 \times 10^{-7}$ Pa. In order to analyze the interface between gate dielectric and Ge substrate, the thickness of the gate dielectric layer was thinned to ~ 5 nm from the Ge surface with a slow etching rate of ~1 nm/min by using an in-situ $Ar^+$ ion beam. And meanwhile, Atomic force microscopy (AFM) was used to check the $HfO_2$ surface morphology, which was influenced by the out-diffused Ge from the interface into the $HfO_2$ dielectric.

## 3 Results and discussion

Figure 2 shows the typical high-frequency (1-MHz) *C-V* curves of the three samples. Obviously, the control sample without IPL exhibits poor *C-V* behavior with the largest stretch out along the horizontal axis and even a large distortion in the weak inversion region, indicating the poor interface quality between $HfO_2$ dielectric and Ge substrate due to the formation of $GeO_x$ and defective $HfGeO_x$. The reduction of accumulation capacitance is understood by considering the formation of this low-*k* $GeO_x$ (~5) [13, 14]. With YON IPL inserted for the YON sample, the stretch out and distortion of *C-V* curve is reduced significantly, indicating much improvement of the interface quality. This is attributed to the formation of stable $YGeO_xN_y$ IPL in the interface, which could effectively suppress the formation of $GeO_x$ and $HfGeO_x$, and introduce the so-called "valency passivation" [8]. Furthermore, the YN sample exhibits even better *C-V* behavior than the YON sample, which has smaller stretch out and eliminated distortion in the weak inversion region. As a result, the YON IPL of YN sample could improve the interface quality more effectively. Both of the three samples exhibit positive flat-band voltage ($V_{fb}$), due to the negative oxide-charge density ($Q_{ox}$) caused by the diffusion of Ge (or $GeO_x$) into the $HfO_2$ dielectric [15]. The YN sample exhibits the smallest $V_{fb}$, indicating the most effective suppression of the diffusion of Ge (or $GeO_x$).

The typical gate leakage properties of the samples are shown in the inset of figure 2. Due to the poor interface quality, large $Q_{ox}$ and reduced conduction-band offset caused by the formation of Ge oxides [16], the control sample without IPL exhibits the largest gate leakage current. With YON IPL inserted, the gate leakage current is greatly reduced for the YON sample. Moreover, the YN sample achieves the smallest gate leakage current due to its best interface quality and smallest $Q_{ox}$.

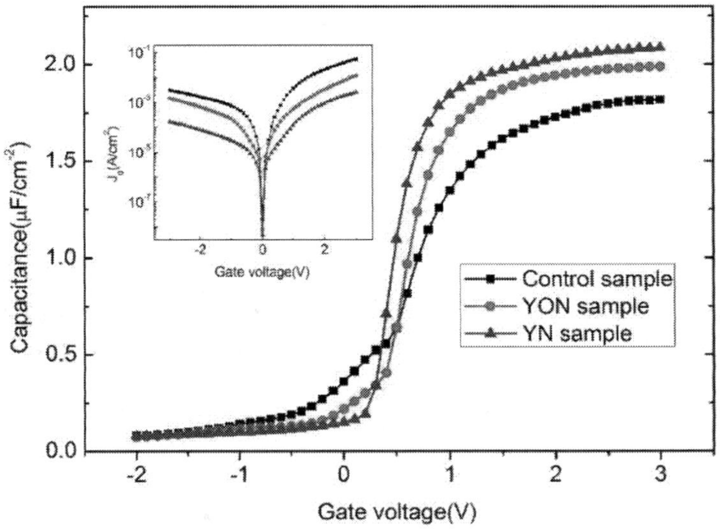


**Figure 2.** High-frequency (1-MHz) *C-V* curves of the three samples. The inset shows the $J_g$-$V_g$ curves of the three samples.

The frequency dispersion of *C-V* curves for the three samples is measured and shown in figure 3 to further compare the interface quality of the samples. The dispersive behavior of *C-V* curves from depletion to accumulation region and from weak to strong inversion region is attributed to the interface traps near the Ge valance band and conduction band, respectively [17]. Obviously, the control sample without IPL exhibits the largest frequency dispersion in both regions, implying a poor interface quality. With IPL inserted, much smaller frequency dispersion has been obtained. Moreover, the YN sample exhibits the smallest frequency dispersion, indicating the best interface quality.

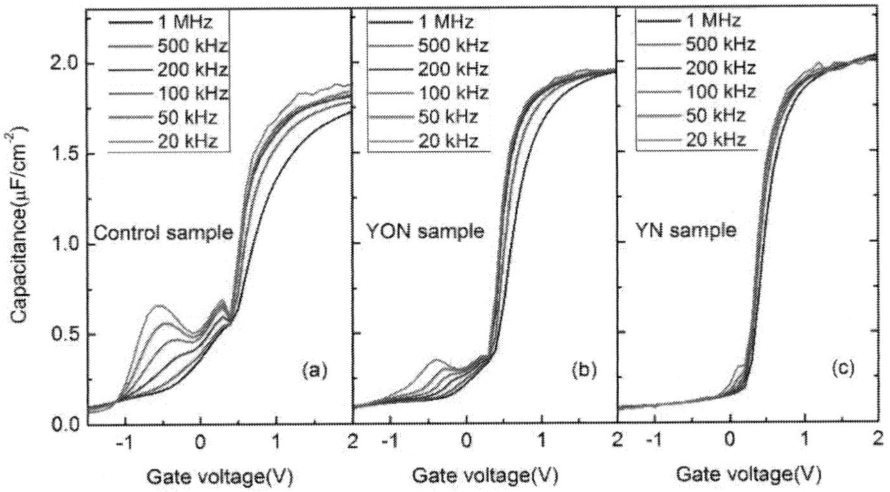


**Figure 3.** Frequency dispersion of *C-V* curves: (a) control sample, (b) YON sample and (c) YN sample

In order to analyse the reason for the different electrical and interfacial properties of the samples, XPS analysis is used to investigate the composition and chemical states of the interface. The energy scale is calibrated by fixing the Ge 3*d* peak of Ge substrate at a binding energy (BE) of 29.4 eV to eliminate the charging effect of samples [18]. Ge $2p_{3/2}$ core level spectra for the samples are shown in figure 4 (a) - (c). For the control sample, large $GeO_x$ and $HfGeO_x$ peaks in figure 4 (a) indicate the formation of Ge sub-oxide and $HfGeO_x$ in the interface respectively, both of which will degrade the interface quality and cause poor electrical properties. With YON IPL inserted for the YON and YN samples, the defective $HfGeO_x$ is replaced by stable $YGeO_xN_y$, which could significantly suppress the formation of Ge oxide and thus less $GeO_x$ is observed in figure 4(b) and (c). Compared to the YN sample, there is still an amount of $GeO_x$ in the interface for the YON sample, which is 11.33% described as the $GeO_x$/Ge-Sub peak-area ratio. This may be formed during the formation of $YGeO_xN_y$, due to the oxidation of Ge surface by the in-diffused O [4]. The YN sample exhibits much less $GeO_x$ (1.5%) than the YON sample, due to the absorption of in-diffused O by YN IPL prior to Ge substrate, which could effectively suppress the diffusion of O to the Ge surface and thus the formation of Ge oxide. The defective $GeO_x$ is known to be an obstacle to improve the interface quality of Ge substrate. Therefore, the YN sample with the least $GeO_x$ leads to the best interface quality and consequently exhibits the best electrical properties.

To further verify the conclusion above, O 1*s* core level spectra for the samples are also shown in figure 4 (d) - (f). The peak with the BE of 529.5 eV is attributed to the O-Hf bond, which corresponds to the $HfO_2$ dielectric. The peak of Ge-O-Hf, Ge-O-Y and O-Ge indicates the formation of $HfGeO_x$, $YGeO_xN_y$ and Ge oxide in the interface, respectively. The amount of Ge oxide for the control, YON and YN samples, described as the O-Ge/O 1*s* peak-area ratio, is 20.9%, 13.1% and 4.6% respectively. This indicates that the YON IPL could effectively suppress the formation of Ge oxide and the YN sample exhibits better efficiency than the YON sample, which is consistent with the result from Ge $2p_{3/2}$ XPS spectra.

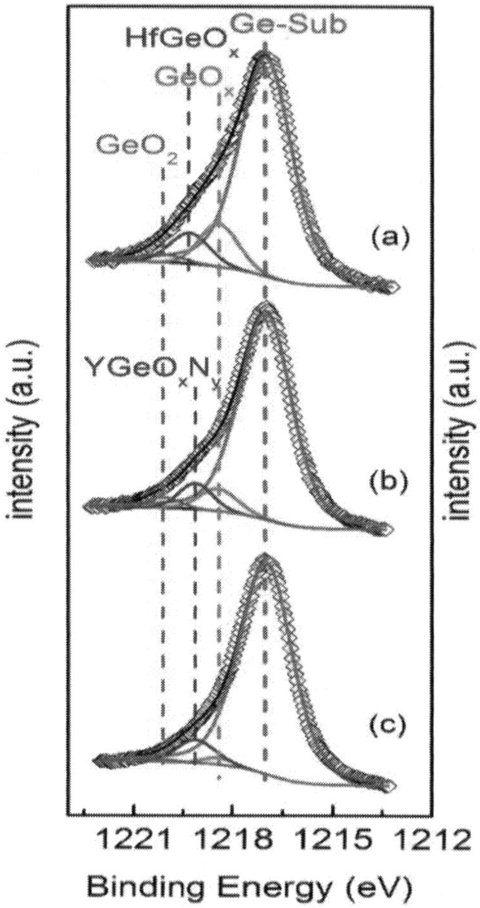


**Figure 4.** Ge $2p_{3/2}$ XPS spectrum: (a) control sample, (b) YON sample and (c) YN sample, and O 1*s* XPS spectrum: (d) control sample, (e) YON sample and (f) YN sample

The AFM is used to check the $HfO_2$ surface morphology of the samples, as shown in figure 5. The control sample exhibits rough surface with a large root mean square (rms) roughness of 0.467nm, due to the poor interface and formation of $GeO_x$, which could diffuse into the $HfO_2$ dielectric and worsen the film property. With YON IPL inserted, the improved interface quality and suppressed formation of $GeO_x$ lead to a smoother surface for both the YON and YN samples. Moreover, the rms roughness of YN sample is 0.158 nm, which is smaller than that of the YON sample (0.233nm). This is attributed to the more effective suppression of the formation of $GeO_x$ for the YN sample, as confirmed by the XPS analysis.

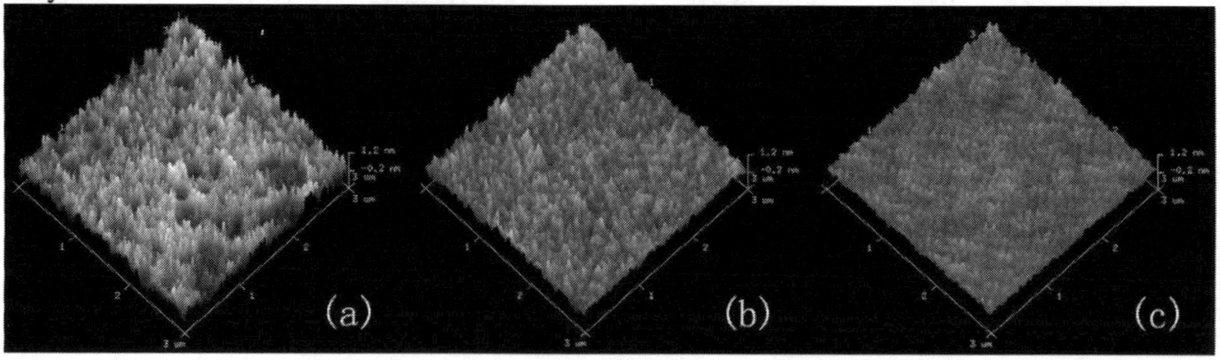


**Figure 5.** AFM image: (a) control sample, (b) YON sample and (c) YN sample

## 4 Conclusion

With YON IPL inserted between $HfO_2$ dielectric and Ge substrate, much improvement of the interface quality could been obtained. Compared to the YON IPL fabricated by sputtering $Y_2O_3$ target in $Ar+N_2$, the YON IPL, which is fabricated by sputtering Y target in $Ar+N_2$ to deposit YN first followed by annealing in $N_2+O_2$ to convert YN to YON, could suppress the formation of Ge oxide in the interface more efficiently to achieve more excellent interfacial and electrical properties.

## Acknowledgments

This work is financially supported by the National Natural Science Foundation of China (Grant No. 61274112, 61176100, 61404055).

## References

[1] Iwai H 2015 *Solid State Electron.* **112** 56-67
[2] Takagi S, Kim S H, Yokoyama M, Zhang R, Taoka N, Urabe Y, Yasuda T, Yamada H, Ichikawa O, Fukuhara N, Hata M, Takenaka M 2013 *Solid State Electron.* **88** 2-8
[3] Kamata Y 2008 *Mater. Today* **11** 30-38
[4] Jung H, Yu I, Kim H K, Lee S Y, Lee J, Choi Y, Chung Y J, Lee N, Park T J, Choi J, Hwang C S 2012 *IEEE T. Electron Dev.* **59** 2350-2356
[5] Houssa M, Pourtois G, Caymax M, Meuris M, Heyns M M 2008 *Appl. Phys. Lett.* **92** 242101
[6] Li C X, Lai P T 2009 *Appl. Phys. Lett.* **95** 22910
[7] Chu L K, Lee W C, Huang M L, Chang Y H, Tung L T, Chang C C, Lee Y J, Kwo J, Hong M 2009 *J. Cryst. Growth* **311** 2195-2198
[8] Wu M, Wu Y, Lyu R, Chao C, Wu C, Lin C, Chen L 2013 *Microelectron. Eng.* **109** 216-219
[9] Rouffignac P D, Park J S, Gordon R G 2005 *Chem. Mater.* **17** 4808-4814
[10] Rastogi A C, Sharma R N 1992 *J. Appl. Phys.* **71** 5041
[11] Liu Z, Liang L, Yu Z, He S, Ye X 2011 *Journal of Physics D Applied Physics* **44** 155403
[12] Bethge O, Zimmermann C, Lutzer B, Simsek S, Smoliner J, Stöger-Pollach M, Henkel C, Bertagnolli E 2014 *J. Appl. Phys.* **116** 214111
[13] Murad S N A, Baine P T, Mcneill D W, Mitchell S J N, Armstrong B M, Modreanu M, Hughes G, Chellappan R K 2012 *Solid State Electron.* **78** 136-140
[14] Zhang R, Iwasaki T, Taoka N, Takenaka M, Takagi S 2011 *Microelectron. Eng.* **88** 1533-1536
[15] Sun Q Q, Shi Y, Dong L, Liu H, Ding S J, Zhang D W 2008 *Appl. Phys. Lett.* **92** 102908
[16] Li X F, Liu X J, Cao Y Q, Li A D, Li H, Wu D 2013 *Appl. Surf. Sci.* **264** 783-786
[17] Zhang L L, Li H L, Guo Y Z, Tang K C, Woicik J, Robertson J, Mcintyre P C 2015 *ACS Appl. Mater. Inter.* **7** 1173044747
[18] Dimoulas A, Tsoutsou D, Panayiotatos Y, Sotiropoulos A, Mavrou G, Galata S F, Golias E 2010 *Appl. Phys. Lett.* **96** 12902

MATEC Web of Conferences **104**, 02001 (2017) DOI: 10.1051/matecconf/201710402001
*IC4M & ICDES 2017*

# Variable Displacement Control of the Concrete Pumping System Based on Dynamic Programming

Min Ye[1,a], Limin Qin[1]

[1]*National Engineering Laboratory for Highway Maintenance Equipment, Chang'an University, Shaanxi Xi'an, China*

**Abstract.** To solve the problems of cylinder piston striking cylinder and the hydraulic shocking of the main pump, and causing energy waste problem, the method of variable displacement control of piston stroke was proposed. In order to achieve effective control of the piston stroke, variable displacement control model was established under the physical constraint condition of non-collision between piston and cylinder. And the control process was realized by Dynamic Programming(DP), the simulation and test results show that piston of concrete pumping system don't strike cylinder and reduce the hydraulic shock of the main pump outlet, meanwhile improve the response speed of the cylinder and achieve energy-saving purposes under varying loads. This control model built in the integration design space of structure variable and control variable is of guiding significance for solving open-loop system's engineering problems.

## 1 Introduction

Concrete pump is widely used in high buildings, highways and other construction projects. It can not only speed up the construction schedule and reduce labor intensity, but also can improve economic efficiency significantly. Since the concrete pump hydraulic systems often work under high pressure condition, and the two main cylinders work alternately, so there will be piston striking and shaking between the changing and increase energy consumption [1-3]. Therefore, it is necessary to research the dynamic characteristics of concrete pump hydraulic system, and find ways to solve reducing the hydraulic shaking. There will be important engineering value to solve the concrete problems.

To solve the foresaid problems engineering and technical personnel have made great efforts. Reference [4] suggests that accumulator or unloading valve should be installed to change structure of valve spool, etc; but without considering variation of the load, this problem cannot be solved completely. Reference [5] proposed that the reason of striking caused by changing was the mismatch oil flow between provided by main pump and the demand of it, and proposed that the main pump and the reversing should be commended synthetically, but no specific control method was given.

This paper discussed the principle of pumping displacement control, and proposed the method of variable displacement control. Based on discussing pump open-loop characteristics, variable displacement piston stroke control model was established on varying load conditions. The position of changing signal point, variable signal point and the value of each variable displacement control period will be determined by Dynamic Programming arithmetic. Finally, concrete pump for pumping process is simulated and analyzed by a model concrete pump and achieve energy saving purposes.

[a]Corresponding author : mingye@chd.edu.cn

DOI: 10.1051/matecconf/201710402001

## 2 Concrete pump working principle and open-loop control characteristics

Concrete pumping hydraulic system is an open-loop, consists of hydraulic pump, reversing valve, hydraulic cylinder, water tanks and delivery tanks and other components, as shown in the **Figure 1**. The cylinder piston is connected with concrete delivery cylinder piston by a rigid axle and the reversing valve is electro-hydraulic valve [6]. Sensor sends a changing signal to reversing valve when No.2 cylinder piston move into the changing control signal point *a*. After some delay time, reversing valve will be triggered to change. No.1 cylinder piston rod will become retracted from extending, and No.2 cylinder rod is retracted into the extended and vice versa. Therefore cylinder to achieve changing, the corresponding delivery cylinder complete feed and suction state switch.

The pump is constant power variable pump, the $p-Q$ characteristic curve shown as **Figure 2**[7].

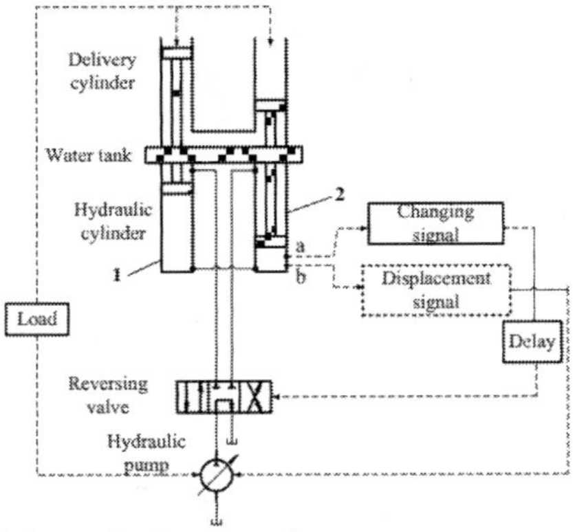


**Figure 1.** Schematic diagram of concrete pump

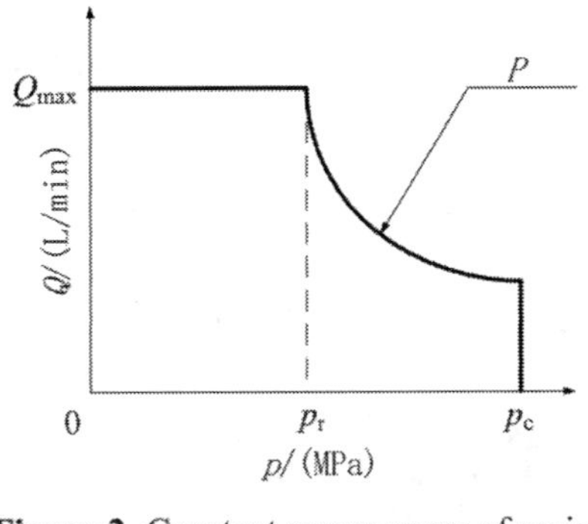


**Figure 2.** Constant power curve of main pump

As Pascal Theorem known that hydraulic pump outlet pressure is determined by the pumping load. Therefore, we can get the function of relationship between its cylinder piston velocity and the pumping load as follows:

$$v=\begin{cases}\dfrac{n\cdot V_{g\max}}{60A_s} & (0\le p_p<p_r)\\ \dfrac{1000P}{p_p\cdot A_s} & (p_r\le p_p<p_c)\\ 0 & (p_p\ge p_c)\end{cases} \tag{1}$$

Wherein, $v$ is velocity (m/s) of the piston; $n$ is rated speed(r/min) of the pump; $V_{g\max}$ is the maximum displacement of the pump(mL/r); $A_s$ is piston rod chamber area(mm$^2$) of hydraulic cylinder; $P$ is rated power(kW) of the pump; $p_p$ is outlet pressure(MPa) of the pump; $p_r$ is the pressure(MPa) of the pump when it start control constant; $p_c$ is deadline pressure (MPa) of the pump.

By Eq.1, we know hydraulic cylinder piston stroke depends on pumping load ultimately. At present, the commonly used formula to calculate the pumping load as:

$$p_L=\left\{\frac{2}{r}[(3-0.1S)\times10^2+(4-0.1S)\times10^2\cdot(1+\frac{t_2}{t_1})\bar{v}]+\rho g\sin\varphi\right\}x \tag{2}$$

## 3 The variable displacement control model of concrete pump piston stroke

### 3.1 Principle of Variable Displacement Control.

The basic idea of variable displacement control method of the cylinder piston stroke is to control the main hydraulic cylinder piston stroke by changing the pump displacement. First, the variable displacement signal point was added to the cylinder. Then pump variable displacement control model was built under some constraints. Finally, the position of variable displacement signal point and the each time value of variable displacement control stage were determined by DP arithmetic. The structure parameters of the cylinder are shown in **Figure 3**. Point *a* is the location of the reversing valve sensor, and the location is fixed; *b* is the variable displacement signal point, and different loads with different *b*; *c* is the point at which piston velocity is zero, and different loads with different *c*; *d* is the bottom of the hydraulic cylinder; *X* is the location where piston is after the changing signal point; *Y* is the distance between displacement signal point and point at which piston velocity is zero.

When the cylinder piston motion to point *a*, the sensor sends a signal to the reversing valve. After a certain delay the valve completed reversing. The cylinder piston is still at greater speed during this time, and it will hit bottom before completed changing causing the pump outlet hydraulic shock. Thus, not only the reversing valve should be controlled, but also the cylinder piston speed in the process should be controlled. So point *b* at which displacement of the variable pump was added to control the speed of cylinder piston. When the cylinder piston move to point *a* and send out signal to changing, the speed remains unchanged. When the cylinder move to point *b*, the displacement of the variable pump is start to be adjusted. In order to avoid hitting the cylinder under all load condition, there is a certain distance between the piston reaching maximum stroke and the cylinder endpoint bottom, set *e*.

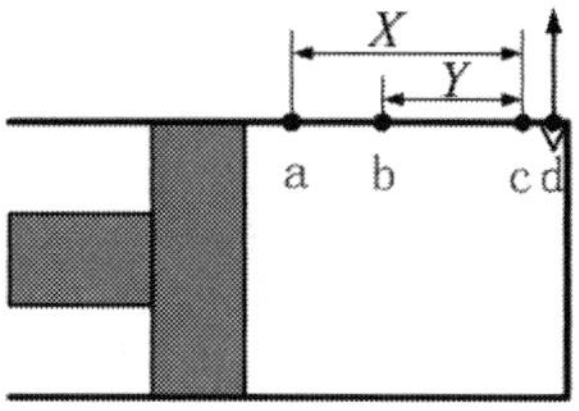


**Figure 3.** Position of signals on main cylinder

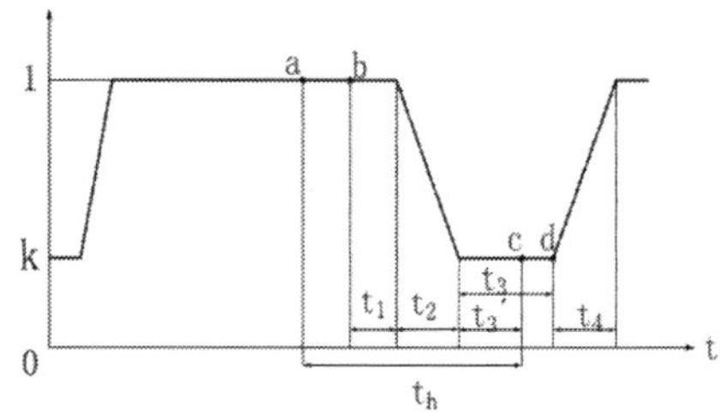


**Figure 4.** Parameters of changing displacement control

## 3.2 Model of Variable Displacement Control.

As shown in **Figure 3**, it is parameters of variable displacement control model. The time is set 0 when piston is at point *b*. According to **Figure 3** and **Figure 4**, we could get the distance of piston's moving $s$ from starting variable displacement signal to completing piston's moving. The distance $s$ is:

$$s=\int_0^{t_1} v dt+\int_{t_1}^{t_1+t_2}[1-\frac{(1-k)(t-t_1)}{t_2}]v dt+\int_{t_1+t_2}^{t_1+t_2+t_3'} kv dt=v(t_1+\frac{1+k}{2}t_2+kt_3') \tag{3}$$

Where: $t_1$ is the time between sending variable displacement signal and pump displacement being adjusted; $t_2$ is the time between the beginning of the working of displacement control signal and the stable displacement period; $t_3'$ is the time between the stable displacement and beginning of reverse direction; $k$ is the parameters of displacement control, and in engineering, $k=0.25$.

As shown in **Figure4**, the time $t_h$ between sending changing signal and completing changing is [8]:

$$\frac{X-Y}{v}+t_1+t_2+t_3'=t_h \tag{4}$$

By Eq.3 and Eq.4 we reach:

$$s=v\cdot[t_1+\frac{1+k}{2}t_2+k\cdot(t_h-\frac{X-Y}{v}-t_1-t_2)] \tag{5}$$

Putting $v_{max}$, $v_{min}$ into Eq.5 we reach:

$$s_{max}-s_{min}=(v_{max}-v_{min})\cdot[t_1+\frac{1}{2}(1+k)t_2+k\cdot(t_h-t_1-t_2)] \tag{6}$$

According to the pumping requirement, the piston stroke must be controlled within a certain range under different load conditions. So the difference value between the maximum of the piston stroke and minimum of the piston stroke must be an upper limit value, setting $a$. Then:

$$s_{\max} - s_{\min} = (v_{\max} - v_{\min}) \cdot [t_1 + \frac{1}{2}(1+k)t_2 + k \cdot (t_h - t_1 - t_2)] \le a \tag{7}$$

In order to reduce the hydraulic pump outlet hydraulic shock changing, the hydraulic cylinder changing must be finished before the hydraulic pump displacement control completed. Namely the hydraulic cylinder changing must be completed in within the time $t_h$ period, so $t_h$ must satisfy the following relationship:

$$\frac{X-Y}{v_{\min}} + t_1 + t_2 < t_h < \frac{X-Y}{v_{\max}} + t_1 + t_2 + t_3 \tag{8}$$

Where: $t_3$ is continuous working time after the value of the pump displacement is the minimum.

In order to avoid hitting the cylinder under all load condition, then $Y \ge s_{\max}$. And in order to make effective stroke of the piston to be maximum, then:

$$Y = s_{\max} \tag{9}$$

Combine with Eq.3, Eq.8, Eq.9, we reach:

$$\frac{X - v_{\max}(t_1 + \frac{1+k}{2}t_2 + kt_3')}{v_{\min}} + t_1 + t_2 < t_h < \frac{X - v_{\max}(t_1 + \frac{1+k}{2}t_2 + kt_3')}{v_{\max}} + t_1 + t_2 + t_3 \tag{10}$$

When pumping load is the maximum, speed of the piston is the minimum and time of the all piston stroke completed is the maximum, we get:

$$f(t) = \frac{2 - v_{\max}(t_1 + \frac{1+k}{2}t_2 + k \cdot t_3') - b}{v_{\min}} + t_1 + t_2 + t_3' \tag{11}$$

In order to improve the pumping efficiency, the time should be the minimum, and the objective function is $\min f(t)$.

### 3.3 Dynamic Programming (DP).

According to the principle of variable displacement pump control, reverse form of Dynamic Programming arithmetic was used to control the adjustment process [9 -11], we get:

$$f_k(x_k) = \underset{u_k \in D_k(x_k)}{opt} \{v_k(x_k, u_k) + f_{k+1}(x_{k+1})\}, k = n, n-1, \ldots, 2, 1 \tag{12}$$

Wherein the boundary condition is:

$$f_{n+1}(x_{n+1}) = 0 \quad or \quad f_n(x_n) = v_n(x_n, u_n) \tag{13}$$

Parameter $x_k$ is the state of $k$ th stage; decision variable $u_k$ is the decision-making in the state $x_k$; $x_{k+1} = T_k(x_k, u_k)$ is state transition equation; $D_k(x_k)$ is the set of decision allowable in $k$ th stage; $v_k(x_k, u_k)$ is index function. In the solution process, optimal decision and optimal value of the process were found at each stage starting from the boundary condition $k = n$. Until $f_1(x_1)$ was determined finally, that is the optimal solution of the problem. In the present paper, the position of point $b$ was determined by Dynamic Programming arithmetic meeting the conditions of the piston don't strike cylinder. Notably $b$ is the variable displacement signal point, and different loads with different $b$.

## 4 Analysis of the simulation and experimental results

A concrete pump simulation model was built using MATLAB toolbox Simulink, and the model was shown as **Figure 5**.

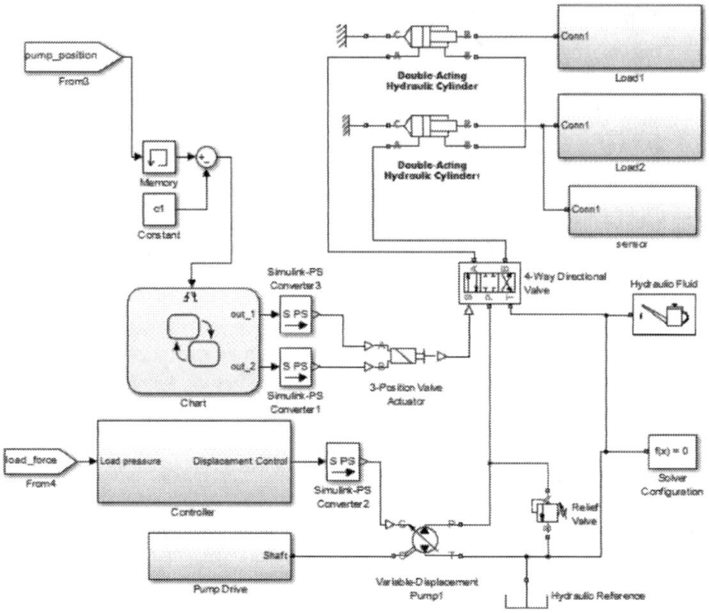


**Figure 5.** Simulation model of concrete pump pumping

During pumping, the engine and the hydraulic pump are working under power match control. Wherein the main simulation parameters were shown as **Table 1**:

**Table 1.** Main parameters of simulation model

| Parameters | Symbol | Value | Parameters | Symbol | Value |
|---|---|---|---|---|---|
| Revolving speed of main pump /(r/min) | n | 2300 | Motion Viscosity of oil/($m^2.s^{-1}$) | $v$ | $3\times10^{-5}$ |
| Outer diameter of the | $D_t$ | 0.065 | Oil density /($kg/m^{-3}$) | $\rho$ | 900 |
| Inner diameter of the tube /(m) | $d_t$ | 0.049 | Piston diameter of main cylinder /(m) | D | 0.14 |
| Tube length /(m) | $L_t$ | 1.5 | Piston rod diameter of main cylinder /(m) | d | 0.095 |
| Young's modulus of tube /(Pa) | $E_t$ | $2\times10^{11}$ | Inertia of piston and piston rod /(kg) | I | 246 |
| Poisson's ratio of tube | $\lambda_t$ | 0.3 | coefficient /(N.m1.s-1) | $R_f$ | $3.71\times10^4$ |
| Oil bulk modulus /(Pa) | E | $1.6\times10^9$ | Piston stroke of main cylinder /(m) | L | 2 |

A certain type of concrete pump was tested based on the variable displacement control theory, and the main parameters of concrete pump were shown as **Table 2**.

**Table 2.** Main parameters of concrete pump

| Parameters/unit | Symbol | Value |
|---|---|---|
| Maximum displacement of the main pump/($m^3.s^{-1}$) | $V_{g\max}$ | $3.167\times10^{-6}$ |
| Rated revolving speed of the main pump/($r.s^{-1}$) | $n$ | 240.855 |
| Rated power of the main pump/W | $P$ | $1.15\times10^5$ |
| Control pressure of main pump working in constant power/W | $p_r$ | $1.58\times10^7$ |
| Outlet pressure of the main pump/(Pa) | $p_p$ | $(6\sim30)\times10^6$ |
| Rod chamber piston area of the main cylinder/ $m^2$ | $A_s$ | $8.31\times10^{-3}$ |
| Piston stroke of the main cylinder/ m | $s_p$ | 2 |
| Changing time of hydraulic system/ s | $t_h$ | 0.2 |

The model of the control system was simulated with DP arithmetic and experimental data also imported in, then we reached the following results.

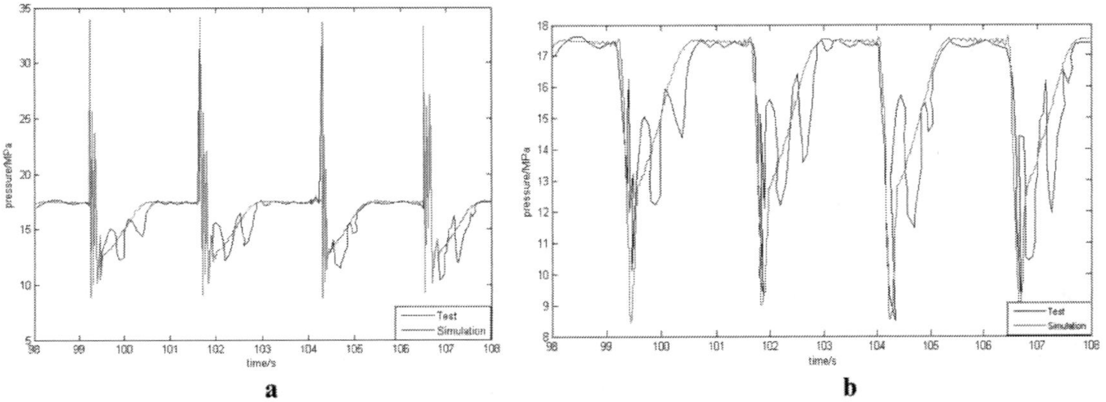


**Figure 6.** Pressure of main pump with changing displacement control

The simulation and the experiment curve of the outlet pressure of hydraulic pump without control are shown individually in **Figure 6 a**. The pump outlet pressure hold steady after slowly increasing to 17.5MPa, and there has hydraulic shocking when the cylinder changing, peak pressure up to 33MPa about two times the normal working pressure. The simulation and the experiment curve of the outlet pressure of hydraulic pump with variable displacement control are shown individually in **Figure 6 b**. The outlet pressure of pump is gradually increased and stabilized, and there is no peak value of the outlet pressure after changing. Also there is no crash in test site. It can avoid striking the cylinder and reduce the outlet shocking of the pump by controlling variable displacement of piston stroke. Experiment shows that the fluctuation rang of engine speed are lower under variable displacement control than without control, and the fuel consumption of the engine is much lower.

## 5 Conclusion

To solve the problems of cylinder piston striking and the hydraulic shocking of the pump when the cylinder changing, the method of variable displacement control of piston stroke was proposed. The control model of concrete pump piston stroke was built. The position of variable displacement signal point and the value of each time period were decided by DP arithmetic under different load. And the control process was realized by DP, the simulation and test results show that piston of concrete pumping system don't strike cylinder and reduce shock of the pump outlet, meanwhile improve the response speed of the cylinder and achieve energy-saving purposes under varying loads. Since the variable displacement piston stroke control method is based on the general concrete pump, it has some versatility. There has some reference value for similar problem of automotive pumps or trailer pump.

## References

1. L. F. Jie, L. Liu, Y. Li, Zh. Sh. Cheng. Hydraulics Pneumatics & Seals **10**, 56, (2010).
2. Zh. X. Li, Sh. X. Wang, H. Jiang. Machinery Design & Manufacture **6**, 76, (2011).
3. M. Ye, X. G. Yi, Sh. J. Jiao, Energy Procedia **88**,574, (2016).
4. B. X. Cao, G. A. Chen, T. J. Fan. *Proceeding of the sixth international conference on fluid power transmission and control* (International Academic Publisher, Beijing, 2005).
5. G. A. Chen, T. J. Fan, B. X. Cao, Construction Machinery and Equipment **39**, 10, (2008).
6. W. R. Wu, Zh. W. Mao, X. Y. Huang, Computer Simulation **31**, 222,(2014).
7. R. K. Guo, J. Li, G. X. Chen. Computer Simulation **32**, 239-244, (2015).
8. A. L. Wang, X. H. Yu, B. Shi. Journal of Shanghai Jiaotong University **43**, 114, (2009).
9. Miguel Areias, Ricardo Rocha, The Journal of System & Software **8**, 1, (2016).
10. W. F. Lan, Z. Y. Wu, Q. Li, X. Li. Qiang, Journal of South-Central University for Nationalities **35**, 135, (2016).
11. J. Chen, Computer and Information Technology **21**, 23, (2013).

# Energy dissipation mechanism of a non-obstructive particle damper based on DEM simulation

Fan Su[1], Hang Zhang[1] and Zhongjun Yin [1, a]

[1]*School of mechanical engineering, University of science and technology Beijing, Beijing 100083, China*

**Abstract.** The Non-obstructive Particle Damping (NOPD) has been widely used as an effective passive damping technique. However, the understanding of energy dissipation mechanism and meso-scale behaviors of granular system in NOPD is not in-depth enough. In this paper, the damping effectiveness of NOPD was analyzed combining with the motion modes of granular system and the energy dissipation mechanism was illustrated at the meso-scale. The results shows that the different motion states and meso-structure leading to the variations of the damping effectiveness. Generally, the NOPD has the optimal damping effectiveness in the buoyancy convection state. The purpose of this paper is to provide better understanding and theory support for the design and application of NOPD.

## 1 Introduction

The Non-obstructive Particle Damping (NOPD) has been widely used as an effective passive damping technique in aerospace [1, 2], mechanical engineering [3-5] and construction[6-8] fields for its simplicity and high efficiency. The mechanism of NOPDs for reducing the vibrations is based on the dissipative nature of the combined effects of inelastic collisions and frictional losses when the particles in the cavity collide and rub with themselves and the cavity walls [9, 10].

Recent studies on NOPD mainly focused on the vibration damping effectiveness and the contributions to the modal damping of structures based on experimental investigation and numerical simulation [5, 9, 11]. However, the understanding of energy dissipation mechanism and meso-scale behaviors of granular system in NOPD is not in-depth enough. In fact, granular matter has various and special rheology behaviors, just as the solid state, liquid state and gas state [12]. Only a few researches concerned the characteristic of energy dissipation. These impede the further development and application of NOPD.

Discrete element method (DEM) has gradually become an important tool in the research of granular system with the continuous improvement of DEM and development of computer technology [13, 14], which can track the stresses and motions of individual particles in time, providing sufficient information for analyzing the principles of energy dissipation in the granular system of NOPD.

In this paper, DEM was applied in the simulation of NOPD. The damping effectiveness of NOPD was analyzed combining with the motion modes of granular system to obtain the influence of different vibration conditions on damping property and demonstrate the energy dissipation mechanism of

[a] Corresponding author : yinzhongjun@ustb.edu.cn

granular system at the meso-scale. The purpose of this paper is to provide better understanding and theory support for the design and application of NOPD.

## 2 Simulation model and parameters

In this work, a 3D scale granular system of NOPD model (Fig. 1) was built. It consisted of a closed Perspex cylindrical container (D×H=26×78 mm$^2$) filled with 500 (d=3.5mm) steel spheres. The material properties used in the simulation are based on the references [12], which is listed in Table 1.

The container was subjected to vertical sinusoidal excitation. The two control parameters were excitation frequency $f$(15-90 Hz) and dimensionless excitation acceleration $\Gamma$ (1-30g).

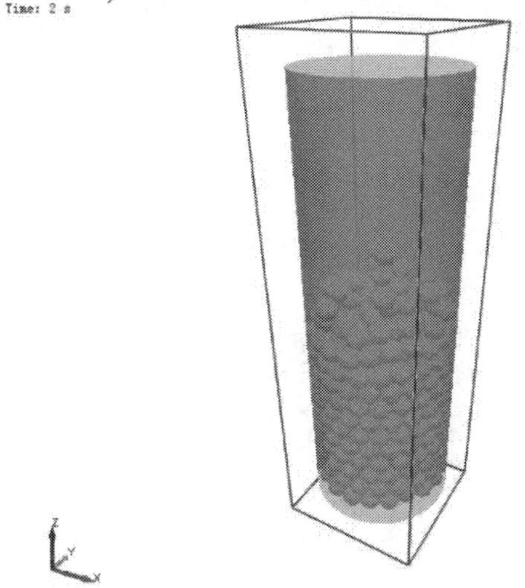

**Figure 1.** DEM model of the NOPD.

**Table 1.** Physical properties and parameters used in the DEM simulation.

| Parameter | Value | Parameter | Value |
|---|---|---|---|
| Density of particle, $\rho_p$ (kg/m$^3$) | 7800 | Density of container, $\rho_c$ (kg/m$^3$) | 1190 |
| Elasticity modulus of particle, $E_p$ (Pa) | 206e+09 | Elasticity modulus of container, $E_c$ (Pa) | 3.3e+09 |
| Poisson ratio of particle, $\nu_p$ | 0.3 | Poisson ratio of container, $\nu_c$ | 0.37 |
| Diameter of particle, $d$ (m) | 0.0035 | Coefficient of restitution, $e$ | 0.92 |
| Total quantity of particles | 500 | Coefficient of static friction, $\mu_s$ | 0.3 |
| Time step, $\Delta t(s)$ | 2.07e-07 | Coefficient of rolling friction, $\mu_r$ | 0.01 |
| Total time of each simulation, $t(s)$ | 2 | | |

## 3 Results and Discussion

### 3.1 Phase diagrams

When containers filled with granular material are subjected to vertical vibrational motion, a series of rheology behaviors were observed. Based on the analysis of the simulation results, the motion modes of granular system were divided into seven different ones including solid-like state, local fluidization, global fluidization, convection [15], Leidenfrost effect [16], buoyancy convection and bouncing bed, which are shown in Fig. 2.

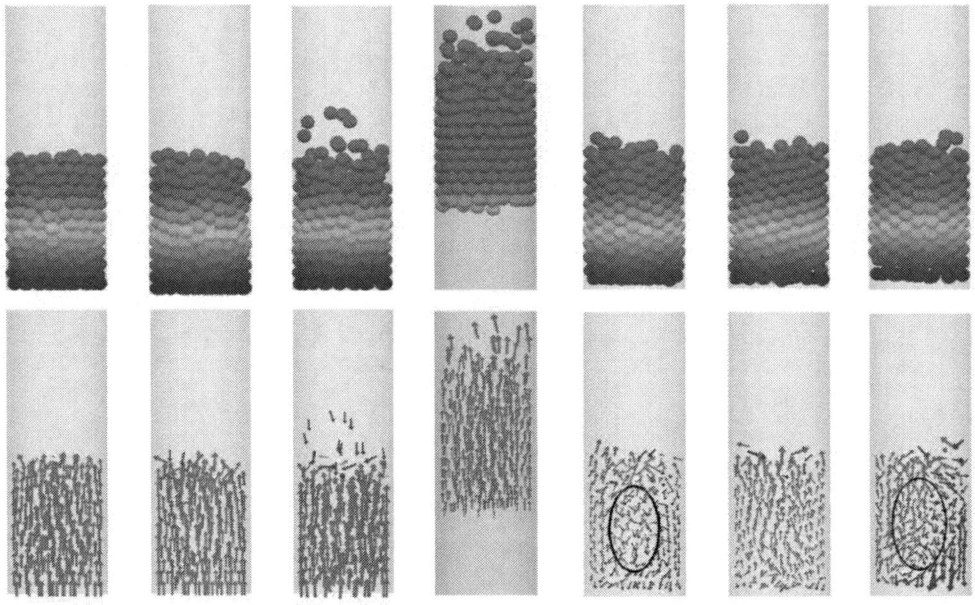


**Figure 2.** Snapshots of particle motion modes and the corresponding velocity vector: a) solid-like state, b) local fluidization, c) global fluidization, d) bouncing bed, e) convection, f) Leidenfrost effect, g) buoyancy convection

When the excitation acceleration is below 1.2g, damping particles are in solid-like state (Fig.2 a)), there is almost no relative motion between particles and between particles and the wall, and the whole granular system looks like an added static mass.

When the excitation acceleration reached 1.2g, the phase transition of granular materials from solid to liquid happened. Particles which are closer from the top surface of the granular bed become fluidized while particles in the lower granular layer are still relatively motionless. Fig. 2 b) shows the slight relative motions between particles in the top granular layers.

With the increasing of amplitude or frequency, the fluidized granular layers were expanding to the whole granular system, and the relative motions between particles occurred in all granular layers. The particles are loose and reached global fluidization state (Fig. 2 c)).

Under the condition of large amplitudes, particles in the granular system behave as inelastic bouncing balls impacting chaotically and show series of phenomena of subharmonic bifurcation [17]. It should be noted that the bouncing bed state in this paper is defined as the condition for which the top granular layer completely contacts with the top of container (Fig. 2 d)).

When convection occurs, there is not only relative motion between particles but also the exchange of the particle position between granular layers. As can be seen from Fig. 2 e), the velocity vector distribution of granular system when $f= 65$ Hz, there was a flowing convection rolls in granular layers.

Above a critical shaking strength, a cluster of particles is elevated and supported by a few fast particles underneath the cluster. This cluster exhibits an almost perfect hexagonal close packing, causing the density to be maximal at the floating cluster and distinctly lower above and below this cluster (Fig. 2 f)). The thickness of the dilute layer oscillates in time (never vanishing) due to the motion of the bottom, while the cluster floats steadily in the vicinities of their original positions [16]. This state appears on condition that both the frequency and vibration intensity are high.

The buoyancy convection (Fig. 2 g)) can be seen as a superpositional state of granular Leidenfrost effect and convection, i.e., convection occurs in suspended granular layers which are in Leidenfrost effect state [12]. Just like the convection, there is also a flowing convection rolls in suspended granular layers.

To directly observe the occurrence conditions of different motion modes in granular system, the phase diagram of damping particles in the NOPD under different excitation conditions was obtained, as depicted in Fig. 3.

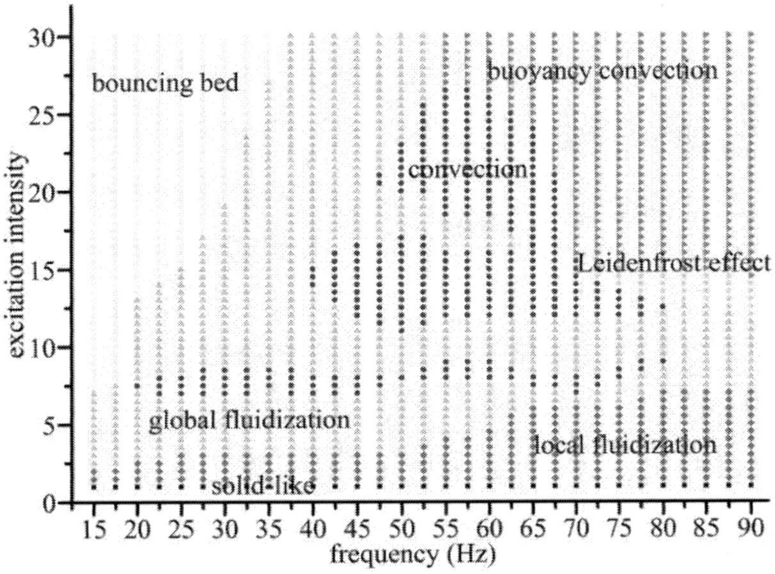


**Figure 3.** Phase diagrams of damping particles in NOPD

The transitions of different motion modes can be seen from the phase diagram obviously. It is notable that global fluidization state is the basis for transitions to other states and convection is a relatively unsteady state compared with other states, and the phase boundary of global fluidization state and convection is not as clear as others.

### 3.2 Loss factor

In this work, the damping efficiency is measured by the damping loss factor [18], which is given as

$$\eta = \frac{E_{dissipated}}{E_{\max imum}} \tag{1}$$

where $E_{dissipated}$ is the dissipated energy in a cycle per radian, and $E_{\max imum}$ is the maximum energy stored in a cycle, which is approximately equal to the maximum kinetic energy in a cycle.

Four cases in Fig. 3 covering different processes of transition were selected, and the energy dissipation and the maximum kinetic energy along the excitation intensity for each case are plotted in Fig. 4 a) and b). As it can be seen, generally both the energy dissipation and kinetic energy increase with the increase of excitation intensity, and higher frequencies have a lower increase.

Fig. 5 shows the loss factor versus excitation intensity under different frequencies. Comparing the different curves, it can be observed that the particles under different frequencies offer best damping in different excitation condition. When the frequency is low, the loss factor presents the maximum value under the condition of low excitation intensity, the loss factor peaks around 3g and then decreases shapely with the increase of the excitation intensity. After 20g, the loss factor gradually stabilizes at a constant value. When the frequency is high, the loss factor reaches another peak (30g) and presents the maximum value under the condition of high excitation intensity.

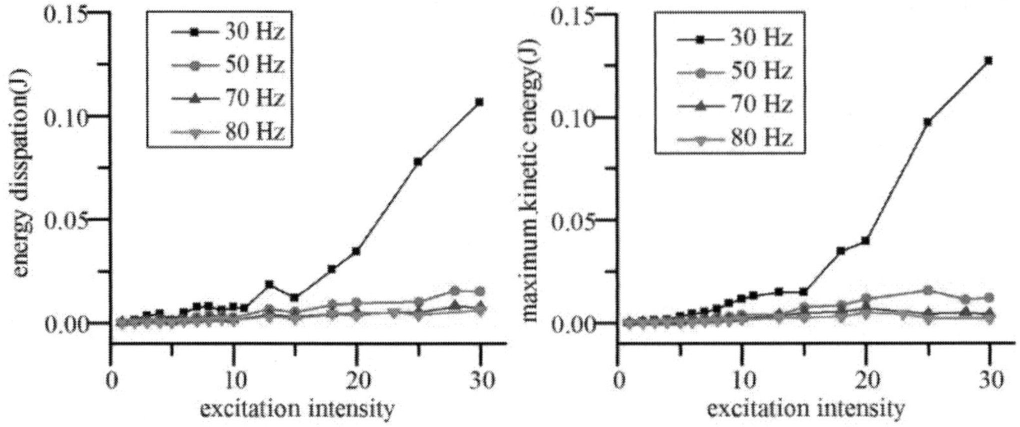


**Figure 4.** a) The energy dissipation in a cycle, b) the maximum kinetic energy in a cycle

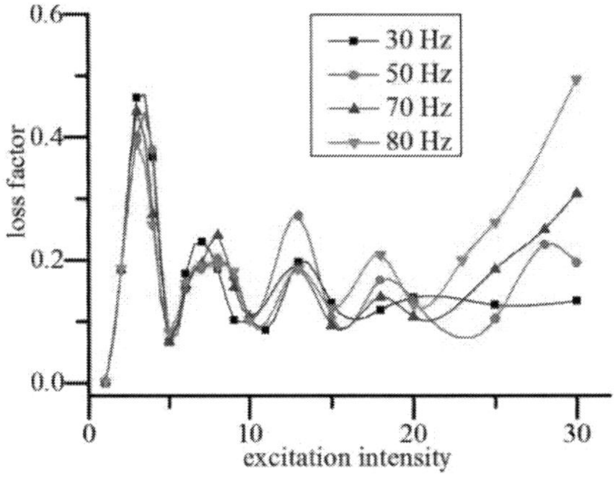


**Figure 5.** Loss factor versus excitation intensity under different frequency

Combined with the phase diagram, it can be concluded that the variations of loss factor are due to the motion modes transitions of the granular system.

When the solid-like state transits to local fluidization state, the higher the amplitude or frequency, the more violent the particles' motion, especially the particles which are closer from the top surface of the granular system, and the more active motion of particles in the local fluidization state results in more energy dissipation because the collisions between particles and the walls become more frequent. However, the maximum kinetic energy is still low in comparison with the energy dissipation, thus a sharp increase on the value of loss factor can be seen at about 3g.

The fluidization state will transit to bouncing bed state with the increase of the excitation intensity under low frequencies, where the increased opportunity of collisions and the severity of impacts dissipate a lot of energy. However, the loss factor would not increase infinitely along with the increase of excitation intensity. Most of the collisions happen between the particles and the walls of the container, and the impacts and motions of particles are too chaotic to reduce vibration because of the subharmonic bifurcation [17]. Therefore, the value of loss factor at $f$= 30 Hz is small and staying near 0.12 at large excitation intensity.

Compared to other states, convection is an unstable state, the phase boundary between it and the global fluidization state is not as obvious as other boundaries, and the convection shows period-like properties with excitation intensity and frequency, which occurs in a higher frequency with the increases of excitation intensity. The value of loss factor of convection has less difference with the global fluidization state, but the periodic switch of directivity of convective motion in granular system causes more collision chances, which means that in convection state, the loss factor of NOPD is slightly larger than the value in the global fluidization state.

The Leidenfrost effect and the buoyancy convection state appear under the condition of high frequency and large vibration intensity, and a cluster of particles is elevated and supported by a few fast particles underneath the cluster. There will be an extra energy dissipation to stabilize the suspension of the granular layers except the collisions and frictions between particle–particle and particle–wall. Therefore, these two states have relatively large loss factor. But the meso-structure of suspended layer in the Leidenfrost effect and the buoyancy convection state is different. The suspended layer is in the solid-like state for the Leidenfrost effect that the meso-structure is an almost perfect hexagonal close packing, while is in the convection state for the buoyancy convection state. In analogy to the difference between the solid-like state and convection state, the energy disspation of collisions and frictions in the particle suspension layer in buoyancy convection is more than in the Leidenfrost effect state. Therefore, when NOPD plays the optimal damping effectiveness, particles in NOPD are in the buoyancy convection state.

## 4 Conclusion

In this work, the loss factor of particle damping was analyzed combing with the motion modes of granular system based on DEM, and the following conclusions are obtained:

1. The NOPD offers best damping in different vibration parameters. When frequency is low, the value of loss factor peaks at low excitation intensity. When frequency is high, the loss factor presents the maximum value at high excitation intensity.
2. The different motion states and meso-structure leading to the variations of the loss factor of granular system. Generally, the NOPD has the optimal damping effectiveness in the buoyancy convection state.

The purpose of this work is to provide a better understanding of the particle damping effectiveness and theory support for the design of the particle damper.

## References

1. P. Lieber, D. Jensen, Trans. ASME, **67** (1945)
2. J.J. Moore, A. Palazzolo, R. Gadangi, T. Nale, S. Klusman, G. Brown, A. Kascak, J. Vib. Acoust., **117** (1995)
3. A. Paget, Engineering, **143** (1937)
4. E. Skipor, L. Bain, J. Mech. Des, **102** (1980)
5. W. Xiao, J. Li, S. Wang, X. Fang, J. Sound Vib., **366** (2016)
6. K. Ogawa, T. Ide, T. Saitou, J. Wind Eng. Ind. Aerodyn., **72** (1997)
7. J. Yao, J. Struct. Div., **98** (1972)
8. J.M. Kelly, R. Skinner, A. Heine, Bull. New Zealand Soc. Earthq. Eng., **5** (1972)
9. J.M. Bajkowski, B. Dyniewicz, C.I. Bajer, J. Sound Vib., **341** (2015)
10. H. Panossian, J. Vib. Acoust., **114** (1992)
11. B. Yao, Q. Chen, H. Xiang, X. Gao, Int. J. Mech. Sci., **80** (2014)
12. K. Zhang, T. Chen, X. Wang, J. Fang, J. Sound Vib., **364** (2016)
13. K. Mao, M.Y. Wang, Z. Xu, T. Chen, Powder Technol., **142** (2004)
14. M. Saeki, J. Sound Vib., **281** (2005)
15. E. Ehrichs, H. Jaeger, G.S. Karczmar, J.B. Knight, Science, **267** (1995)
16. P. Eshuis, K. van der Weele, D. van der Meer, D. Lohse, Phys. Rev. Lett., **95** (2005)
17. N. Mujica, F. Melo, Phys. Rev. E, **63** (2000)
18. M.B. Romdhane, N. Bouhaddi, M. Trigui, E. Foltête, M. Haddar, MSSP, **38** (2013)

MATEC Web of Conferences **104**, 02003 (2017) DOI: 10.1051/matecconf/201710402003
*IC4M & ICDES 2017*

# The Transfer efficiency analysis and modeling technology of new non - contact power transmission equipment

Shi Cao[1], NianSong Zhang[1,a] and AiMin Wang[2]

[1]*School of Mechanical Engineering, Nanjing University of Science and Technology, 210094 Nan Jing, China*
[2]*School of Mechanical Engineering, Beijing Institute of Technology, 100081 Bei Jing, China*

**Abstract.** Due to the shortcomings of current power transmission which is used in ultrasound - assisted machining and the different transfer efficiency caused by the related parameters of the electromagnetic converter, this paper proposes an analysis model of the new non-contact power transmission device with more stable output and higher transmission efficiency. Then By utilizing Maxwell finite element analysis software, this paper studies the law of the transfer efficiency of the new non-contact transformer and compares new type with traditional type with the method of setting the boundary conditions of non-contact power supply device. At last, combining with the practical application, the relevant requirements which have a certain reference value in the application are put forward in the actual processing.

## 1 Introduction

Traditional energy transmission uses carbon brush and other physical contact approaches, these exist some shortcomings such as low speed, heating serious, wearing too fast , easy to produce EDM and so on[1-3]. In response to these problems, many scholars have proposed non-contact power supply technology. At present, non-contact power transmission devices are divided into two categories: sliding type and rotary type[4]. In the actual processing applications, rotating type occupies a dominant position. The non-contact power transmission devices of traditional rotary type up and down are 360°core, for the new rotary non-contact power transmission device, its first half is the 120° core, the second half is the 360°core. Some scholars have improved non-contact power transmission efficiency through the finite element method, however, these studies are only based on the traditional non-contact power transmission device, the new non-contact power transmission device is rarely involved[5-6].

This paper presents an analysis model of the new non-contact power transmission device, through researching the influence of three parameters which are core gap, core offset and core thickness having effect on non-contact power transmission efficiency, the curve whose relationship is between the relevant parameters and transfer efficiency is drawing in the text so that it can give the specific transfer efficiency. The three machining parameters are validated by finite element analysis, which provides the theoretical basis for the subsequent parameter optimization.

[a]Corresponding author :zns@ustc.edu.cn

## 2 Maxwell simulation modeling

Maxwell is a powerful electromagnetic simulation tool, which is mainly used to analyze two-dimensional and three-dimensional electromagnetic components, such as motors, transformers, actuators and other electrical and mechanical and electrical equipment, where applies in the automotive, military, aerospace and industrial.

Magnetic materials plays an important part in non-contact power transmission devices, today's magnetic components have a very wide operating range where can operate over the audio range up to the megahertz range. The magnetic permeability of a magnetic material is the standard to measure how easy it is to be magnetized. The value can be expressed as:

$$\mu = \frac{B}{H} \tag{1}$$

Among them, $B$ represents the magnetic flux density and $H$ is the magnetic field strength.
Nowadays, the general choice is the ferrite core material. The ferrite is a homogeneous ceramic material consisting of iron oxide and other oxides. A soft ferrite can be divided into two main categories: manganese-zinc and nickel-zinc. In each category, changing the chemical composition or manufacturing method can create more different categories. Among them, the power ferrite is a kind of the manganese-zinc ferrite, which mainly uses in high-power transmission, where operates from audio to hundreds of megahertz. So that in this paper, the power ferrite is selected in the core material. For the sake of calculating simply, a method that the window area of the core is divided by an equal current density according to the power handling capability of the coil is used in this paper. The processing power of the primary coil is $P_{in}$, and the power of the secondary coil to the load is $P_0$.

The total apparent power is calculated as:

$$P_t = P_{tp} + P \tag{2}$$

In the above formula, $P_{in} = \frac{P_0}{\eta}$, $P_{tp} = P_{in} P_a$, In addition, through calculation formula of the electric state and the core geometry constant:

$$K_\varepsilon = 0.145 K_f^2 f^2 B_{AC}^2 \times 10^{-4} \tag{3}$$

$$K_g = \frac{P_t}{2K_\varepsilon \alpha} \tag{4}$$

Selecting the core which has a similar geometric constants $K_g$ to obtain the number of winding turns $N_p$, current density $J$, input current $I_{in}$ and coil resistance $R_p$ respectively:

$$N_p = \frac{V_p \times 10^4}{A_C B_{AC} f K_f} \tag{5}$$

$$J = \frac{P_t \times 10^4}{K_f K_u B_{AC} f A_p} \tag{6}$$

$$I_{in} = \frac{P_0}{V_{in} \eta} \tag{7}$$

$$R_p = MLT(N_p) \frac{\mu\Omega}{cm} \times 10^{-6} \tag{8}$$

The finite element model of the new non-contact power transmission device is built on the Maxwell software platform. The transfer efficiency of the non-contact power transmission device is mainly analyzed from the variation of the core gap, the thickness of the magnetic core and the thickness of the magnetic core. The transient field solver module was first selected and then the 3D model of the non-contact power transmission device was built according to Figures 1 and the dimensions listed in Table 1. Afterwards, setting the boundary conditions, 20KHz ultrasonic frequency AC signal is applied to the primary coil. The number of coil turns are 100.Establishing a three-dimensional model shown in Figure 2.

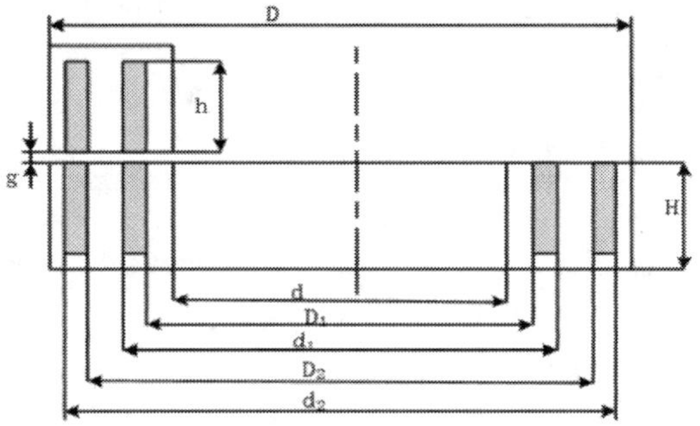


**Figure 1.** Sectional view of new non-contact power transmission device

**Table 1.** Dimensions of non - contact power transmission devices

| Parameter name | symbol | Parameter size (mm) |
|---|---|---|
| Core outer diameter | D | 100 |
| Core inside diameter | d | 65 |
| Core thickness | H | 20 |
| First winding outer diameter | $D_1$ | 70 |
| First winding inside diameter | $d_1$ | 78 |
| Second winding outer diameter | $D_2$ | 88 |
| Second winding inside diameter | $d_2$ | 96 |
| Winding thickness | h | 15 |
| Core clearance | g | 1 |

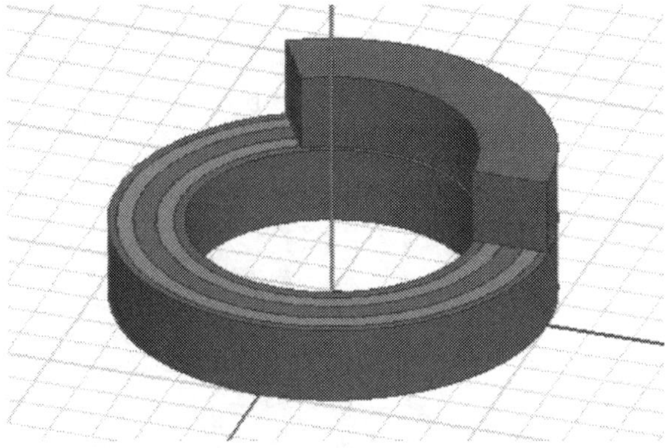

**Figure 2.** Three - dimensional model of non - contact power transmission device

When the secondary winding speed gradually increases, the current amplitude of the primary coil and secondary coil remain unchanged, the rotational speed of the secondary coil has little effect on the efficiency of the non-contact power transmission device. Because of the characteristics of the ferrite core, it can react instantaneously to the changes in the tens or even hundreds of hertz, moreover at the speed of 2000 rev / min. Therefore, in this paper, you can ignore the impact of the speed on transmission efficiency.

As can be seen in Figure 3, the input current of the primary coil and the induced current of the secondary coil have the same frequency and opposite phase angle. When the input current amplitude of the primary coil reaches 0.8A, the first secondary coil output current is 0.34A and the second secondary coil output current is 0.28A. This is due to the presence of magnetic flux leakage, resulting in that the induction current of secondary coil is smaller than the input current of primary coil.

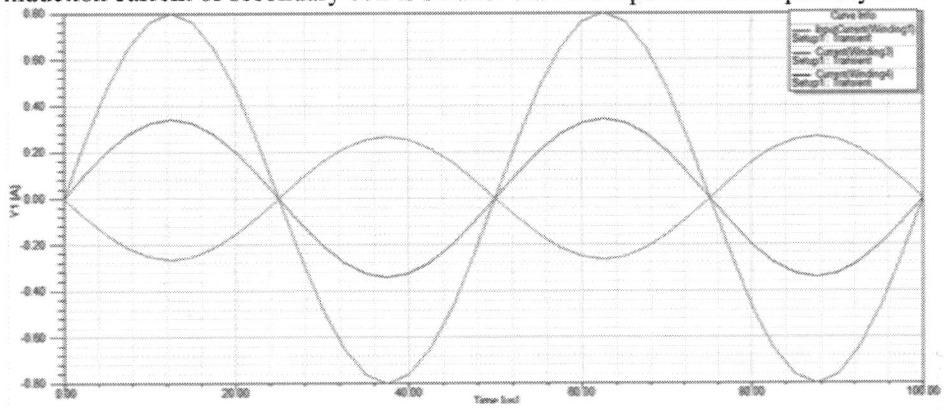

**Figure 3.** Primary and Secondary Coil Current

## 3 Effect of magnetic core spacing

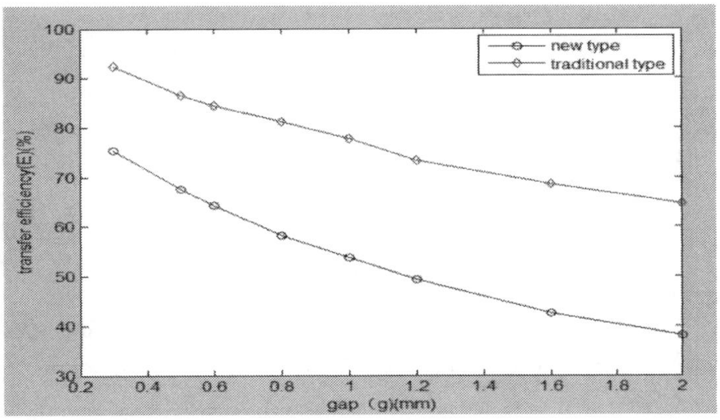


**Figure 4.** Transmission Efficiency with the change of Core Spacing

As shown in Figure 4, when the other parameters remain unchanged, When the two core spacing of the non-contact power transmission device are changed in pitch from 0.3 mm to 2 mm, whether it is a new non-contact power transmission device or a traditional non-contact power transmission device, the transmission efficiency reduces along with the growth of the core spacing. This is because with the increasing of magnetic core spacing, reluctance and magnetic leakage increases, mutual inductance decreases, resulting in the decreasing of the coupling coefficient between original and secondary winding, making the transfer efficiency decrease. In the same core spacing, the transfer efficiency of new non-contact power transmission device is smaller than that of the traditional type, moreover, when the two non-contact power transmission device core spacing are in the 0mm to 1mm, the transmission efficiency changes mostly severe, the change of the transmission efficiency begins to slow down as it continues to increase. When the core spacing increases to a certain extent, the mutual inductance may be very weak, which can't produce the induced current, the transmission efficiency of the non-contact power transmission device will be 0.

Therefore, the transmission effect is the best when the distance is 0 between the original and the secondary winding gap theoretically. However, in practice, due to processing and installed problems, the original, the secondary winding spacing of the non-contact power transmission device can't be infinitely small, the selected core spacing is generally within 0.6mm.

## 4 Effect of core offset

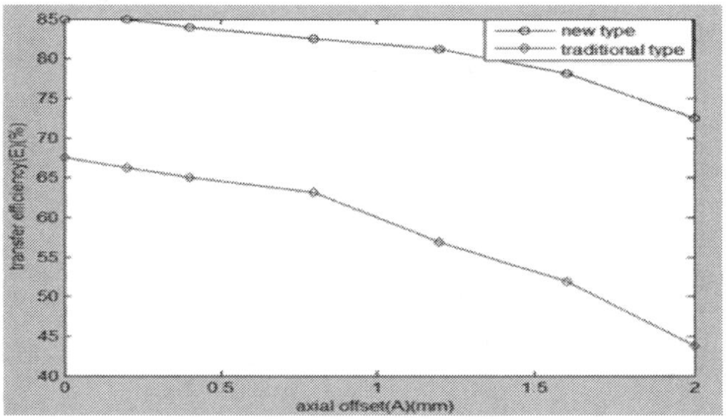


**Figure 5.** Transmission Efficiency with the change of core offset

In the actual processing, as in the processing and assembly of the handle, core, the core coat and some other parts, there is likely to happen coaxial error, leading that the primary winding and secondary winding are not aligned. Whether the non-contact power transmission system working conditions will change in case of this axis offset, and the impact is how much, these are all obtained by Maxwell simulation.

In the above figure, the transfer efficiency of the two non-contact power transmission devices is reduced when the two core offset of the non-contact power transmission device changes from 0mm to 2mm while the other parameters are kept unchanged. This is because with the increasing of core offset, reluctance and magnetic leakage increases, mutual inductance decreases, resulting in the decreasing of the coupling coefficient between original and secondary winding, making the transfer efficiency decrease. In addition, it can be seen that the traditional non-contact power transmission device has little change in the transfer efficiency when the offset is small. As the offset continues to increase, the transfer efficiency decreases slowly, and later becomes fast, which is similar to the new non-contact power transmission device.

In the actual processing of parts, it needs to ensure the core offset that the error of concentricity isn't more than 0.3mm so that a higher energy transfer can be achieved.

## 5 Effect of core thickness

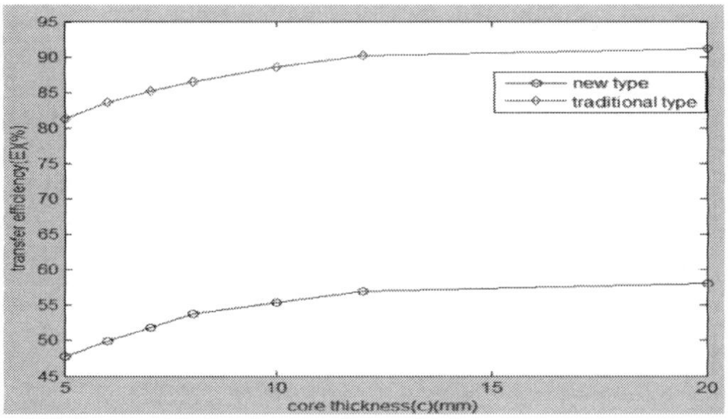


**Figure 6.** Transmission Efficiency with the change of core thickness

It can be seen from Figure.6 that the increase of the core thickness can effectively improve the transfer efficiency when the other parameters are unchanged whether traditional or new non-contact power transmission devices. This is because the thicker core has a good poly-magnetic effect, thereby reducing the leakage of magnetic energy into the air. According to the fitting curve in the figure, when the thickness is less than 10mm, the increase rate of transmission efficiency is relatively large. As the thickness continues to increase, the increase rate of transmission efficiency begins to slow down gradually, which indicates that the variation of transmission efficiency and core thickness is convergent.

In the actual processing of parts, the core is generally maintained at a thickness of 10mm or more appropriate.

## 6 Conclusion

In this paper, the non-contact power transmission device is used as a model, utilizing Maxwell software to obtain relationship curve between the three parameters and transmission efficiency belong to the core. By comparing the new and traditional non-contact power transmission devices, It analyzes the factors affecting the transfer efficiency in the actual processing in order to get the maximum energy transferred in the ultrasonic vibration processing.

In the same conditions, although the transfer efficiency of the traditional non-contact power transmission device is higher than that of the new non-contact power transmission device, however, on the one hand, the transfer efficiency of new type has been basically meet the production requirements in the industrial application, on the other hand, it isn't convenient to change cutleries in traditional type, which will reduce the production efficiency greatly.

So the non-contact power transmission device is currently in the direction of changeable cutlery. The reasonable matching network with experiments will be implemented in future work.

## References


1. Wen Yuan, Zhili Long, Jianguo Zhang, et al. The application of new non-contact power transducer in ultrasonic machining, Transactions of China Electrotechnical Society[J], **30**(10):53-57(2015)
2. Valtchev S, Borges B, Brandisky K, et al. Resonant contactless energy transfer with improved efficiency[J]. IEEE Transactions on Power Electronics, **24**(3):685-699(2009)
3. Yajun Li, Xinquan Lai, Qiang Ye, et al. High efficiency and low electromagnetic interference boost DC-DC converter. Journal of Semiconductors[J], **35**(4):1-8(2014)
4. Qiang Ye, Jie Liu, Bing Yuan, et al. On-ship frequency compensation with a dual signal path operational transconductance amplifier for a voltage mode control DC/DC converters. Journal of Semiconductors, **33**(4):045006(2012)
5. Reiter T, Polenov D, Probstle H, et al. PWM dead time optimization method for automotive multiphase DC/DC-converters. IEEE Trans Power Electron, **25**(6):1604(2010)
6. M Du, H Lee, J Lin. A 5-MHz 91% peak-power-efficiency buck regulator with auto-selectable peak and valley-current control IEEE J Solid-State Circuit, **46**(8):1928(2011)

# Simulating fiber motion in the airflow in a rotor spinning unit

Huiting Lin[1], Nicholus Tayari Akankwasa[1] and Jun Wang[1,2,a]

[1] *College of Textiles, Donghua University, Shanghai 201620, China*
[2] *Key Laboratory of Textile Science & Technology, Ministry of Education, Donghua University, Shanghai 201620, China*

**Abstract.** In this study, a numerical model for simulation of fiber motion in the transport channel of a rotor spinning unit has been developed. A chain represents the fiber consisting of a series of cylindrical rods connected with beads. The fiber model is applied on the turbulent airflow field of the transport channel to present fiber configuration at different times. Results show that the airflow field in the transport channel has a great impact on fiber configuration. The vortex generated in the channel inlet tends to cause a hook on the fiber leading end. As the leading end escapes from the vortex and moves downwards the transport channel, the accelerating air will facilitate fiber straightening.

## 1 Introduction

Air technology is widely used in the manufacture of textile. Especially in the rotor spinning process, it is used to transport and orientate fibers, which is a complex two-phase flow problem. In the study of two-phase (fluid/solid) problem, two basic approaches are used. One is the Eulerian approach, in which the solid phase is regarded as a continuous phase as the fluid phase. In this case, the fibers need to be treated as particles. However, as the fiber is elastic and flexible with a large aspect ratio, it cannot be considered as particles. Another is the Lagrangian approach, which treats the solid phase as discrete phase and the fluid phase as a continuous phase. The latter is usually used to study fluid/fiber interaction.

The key to study air/fiber interaction is to construct fiber mechanical model. Smith and Roberts [1] assumed the fiber as a chain composed of both spherical and cylindrical elements. The fiber model considering fiber stiffness was then applied on a laminar airflow. Reported by Bangert and Sagdeo [2], the fiber was represented by a rigid cylinder. However, their model cannot present fiber flexibility. Yamamoto and Matsuoka [3] modelled the motion of rigid and flexible fibers in a simple shear flow, and the fiber was composed of a series of contiguous spheres. The fiber could stretch, bend and twist by changing the bond distance, bond angle and torsion angle respectively between the neighbour spheres. However, it was computationally expensive when modelling a long fiber. Kong and Platfoot [4, 5] proposed a model represented by discrete nodes held together by weightless chains. Nevertheless, this model would not give a clear picture of elastic deformation of the fiber. Zeng et. al [6] established a bead-elastic rod model for simulating the fiber motion in high-speed air flow.

In rotor spun yarn spinning process, the air phase flow characteristics inside the rotor spinning unit was affected by the variations of structure and spinning parameters [7-9]. The airflow has a critical

[a] Corresponding author: junwang@dhu.edu.cn

impact on yarn properties through influencing the configuration of fibers in yarns. Kong and Platfoot [8] simulated the two-dimensional airflow field in the transport channel. Affected by the channel geometry, a vortex was generated at the channel inlet. The vortex may cause fiber bending and hence deteriorate yarn properties. However, there is still a lack of studies on the relationship between the airflow and fiber configuration in rotor spinning. Therefore, the present paper will focus on studying the air/fiber interaction in the rotor spinning unit. The aim is to provide a theoretical guidance for rotor spun yarn production.

## 2 Theoretical model

The fiber/air interaction is assumed to be a one-way coupling model in which the presence of the fiber is assumed to have a negligible effect on the airflow field. Fiber trajectories can be obtained by applying the fiber model on the airflow field. Therefore, the airflow and fiber models will be constructed separately.

### 2.1 Construction of the airflow model

In rotor spinning, fiber slivers fed by the feeding roller are combed by the rotating opening roller into individual fibers. As the air inside the rotor is sucked by a pump, the pressure differential between the rotor and the trash box and the yarn guiding mouth forces the ambient air flowing into the rotor through the transport channel. Therefore, the individual fibers are transported into the rotor groove through the transport channel by the airflow, which will greatly influence fiber straightness. The fibers are then collected in the rotor groove and finally twisted into yarn by the rotating rotor.

Figure 1 presents the rotor spinning unit model and its main dimensions, including the rotor, the yarn guiding mouth, the transport channel, the opening roller and the trash box. As the fiber feeding channel is always blocked with fibers so that the air could hardly enter, this channel is not considered.

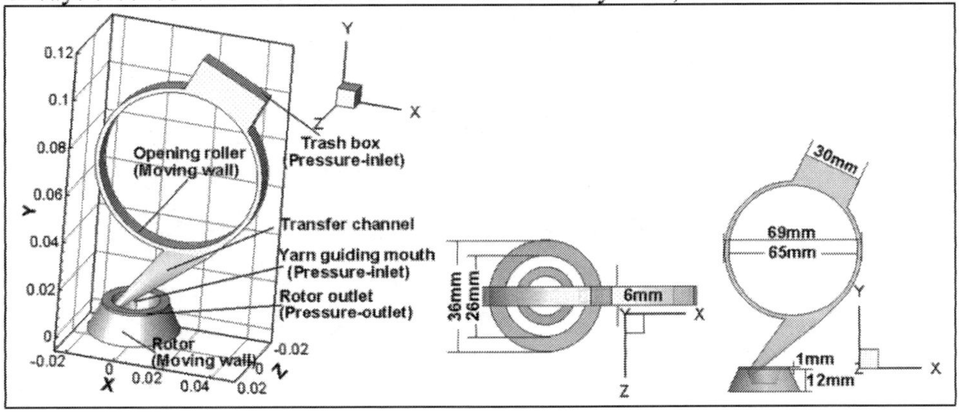


**Figure 1.** Geometry and dimensions of the rotor spinning unit.

Since the Reynolds number in the rotor spinning unit is high (up to 6000) [5], the airflow needs to be assumed as turbulent. As a complete spinning process is finished in a short time, a three-dimensional, steady, viscous turbulence flow is considered. When considering $k-\varepsilon$ turbulence model, the general governing equation is:

$$\frac{\partial(\rho\varphi)}{\partial t}+\operatorname{div}(\rho\mathbf{u}\varphi)=\operatorname{div}(\Gamma\operatorname{grad}\varphi)+S \quad (1)$$

where $\rho$ is the fluid density, and $\mathbf{u}$ is a velocity vector. $\varphi$ represents a generic variable. $\Gamma$ is the diffusion coefficient and $S$ is the general source term. The different terms of the governing equations are illustrated in Table 1. $p$ is pressure. $u$, $v$ and $w$ are the velocity components in $x$, $y$ and $z$-direction, respectively. $\mu_{eff}$ is the effective viscosity, which is the sum of the fluid viscosity $\mu$ and the Reynolds apparent viscosity $\mu_t$. The turbulent kinetic energy $k$ and dissipation rate $\varepsilon$ are related to $\mu_t$ by the expression $\mu_t = \rho C_\mu k^2/\varepsilon$. $G_k$ is the generation of turbulent kinetic energy $k$ due to the mean velocity gradients. $\sigma_k$ and $\sigma_\varepsilon$ represent the turbulent Prandtl numbers corresponding to $k$ and $\varepsilon$. $C_\mu$, $C_{1\varepsilon}$ and $C_{2\varepsilon}$ are empirical constants.

A mix of structured hexahedral and unstructured tetrahedral grids are generated in the whole computational domain. The simulation is conducted using a commercial CFD package FLUENT. The finite volume discretization method is used and the second-order-upwind scheme is adopted due to its high accuracy to solve the governing equations.

**Table 1.** Terms of the governing equations considering $k-\varepsilon$ model.

| Equation | $\psi$ | Diffusivity $\Gamma$ | Source term $S$ |
|---|---|---|---|
| Continuity | 1 | 0 | 0 |
| *x*-momentum | $u$ | $\mu_{eff} = \mu + \mu_t$ | $-\frac{\partial p}{\partial x} + \frac{\partial}{\partial x}\left(\mu_{eff}\frac{\partial u}{\partial x}\right) + \frac{\partial}{\partial y}\left(\mu_{eff}\frac{\partial v}{\partial x}\right) + \frac{\partial}{\partial z}\left(\mu_{eff}\frac{\partial w}{\partial x}\right) + S_u$ |
| *y*-momentum | $v$ | $\mu_{eff} = \mu + \mu_t$ | $-\frac{\partial p}{\partial y} + \frac{\partial}{\partial x}\left(\mu_{eff}\frac{\partial u}{\partial y}\right) + \frac{\partial}{\partial y}\left(\mu_{eff}\frac{\partial v}{\partial y}\right) + \frac{\partial}{\partial z}\left(\mu_{eff}\frac{\partial w}{\partial y}\right) + S_v$ |
| *z*-momentum | $w$ | $\mu_{eff} = \mu + \mu_t$ | $-\frac{\partial p}{\partial z} + \frac{\partial}{\partial x}\left(\mu_{eff}\frac{\partial u}{\partial z}\right) + \frac{\partial}{\partial y}\left(\mu_{eff}\frac{\partial v}{\partial z}\right) + \frac{\partial}{\partial z}\left(\mu_{eff}\frac{\partial w}{\partial z}\right) + S_w$ |
| Kinetic energy | $\kappa$ | $\mu + \mu_t/\sigma_k$ | $G_k + \rho\varepsilon$ |
| Dissipation rate | $\varepsilon$ | $\mu + \mu_t/\sigma_\varepsilon$ | $(\varepsilon/k)(C_{1\varepsilon}G_k - C_{2\varepsilon}\rho\varepsilon)$ |

## 2.2 Construction of a fiber model

The fiber is assumed to move freely in the air without being stretched, compressed and twisted. The fiber is a chain composed of *n* beads and *n*-1 massless cylindrical rods, see Figure 2. The cylindrical rods have the same diameter as the beads. Only the beads are affected by forces, while the rods maintain the fiber configuration. The spatial positions of the beads represent the trajectory of the fiber.

In the fiber model, bead *i* only connected with bead *i*-1 and *i*+1. Therefore, bead 1 and bead *n* become the first and last beads, respectively. The mass of bead *i* $m_i$ is described as:

$$m_i = 0.5 * \rho_f\left(l_{(i-1,i)} + l_{(i,i+1)}\right) \quad (i\text{=1,2,3......n}) \tag{2}$$

where $\rho_f$ is the fiber density. $l_{(i-1,i)}$ and $l_{(i,i+1)}$ are the length of fiber section (*i*-1, *i*) and(*i*, *i*+1), respectively. For beads 1 and *n*+1, the mirror reflection is used, therefore, $l_{(0,1)} = l_{(1,2)}$ and $l_{(n-1,n)} = l_{(n,n+1)}$ are obtained.

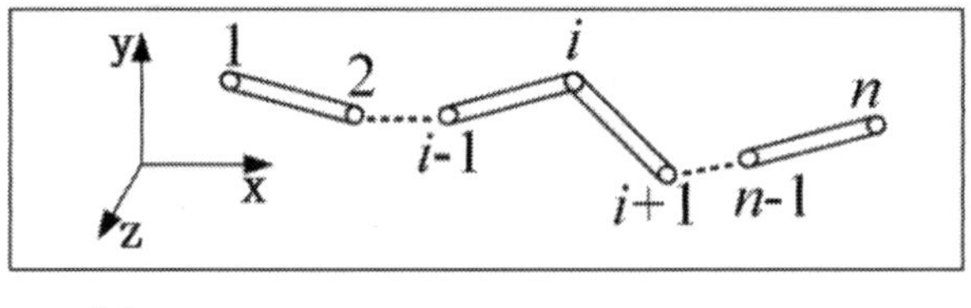


**Figure 2.** Diagram of the fiber model.

According to Newton's second law, the equation for the motion of the beads that constitute the fiber is:

$$m_i\frac{d^2\mathbf{r}_i}{dt} = \mathbf{F}_{bi} + \mathbf{F}_{di} \tag{3}$$

$\mathbf{r}_i$ represents the position of bead *i* in the airflow field, which can be expressed as $\mathbf{r}_i = x_i(t)i + y_i(t)j + z_i(t)k$. $\mathbf{F}_{bi}$ and $\mathbf{F}_{di}$ are respectively the bending force and the drag force exerted on bead *i*.

The angle $\theta_i$ between section (*i*-1, *i*) and (*i*, *i*+1) represents that section (*i*, *i*+1) is subjected to the bending force (see Figure 3). The bending torque $\mathbf{M}_{bi}$ acted on bead *i* is:

$$\mathbf{M}_{bi} = -k_b\theta_i \tag{4}$$

where $k_b$ is the bending constant related to fiber elasticity modulus *E* and inertia moment of the fiber cross section to the central axis $I_b$, which are:

$$k_b = 2EI_b/l_{(i,i+1)} \quad \text{and} \quad I_b = \pi d_f^4/64 \tag{5}$$

where $d_f$ is the fiber diameter. $\theta_i$ can be obtained by the following equation:

$$Cos\theta_i = \mathbf{e}_i\mathbf{e}_{i+1} \tag{6}$$

$\mathbf{e}_i$ and $\mathbf{e}_{i+1}$ are the unit vectors of section (*i*-1, *i*) and (*i*, *i*+1).

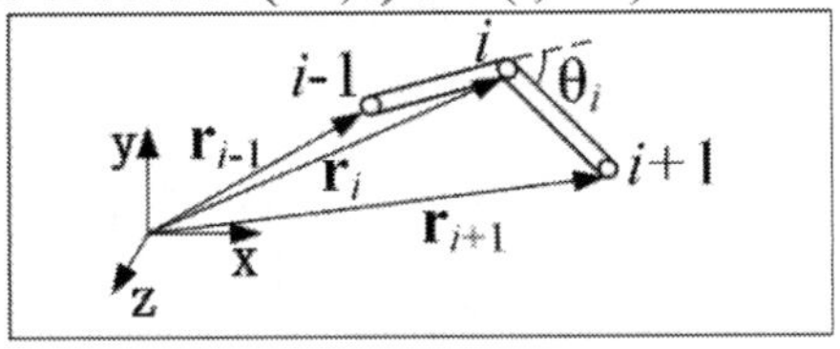


**Figure 3.** Sketch of fiber bending.

The bending force $\mathbf{F}_{bi}$ exerted on bead *i* can be expressed as:

$$\mathbf{F}_{bi} = \mathbf{M}_{bi}/l_{(i,i+1)} \tag{7}$$

$\mathbf{F}_{bi}$ is perpendicular to $\mathbf{e}_{i+1}$ with its direction to the equilibrium position.

When immersed in the airflow, the fiber is subjected to the aerodynamic forces. Only the drag force is considered in this paper. The drag force is related to the air characteristics, fiber shape and fiber orientation. If considering section (*i*, *i*+1), the drag force $\mathbf{F}_{di}$ is:

$$\mathbf{F}_{di} = (\pi/8)\rho_a d_{sphi}^2 C_d(\mathbf{V}_{ai} - \mathbf{V}_{fi})|V_{ai} - V_{fi}| \tag{8}$$

where $\rho_a$ is the air density and $d_{sphi}$ is the equivalent spherical diameter of section (*i*, *i*+1).

$$d_{sphi} = \sqrt[3]{1.5d_f^2 l_{(i,i+1)}} \tag{9}$$

$\mathbf{V}_{fi}$ and $\mathbf{V}_{ai}$ are the velocities of bead *i* and the air in the position of bead *i*. The direction of $\mathbf{F}_{di}$ is the same as the relative velocity $\mathbf{V}_{ai} - \mathbf{V}_{fi}$. $C_d$ is the drag coefficient. The expression to determine $C_d$ proposed by Andrea and Martin [10] is adopted, which is described as:

$$C_d = \frac{8}{Re}\frac{1}{\sqrt{\emptyset_\perp}} + \frac{16}{Re}\frac{1}{\sqrt{\emptyset}} + \frac{3}{Re}\frac{1}{\emptyset^{0.75}} + 0.4210^{0.4(-log\emptyset)^{0.2}}\frac{1}{\emptyset_\perp} \tag{10}$$

where $\emptyset$ and $\emptyset_\perp$ are sphericity and crosswise sphericity. They are given as:

$$\emptyset = s/s_{fi} \quad \text{and} \quad \emptyset_\perp = s_m/s_\perp \tag{11}$$

$s$ is the surface of a sphere having the same volume as section (*i*, *i*+1) and $s_{fi}$ is the surface of section (*i*, *i*+1). $s_m$ is the maximum cross-sectional area of the volume equivalent sphere. $s_\perp$ is the projected cross-sectional area of section (*i*, *i*+1) perpendicular to the airflow. They are expressed as:

$$s = \pi d_{sphi}^2 \quad \text{and} \quad s_{fi} = \pi d_f l_{(i,i+1)} \quad \text{and} \quad s_m = 0.25\pi d_{sphi}^2 \tag{12}$$

$s_\perp$ is related to the relative position between section (*i*, *i*+1) and air velocity vector. Figure 4 illustrates the three possible cases of the relative position.

(a) Air velocity vector is perpendicular to the central axis of section (*i*, *i*+1), then $s_\perp = d_f l_{(i,i+1)}$.

(b) Air velocity vector is parallel to the central axis of section (*i*, *i*+1), then $s_\perp = 0.25\pi d_f^2$.

(c) The angle between air velocity vector and the central axis of section (*i*, *i*+1) is $\alpha_i$, then $s_\perp = d_f l_{(i,i+1)} sin\alpha_i$.

For section (*i*, *i*+1), the Reynolds number $Re$ is:

$$Re = \rho_a(V_{ai} - V_{fi})d_{sphi}/\mu_a \tag{13}$$

where $\mu_a$ is the air dynamic viscosity.

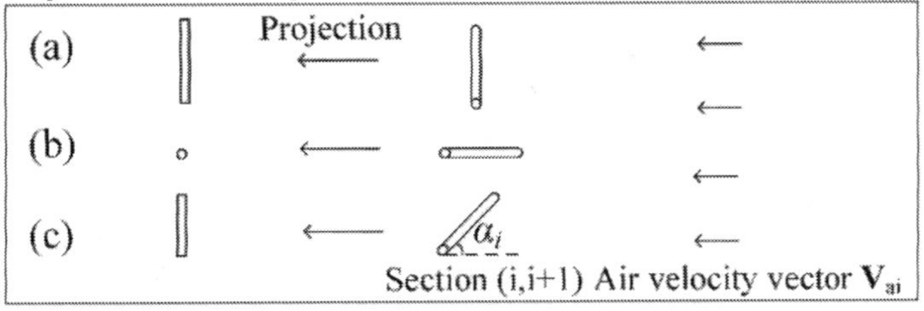


**Figure 4.** Relative position between section (*i*, *i*+1) and the air velocity vector.

### 2.3 Simulation procedure

The following constraint condition needs to be satisfied during calculation:

$$|\mathbf{r}_{i+1} - \mathbf{r}_i| = l_{(i,i+1)} \tag{14}$$

As equation (14) may be broken due to the numerical error, a procedure is conducted for adjustment at each time step. The position of bead $i$ is adjusted from $\mathbf{r}_i$ to $\mathbf{r}_i + \Delta\mathbf{r}_i$ (i=1,2,...,n) . $\Delta\mathbf{r}_i$ can be obtained by solving the quantity as minimum: $\delta = \sum(\Delta\mathbf{r}_i)^2$ [11].

The procedure for the solution is, (a) Calculate the air flow field of the rotor spinning unit without the fiber by equation (1), then the air velocity can be obtained; (b) For a given fiber, calculate equations (4) - (7), $\mathbf{F}_{bi}$ is obtained; (c) Calculate equations (8) - (13), $\mathbf{F}_{di}$ is obtained; (d) Solve equations (2) - (3), the position $\mathbf{r}_i'$ and velocity $\mathbf{V}_{fi}'$ of bead $i$ at $t' = t + \Delta t$ are obtained; (f) If $\mathbf{r}_i'$ does not satisfied the constraint condition (equation (14)), then the procedure is applied to adjust the position; (e) Repeat steps (a) - (f), the trajectory of a fiber as a function of time is obtained.

## 3 Results and discussion

The material parameters for the fiber used are diameter of 0.02 mm, linear density of 1.8 dtex, elastic modulus of 8 GPa and length of 12 mm. Initially, the fiber is assumed to be straight, and the fiber is parallel to the transport channel inlet cross section, see Figure 5 (b). The initial fiber velocity is assumed to be 5 m/s. The calculation ends once the fiber collides with the walls or the leading end of the fiber enters the rotor.

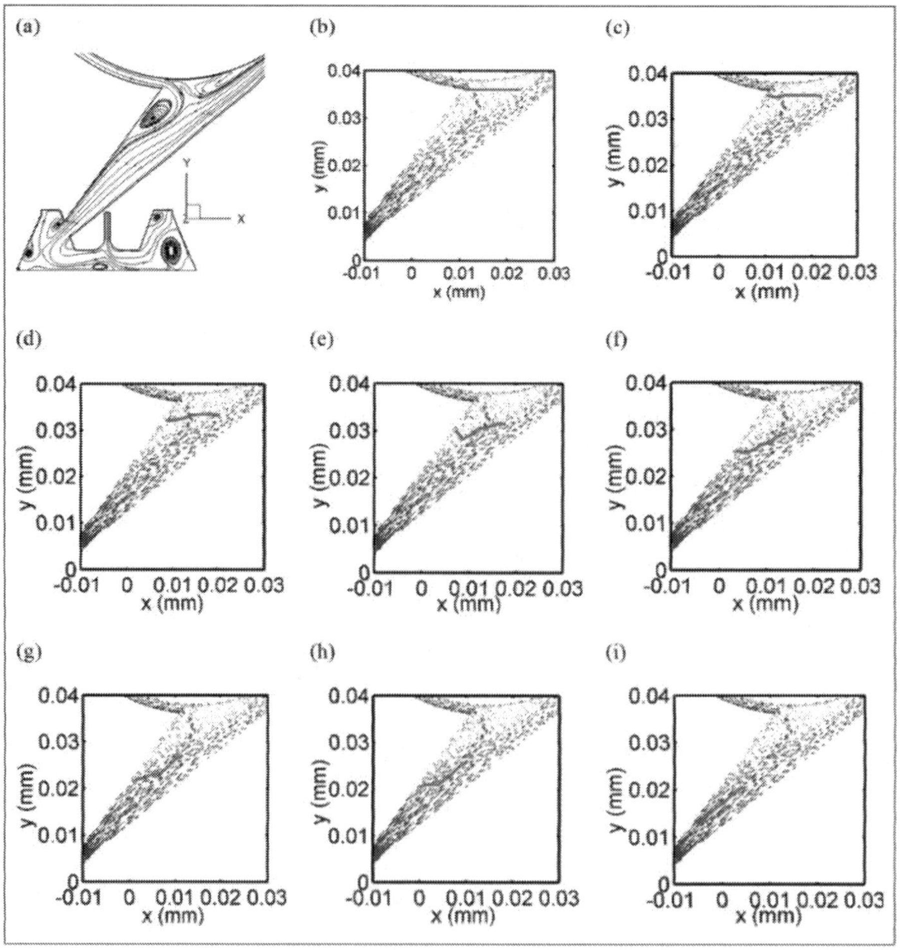


**Figure 5.** (a) Velocity streamline and (b) - (i) velocity vector and fiber configuration in the transport channel at plane z = 0 mm. The time step for (b) - (i) is $4\times10^{-4}$ sec; figure (b) shows the initial position of the fiber ($t = 0$ sec).

Figure 5(a) shows the velocity streamline in the transport channel and rotor. The air from the opening roller chamber enters the rotor through the transport channel, at which a vortex is generated. Figures 5(b) - (i) illustrate the fiber trajectories along the transport channel at different times. The variations of fiber configuration indicate that the airflow inside the rotor spinning unit is critical to the fiber straightness. The fiber moves downwards with its leading end entering the vortex (Figure 5(c)). The fiber leading end is gradually bent due to the vortex. However, the intensity of the vortex is not strong enough to drag the whole fiber into the vortex. The leading end of the fiber moves faster than its trailing end, therefore, as soon as its leading end escapes from the vortex, the fiber moves along the transport channel longitudinal axis. Figure 5(g) - (i) shows that the fiber undergoes straightening gradually as it is moving down, indicating that the converging transport channel along with the inside accelerating air is helpful in straightening the hooked fibers.

## 4 Conclusion

In this paper, the simulation of fiber motion in the turbulent airflow field was carried out. The fiber was modelled as a bead-cylindrical rod chain and was applied in the transport channel airflow field in the rotor spinning machine. The airflow field has a critical influence on fiber configuration. The fiber tends to bend when it runs into a vortex. However, it will escape from the vortex as long as the vortex is not strong enough to trap it. As the fiber moves downwards the transport channel, the accelerating air facilitates fiber-straightening, eliminating the temporary bending caused by the vortex.

## Acknowledgement

This work was supported by the Key Grant Project of Chinese Ministry of Education (Grant 113027A).

## References

1. A.C. Smith, W.W. Roberts, Text. Res. J. **64**, 335 (1994).
2. L.H. Bangert, P.M. Sagdeo, Text. Res. J. **47**, 773 (1977).
3. S. Yamamoto, T. Matsuoka, J. Chem. Phys. **98**, 644 (1993).
4. L.X. Kong, R.A. Platfoot, Text. Res. J. **67**, 269 (1997).
5. L.X. Kong, R.A. Platfoot, Comput. Struct. **78**, 237 (2000).
6. Y.C. Zeng, J.P. Yang, C.W. Yu, Appl. Math. Model. **29**, 253 (2005).
7. H.T. Lin, Y.C. Zeng, J. Wang, Text. Res. J. **86**, 115(2015).
8. L.X. Kong, R.A. Platfoot, Text. Res. J. **66**, 641 (1996).
9. L.H. Zhang, B.X. Zhang, J. China Text. Univ. **17**, 16 (1991).
10. A. Hölzer, M. Sommerfeld, Powder Technol. **184**, 361 (2008).
11. M. Doi, D. Chen, J. Chem. Phys. **90**, 5271 (1989).

# Evaluation of drag and lift in the internal flow field of a dual rotor spinning unit via CFD

Nicholus Tayari Akankwasa[1], Huiting Lin and Jun Wang[1,2,a]

[1]*College of Textile, Donghua University, Shanghai, 201620, China*
[2]*The Key Lab of Textile Science and Technology, Ministry of Education, Shanghai 201620, China*

**Abstract.** In the present study, we evaluate the drag and lift magnitude in the new dual-feed rotor spinning unit using computational fluid dynamics technique. We adopt theoretical and numerical approach based on FLUENT to investigate the influence of new design on the drag and lift in the rotor interior. Results reveal that the drag and lift inside the rotor of the proposed model are reduced by 60-80% and 50-66% respectively as compared to the conventional unit. The velocity and pressure profiles become evenly distributed in the dual-feed rotor interior as opposed to the conventional rotor spinning unit and this modification is anticipated to improve fiber configurations. This phenomenon can be utilized to further optimize the rotor spinning unit and other wall-bounded engineering problems.

## 1 Introduction

In the fluid dynamics concept, previous studies have focused more on the turbulent viscosity, flocculation, eddies and vortices among other flow properties. Growing interest in drag reduction in flows has increased in the recent years. Principally, the drag and lift phenomenon has been widely applied in solving aerodynamics and aeronautics problems. For wall-bounded flows, it is important to note that drag and lift has a significant impact on the processing of fibers and polymers. Witold Brostow[1] reported a comprehensive review on drag reduction in flow covering the applications, mechanisms, and prediction. In his review, Witold Brostow[1] reported an increasing interest of application of drag reduction phenomenon in turbulent flows. Unlike the laminar flows, turbulent flows are rather chaotic, complex and associated with high Reynolds numbers. For effective evaluation of turbulent flows, drag and lift assessment can be helpful in benchmarking several engineering problems. Roberto Benzi[2] reported a well-articulated review on drag reduction by polymers in wall-bounded turbulence. The review reveals that the amount of polymer stretching transversely to mean flow plays a major role in understanding the phenomenon of drag reduction. Study on the behavior of drag and lift forces in the flows is a significant phenomenon that gives a comprehensive understanding of a given processes, which can help to improve the quality and handle costs efficiently.[3–5]

Numerical studies in textile processing have tremendously risen in the past decade with authors focusing on finite element methods to study fiber processing techniques more comprehensively[6–13]. In all industrial rotor spinning machines, the opening of the fibers is performed by the opening roller,

[a] Corresponding author: junwang@dhu.edu.cn

in which sliver is fed from a single point. Increasing the number of feed rollers from one to two may improve fiber opening and trash ejection by two-step loading in the opening zone of the roller. Peyravi et al.[13] and Wang et al.[14] reported the dual-feed concepts that could spin yarns with better yarn properties. Their remarkable findings have significantly contributed to rotor spinning research with critical insights towards design modification, yarn quality improvement and reduction in production costs. Lin et al.[8] applied numerical methods to study rotor spinning process behavior, especially in the rotor interior. Their findings revealed a significant relationship between the geometrical characteristics, spinning parameters and airflow characteristics.

Our recent works on the new spinning dual-feed rotor spinning machine explored the 3D airflow domain with a focus on the effect of varying air velocity on the flow characteristics in the rotor.[12] Results showed that turbulence strength and Re increased with increase in inlet air velocity. Furthermore, pressure profiles and velocity distribution were affected as compared to the findings on the conventional rotor spinning unit by Lin et al.[8] Recent studies on drag and lift in turbulent flows have applauded the significance of the lift and drag in internal flows. This piece of work seeks to evaluate the drag and lift forces associated in the new dual-feed model. The rotor dimensions in this study were based on the model designed by Wang et al.[14] We explore the effects of the second channel on the drag and lift forces using numerical methods based on FLUENT. In the second section, we present a brief mathematical and numerical model. Section 3 provides an account of the airflow regime based on velocity and pressure contours and calculation of drag and lift. Finally, conclusions are drawn.

## 2 Theoretical model and numerical procedure

### 2.1 Dual-feed rotor structure

Figure 1 shows the geometrical model of the proposed dual-feed rotor spinning machine. The new design has two transfer channel inlets as opposed to the conventional unit[8]. The airflow is simulated from the two velocity inlets concurrently to the rotor interior up to the doffing tube. The Cartesian coordinate system is used. The origin of the model is situated at the center of the rotor, and one of the transfer channels lies in the positive x-y plane, and the another one is in the negative x-y plane. The rotor is region shares the three axes.

To demonstrate the effect of the dual-unit principle, we evaluate the drag and lift inside the rotor due to the effect of the second channel in the rotor.

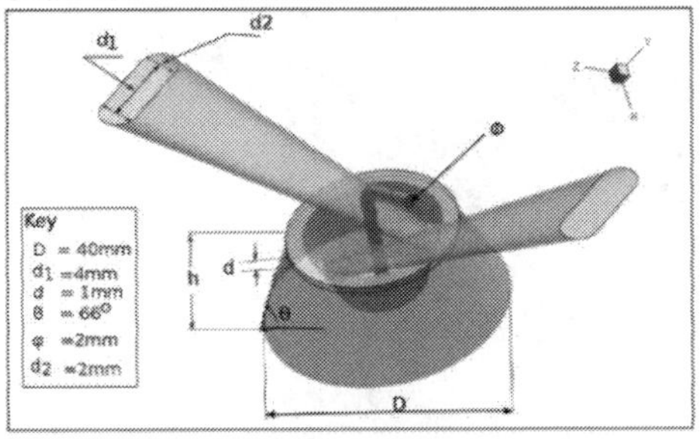


**Figure 1**. Geometrical profiles of a 3-D model of the dual rotor spinning unit.

### 2.2 Governing equations

Considering air flowing through the model domain, the flow direction originates from the channel inlets towards the rotor. The airflow in the rotor unit has Mach number of not more than 0.3[5] so it can be regarded as incompressible. Reynolds number in the rotor spinning box is considerably high, up to 6000 and the flow is turbulent[7,8]. The equations adopted in our simulation intent hypothesize that the airflow in the rotor zone is viscous, incompressible and isentropic. The governing equations are articulate as follows;

$$\frac{\partial v}{\partial t} + \nabla.\rho.\ \vec{v} \tag{1}$$

ρ is the fluid density, v is a velocity vector. Equation 1 implies that, the sum of the time-dependent change of density ρ and the three-dimensional of the current density $\overline{\rho\vec{v}}$ is zero.

$$\frac{\partial(\rho u_i)}{\partial t} + \mathrm{div}(\rho u u_i) = \mathrm{div}(\mu \mathrm{grad} u_i) - \frac{\partial p}{\partial x_i} + S_i \tag{2}$$

The turbulent viscosity term $\mu_t$ expresses the relationship between the turbulent kinetic energy k and the turbulent dissipation rate ε (see equation 3). The vortex viscosity term $\mu_t$ in the flow zone is expressed in terms of k and ε (see equations 3-5).

$$\mu_t = \rho C_\mu \frac{k^2}{\varepsilon} \tag{3}$$

$$\frac{\partial(\rho k)}{\partial t} + \frac{\partial(\rho k u_i)}{\partial x_j} = \frac{\partial}{\partial x_j}\left[\left(\mu + \frac{\mu_t}{\sigma_k}\right)\frac{\partial k}{\partial x_j}\right] + G_k - \rho_\varepsilon \tag{4}$$

$$\frac{\partial(\rho\varepsilon)}{\partial t} + \frac{\partial(\rho\varepsilon u_i)}{\partial x_j} = \frac{\partial}{\partial x_j}\left[\left(\mu + \frac{\mu_t}{\sigma_\varepsilon}\right)\frac{\partial k}{\partial x_j}\right] + \frac{C_{l_\varepsilon}\varepsilon}{k} G_k - C_{2\varepsilon}\rho\frac{\varepsilon^2}{k}, \tag{5}$$

The default values of constants $C_\mu$, $C_{l_\varepsilon}$, $C_{2\varepsilon}$, $\partial k$, and $\sigma_\varepsilon$ is same to literature[8].

## 2.3 Numerical and solution procedure

The investigation is performed using a three-dimensional numerical simulation based on the standard k-ε two-equation turbulence model. The Navier-Stokes equations are numerically integrated using the pressure-velocity coupling and SIMPLE algorithm scheme. In this numerical simulation procedure, the standard pressure spatial discretization is considered, the momentum, kinetic energy, and turbulent dissipation rate are evaluated by the second order upwind method to improve the accuracy of the model. The solution is initialized by hybrid methods and the convergence criterion of volume and surface monitors assigned to $10^{-6}$. Due to incompressible effects, the pressure-velocity approach adopted works on the principle that performs simultaneous solutions of the conservation of mass, energy and turbulent kinetic energy as well as its dissipation rate equations within the physical domain. We used ANSYS ICEM CFD, release 14.5 to develop, design, mesh and simulate our model. Figure 2 shows the mesh topology and boundary conditions. The parameters considered in this model are similar to our recent works on the 3-D airflow field[12]. Mesh independence tests were performed prior final simulation runs. Details of grid independence tests for the reported model are well explained in Akankwasa et al.[12]

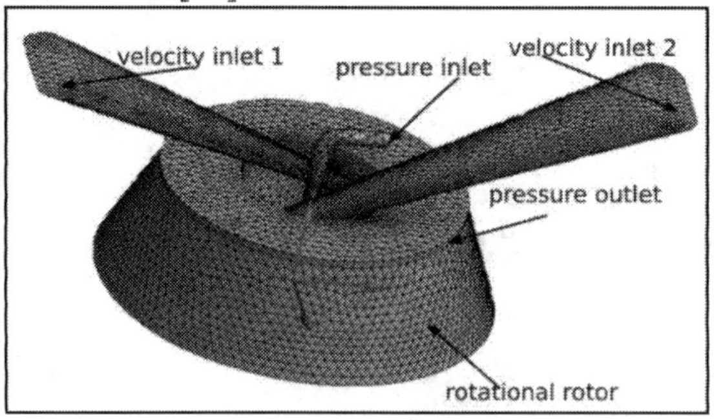


**Figure 2.** Mesh topology and boundary conditions of the dual-feed rotor spinning unit.

## 2.4 Drag and lift calculation

The drag force experienced by a flexible fiber acts in the direction of the flow velocity and is characterized by the drag coefficient. In general, the drag coefficient can be defined by equation 6.

$$C_D = \frac{F_D}{\frac{1}{2}\rho u^2 A_f} \tag{6}$$

Where $F_D$ is the actual drag force, u = $u_\infty - U_f$ is the velocity of the fiber to the local undisturbed fluid velocity $u_\infty$, $\rho$ is the fluid density, and $A_p = \frac{\pi}{4} d_f$ the reference area with $d_p$ being the fiber diameter. To compute the total drag force, equation 7 is considered;

$$F_D = C_f \pi d_f (v - u) l \tag{7}$$

Where $C_f$ is the coefficient of friction between the fiber and air and a given fiber in motion, $\rho$ is the air density, $d_f$ is the fiber diameter, v is the air velocity, u is the fiber velocity and *l* is the fiber length. The numerical simulation data provides the values of parameters such as air density, air velocity through probing specific points in the model.

In the drag equation, $\pi$ and d are theoretically constant, however $C_f$ varies with the relative velocity between air and the fiber. The influence of the coefficient of friction was treated as minimal in this evaluation procedure as proposed by previous studies.[10,11] To simplify the evaluation, we propose treating $C_f$, $\pi$, and d as 1, such that the drag force in the rotor can be evaluated as equation 8 below.

$$F_D = \rho (v - u) d l \tag{8}$$

Considering an element (see Figure 3) inside the rotor with two connecting points, *a* and *b*, where *a* is the is characterized by velocity $v_a$ and density$\rho_a$, and point b presented with different velocity and density as shown in Figure 3.

**Figure 3**: Representation of fiber the rotor

Physical parameters of fiber element such as $\rho$ and $v$ in Figure 3 can be disposed as;

$$\rho = \frac{1}{2}(\rho_a + \rho_b), \qquad v = \frac{1}{2}(v_a + v_b), \tag{9}$$

The total drag is calculated by summation of drag force per unit length (see equation 10).

$$F = \sum F_D / L \tag{10}$$

Lift force on a particle not aligned with the flow velocity is a result of the non-axisymmetric flow field. It acts in the direction perpendicular to the area of velocity its coefficient can be expressed as equation 11.

$$C_L = \frac{F_L}{\frac{1}{2}\rho u^2 A_f} \tag{11}$$

$F_L$ is the exact lift force acting orthogonal to the flow velocity of the fluid. Based on the theoretical modelling and simulation data, we integrate the total drag and lift in the conventional and rotor spinning unit and analysed the results.

## 3 Results and discussions

### 3.1 Evaluation of the airflow regime

To visualize the effect of the second dual channel on velocity profiles, we extracted the flow domain at z=0.000384m perpendicular to the x-y plane. Figure 4 (a) shows the dual-feed rotor spinning unit and (b) shows the conventional rotor unit. As seen in Figure 4, the velocity distribution in the transfer channels exhibits similar patterns in the two models. We notice a difference inside the rotor at the positions near the transfer channel exits. The air velocity of 40-60m/s is observed, and the new dual-feed model reveals a much even distribution compared to the conventional unit. The 3D airflow regime in the dual-feed model has been extensively simulated and verified in our previous works[12]. This section is entirely dedicate on drag and lift evaluation in the proposed dual-feed rotor spinning concept compared to the traditional open-end spinning system(rotor spinning).

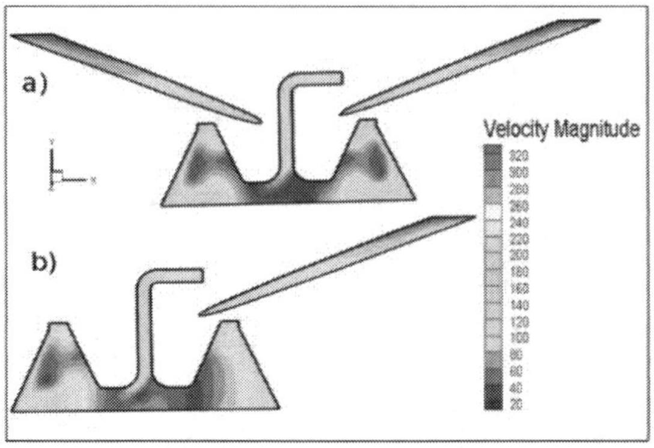


**Figure 4.** Velocity magnitude comparison: (a) modified rotor spinning unit, (b) conventional rotor spinning unit.

The overall velocity magnitude observed, shows that the second feed introduced changes the airflow dynamics. Our results quantitatively agree with previous studies by Kong and Platfoot[7], Lin et al.[8], and Lawrence and Chen[15,16]. The velocity profiles in the dual-feed model predict a significant shift in the way fibers can be aligned inside the rotor. Since both sides, the rotor interior is characterized by balance velocity flow. Our simulation results anticipate improved fiber alignment however, this need to be further proved experimentally.

Figure 5 shows the pressure distribution in the rotor of the conventional and modified rotor spinning process. The region opposite the transfer channel outlet of the conventional model is characterized by pressure ranging from -8000 to -6000pa with the center largely dominated by -8000pa. On the centrally, the pressure distribution in the modified model is evenly distributed with the transfer channel exit positions occupied by traces of positive pressure of up to 6000pa.

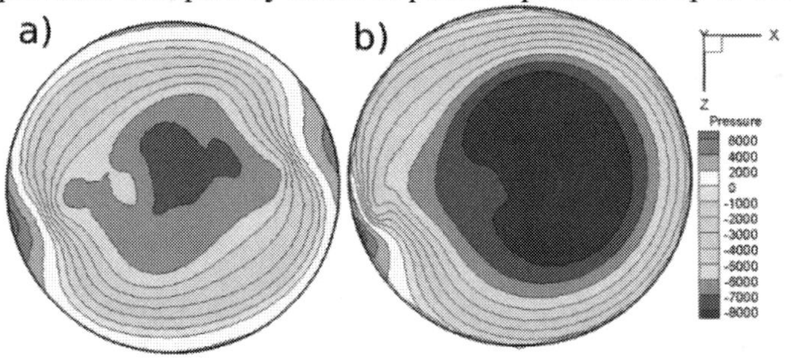


**Figure 5.** Pressure profiles: (a) dual-feed rotor spinning unit (b) conventional rotor spinning system

The pressure difference in internal flow system is a very crucial parameter because it influences the forces and moments such as drag and lift forces. The effect generated by pressure difference eventually influences the airflow dynamics and hence the production process.[11] Even pressure distribution achieved through additional transfer channel will affect the fiber processing mechanics. This observation agrees with Peyravi et al.[13] Findings when they improvised a dual-feed unit to study the yarn characteristics.

## 3.2 Evaluation of drag and lift

Lift in fluid flow is brought about by the pressure difference that exists in the flow regime. In our study, we integrate drag and lift in the two models to evaluate the difference between the models. Figure 6 (a) shows the lift force generated in the conventional unit and the dual-feed with a change in the outlet pressure. Results demonstrate that with the new model reduces the drag force by 50 to 60%. Previous analysis revealed a huge reduction in the drag force as well. Therefore, the new model is anticipated to have a greater positive influence on the yarn quality properties.

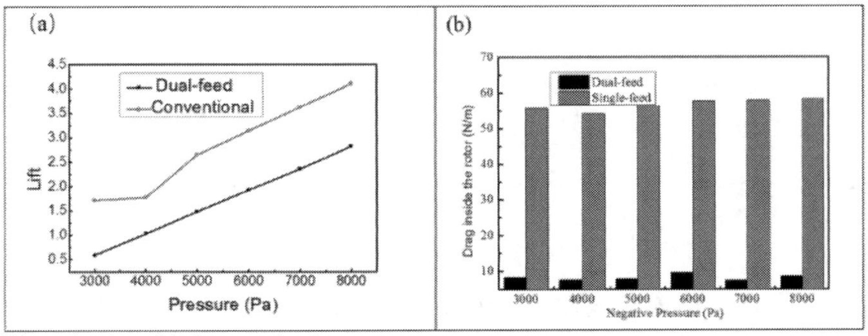


**Figure 6.** (a) lift evaluation, (b) drag evaluation.

Drag is the resistance to flow which acts opposite to the relative motion of any object moving in a fluid or air. Drag exists either between two layers or surfaces or a fluid and a solid surface. In this simulation intent, we evaluated the design of drag that exists in the rotor interior. Even if the fibers are flexible and the rotor walls smooth, for evaluation of physics of flows, form, and frictional drag exists during fiber processing into yarn. Evaluation of the two simulated models revealed drag reduction ranging from 60 to 80% due to the existence of the second feeding unit (see figure 6 (b)).

## 4 Conclusions

A standard k-epsilon model is implemented to evaluate the impact of introducing a second transfer channel in the rotor spinning unit on the drag and lift forces. The flow field is assessed based on the simulation data. Based on results, we conclude that the second channel imparts a significant effect on the flow behavior. The air velocity in the rotor interior of the new model becomes evenly distributed compared to the conventional one. The drag and lift force shows 60-80% and 50-66% reduction respectively. The new model can improve the flow stability in the rotor interior and could improve the fiber alignment and configuration.

## Acknowledgement

This work was supported by the Key Grant Project of the Ministry of Education of the People's Republic of China (Grant No. 113027A)

## References

1. W. Brostow, J. Ind. Eng. Chem. **14**, 409 (2008).
2. R. Benzi, Phys. D Nonlinear Phenom. **239**, 1338 (2010).
3. K. Iwamoto, K. Fukagata, N. Kasagi, and Y. Suzuki, Phys. Fluids **17**, (2005).
4. S. Lardeau and M.A. Leschziner, Phys. Fluids **25**, (2013).
5. S. Tamano and M. Itoh, J. Turbul. **13**, N9 (2012).
6. C.T. DeGroot, C. Wang, and J.M. Floryan, J. Fluids Eng. **138**, 1 (2016).
7. L.X. Kong and R.A. Platfoot, Text. Res. J. **67**, 269 (1997).
8. H.T. Lin, Y.C. Zeng, and J. Wang, Text. Res. J. **86**, 115 (2016).
9. V.C. PATEL, W. RODI, and G. SCHEUERER, AIAA J. **23**, 1308 (1985).
10. Y.-S. Wang, W.-X. Huang, and C.-X. Xu, Fluid Dyn. Res. **48**, 55501 (2016).
11. M. Xiao, H.-S. Dou, and C. Wu, Text. Res. J. (2016).
12. N.T. Akankwasa, H. Lin, Y. Zhang, and J. Wang, Text. Res. J. (2016).
13. A. Peyravi, S. Eskandarnejad, and M.B. Moghadam, J. Text. Inst. **105**, 377 (2014).
14. W. Jun, J.X.Yan, Y.Liang, Z.Yuze, G.Wei, and F.Ting, CN103911696A (2014).
15. C.A. Lawrence and K.Z. Chen, J. Text. Inst. **77**, 201 (1986).
16. C.A. Lawrence and K.Z. Chen, J. Text. Inst. **77**, 139 (1986).

# A Simple Fuzzy PI Control of Dual-Motor Driving Servo System

Haibo Zhao[1,2,a], Chengguang Wang[3]

[1]*Engineering Technology Research Center of Optoelectronic Appliance, Tongling University, Tongling Anhui, 244061, China*
[2]*Department of Electrical Engineering, Tongling University, Tongling Anhui, 244061, China*
[3]*Sichuan Institute of Aerospace System Engineering, Chengdu Sichuan, 610100, China*

**Abstract.** Aiming at the control problem of dual-motor driving servo system, in order to avoid the complicated numerical calculation of complicated control algorithm, we presented a simple fuzzy PI controller for the first time. Firstly, we explained system control principle. In view of the shortcomings of current-error negative feedback, we used speed-error negative feedback synchronous driving control strategy. Then we theoretically analyzed dual-motor synchronous driving system in detail. We proposed the conditions of dual-motor driving servo system synchronous operation for step signal and ramp signal, respectively. Finally, we presented a simple fuzzy PI controller, by combining fuzzy controller with PI controller. We analyzed the fuzzy PI control system and found it was stable. We considered two cases in simulation tests comparing fuzzy PI control with PI control. Simulation results show that fuzzy PI control has better adaptability than conventional PI control, validating the effectiveness and efficiency of the proposed control strategy.

## 1 Introduction

Dual-motor driving servo system is widely used in military, civilian and other occasions. Many scholars have studied it. In [1, 2], the mechanical model of the rotary kiln drive as well as the development of an appropriate laboratory simulating system were presented. Experimental results proved that gear faults can be identified by monitoring the motor phase current as well as the frequency spectrum of this current. In [3], the dynamic model of dual-motor driving system was given, using switching toeque control, simulation results showed that system has good position control precision. In [4], simulation model based on the adaptive full-order flux observer and torque sensor vector control was established to observe or test variety of system dynamic response directly. However, all the above mentioned results have used more complicated intelligent control strategy.

Fuzzy control algorithm has a high maturity and has been widely used in many occasions. Quang Hieu Ngo, et al. [5] presented a fuzzy sliding mode control strategy for container cranes. Simulation results illustrated the efficiency of the proposed control law. Chun-Yan Wang, et al. [6] provided a common adaptive fuzzy smooth dynamic controller for a class of uncertain switched nonlinear systems. Its effectiveness and feasibility were demonstrated by both a numerical example and a chemical system. Do Wan Kim, et al. [7] presented a new direct discrete-time design methodology of

[a] Corresponding author : happyzhaohaibo@126.com

a robust sampled-data fuzzy controller for a class of nonlinear systems. The simulation results for the sampled-data depth control of AUV convincingly illustrated the effectiveness of the developed techniques. Xiaojie Su, et al. [8] presented a T-S model-based fuzzy controller design approach for electromagnetic suspension systems. The T-S fuzzy model was firstly applied to represent the nonlinear electromagnetic suspension systems. Numerical simulation on an electromagnetic suspension system was performed to validate the effectiveness of the proposed approach. From [5-8], we can see that fuzzy control is rarely used alone, and is usually used in conjunction with other control methods.

Fuzzy control is introduced into the logical reasoning, has strong adaptive ability and a good control effect on complex system, but separate fuzzy control cannot fundamentally eliminate steady-state error, control precision is low [9]. Conventional PI control principle is simple, easy to implement, has been widely used in industrial process control. But the coefficient setting of proportion and integral is a very troublesome thing, reasonable PI parameters are usually determined online by experienced technical personnel. In the case of the control object's time variant, a set of fixed PI parameters cannot meet the requirements of system. Fuzzy PI control algorithm is formed by combining fuzzy control with conventional PI control.

In recent years, a simple fuzzy PI control has rarely been used in dual-motor driving servo system. Hence, it is imperative to study the fuzzy PI control of dual-motor driving servo system. This algorithm can not only achieve precise control, but also has strong adaptability. It can automatically modify the control parameters of PI according to the changes of the environment.

This paper is organized as follows: system modeling is given in Section 2. System control principle is presented in Section 3. The theoretic analysis of dual-motor synchronous driving system is given in Section 4. Section 5 presents the design of fuzzy PI controller. Simulation analysis in Section 6 to show the effectiveness of fuzzy PI control strategy. Section 7 concludes this paper.

## 2 System Control Principle

Since the system is a dual-motor driving mode, it must be ensured that the speeds of the two motors are synchronous. In the ideal case, if the two motors are the same, for the same command signal, the output speeds must be the same. However, due to the discrete characteristic of motors and other components, even if the same tag parameters of components, the actual parameters cannot be completely consistent. Therefore, it is necessary to solve the problem of dual-motor synchronous driving, otherwise there will be a so-called "servo fight" phenomenon.

The existing method is to use "current-error negative feedback", which can achieve the synchronous driving of two motors by the balanced control of current. Because the motor parameters cannot be completely consistent, even if the consistent motor current may not be able to achieve the same speed, and the effect of synchronous driving must be based on the speed synchronization. In view of the shortcomings of current-error negative feedback, we use "speed-error negative feedback" synchronous driving control strategy.

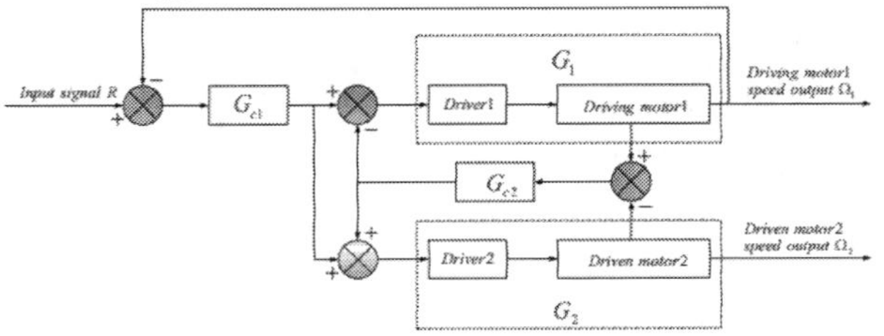


**Figure 1.** Control block diagram of dual-motor synchronous driving.

As is shown in Fig. 1, take one of the motors as driving motor, retaining its speed loop, the other is driven motor, disconnecting the speed loop. For each motor, the input current instruction is a composite signal of the basic current given signal and speed-error feedback signal. And the speed-error negative feedback signals of two motors are equal, but the polarities are opposite.

Where $G_{C1}$ is speed loop controller(type-PI), $G_1$ and $G_2$ are transfer functions between servo motor 1, 2 and respective current loop. $G_{C2}$ is fuzzy PI controller, $R$ is speed input signal. $\Omega_1$ and $\Omega_2$ are speed output of motor 1 and 2, respectively.

In a dynamic process, if the two motor's speeds are not synchronous, after speed-error was linear amplified, feedback to the two motors current given end respectively as auxiliary input. Because the current loop response is much faster than the speed loop response, the speed-error signal is introduced into the current given end to inhibit the two motors speed out of synchronization as soon as possible. So when the speed of motor 1 is faster than the speed of motor 2, the given input of motor 1 will slightly decrease on the basis of basic current, and the given input of motor 2 will increase and vice versa.

## 3 Theoretic Analysis of Dual-Motor Synchronous Driving System

Dual-motor synchronous driving control block diagram, as shown in fig. 1. Using Mason formula, we can solve the transfer function between $\Omega_1$ and $R$

$$G_{Y1-R\_CL} = \frac{W_1(s)}{R(s)} = \frac{G_1 G_{C1} + 2G_1 G_2 G_{C1} G_{C2}}{D} \tag{1}$$

where $D = 1 + G_{C1}G_1 + G_{C2}G_1 + G_{C2}G_2 + 2G_{C1}G_{C2}G_1G_2$

Similarly, the transfer function between $\Omega_2$ and $R$

$$G_{Y2-R\_CL} = \frac{W_2(s)}{R(s)} = \frac{G_{C1}G_2 + 2G_1 G_2 G_{C1} G_{C2}}{D} \tag{2}$$

Then the transfer function between $\Omega_1 - \Omega_2$ and $R$

$$\frac{W_1(s) - W_2(s)}{R(s)} = \frac{G_{C1}(G_1 - G_2)}{D} \tag{3}$$

From Eq. (11), we can see that when the two motors are exactly the same, namely $G_1 = G_2$, regardless of the input signal, regardless of whether there is a speed-error negative feedback or not, $\Omega_1 - \Omega_2 = 0$, the two motors' speed must be synchronized(assume there is no disturbance).

Eqs. (1), (2), (3) are close loop transfer functions, the equivalent open-loop transfer functions of the unit feedback are

$$G_{Y1-R\_OP} = \frac{G_{Y1-R\_CL}}{1 - G_{Y1-R\_CL}} = \frac{G_{C1}G_1(1 + 2G_{C2}G_2)}{1 + G_{C2}G_1 + G_{C2}G_2} \tag{4}$$

$$G_{Y2-R\_OP} = \frac{G_{Y2-R\_CL}}{1 - G_{Y2-R\_CL}} = \frac{G_{C1}G_2(1 + 2G_{C2}G_1)}{1 + G_{C1}G_1 + G_{C2}G_1 + G_{C2}G_2 - G_{C1}G_2} \tag{5}$$

$$G_{DY-R-OP} = \frac{G_{DY-R\_CL}}{1 - G_{DY-R\_CL}} = \frac{G_{c1}(G_1 - G_2)}{1 + G_{C1}G_2 + G_{C2}G_1 + G_{C2}G_2 + 2G_{C1}G_{C2}G_1G_2} \tag{6}$$

From fig.1, we can get the transfer function between current loop and servo motor as follows:

$$G_I'(s) = \frac{K_M K_I T_I s + K_M}{C_1' s^3 + C_2' s^2 + C_3' s + \alpha D_M + \alpha D_L} \tag{7}$$

Where,

$$C_1' = LT_I J_m + LT_I J_L,\ C_2' = RT_I J_m + RT_I J_L + LT_I D_M + LT_I D_L + \alpha K_I T_I J_m + \alpha K_I T_I J_L,$$

$$C_3' = RT_I D_M + RT_I D_L + \alpha K_I T_I D_M + \alpha K_I T_I D_L + \alpha J_m + \alpha J_L + K_M K_\Omega T_I$$

For convenience of analysis, let

$$G_1 = \frac{m_1 s + n_1}{a_1 s^3 + b_1 s^2 + c_1 s + d_1} \quad , \quad G_2 = \frac{m_2 s + n_2}{a_2 s^3 + b_2 s^2 + c_2 s + d_2}$$
$$G_{C1} = \frac{K_{P1} s + K_{I1}}{s} \quad , \quad G_{C2} = \frac{K_{P2} s + K_{I2}}{s} \tag{8}$$

Where, $m_1$ , $m_2$ , $n_1$ , $n_2$ , $a_1$ , $a_2$ , $b_1$ , $b_2$ , $c_1$ , $c_2$ , $d_1$ and $d_2$ are all greater than zero, $K_{I1} > 0$ , when using proportional feedback $K_{I2} = 0$ , when using proportional integral feedback $K_{I2} > 0$ , substitute Eq.(8) into Eqs.(4), (5) and (6) respectively, we get:

$$\begin{cases} G_{Y1-R\_OP} = \dfrac{(m_1 s + n_1)(K_{P1} s + K_{I1})(C_1 s^4 + C_2 s^3 + C_3 s^2 + C_4 s + 2n_2 K_{I2})}{s(C_5 s^7 + C_6 s^6 + C_7 s^5 + C_8 s^4 + C_9 s^3 + C_{10} s^2 + C_{11} s + K_{I2} n_1 d_2 + K_{I2} n_2 d_1)} \\ G_{Y2-R\_OP} = \dfrac{(m_2 s + n_2)(K_{P1} s + K_{I1})(C_{12} s^4 + C_{13} s^3 + C_{14} s^2 + C_{15} s + 2n_1 K_{I2})}{s(C_{16} s^7 + C_{17} s^6 + C_{18} s^5 + C_{19} s^4 + C_{20} s^3 + C_{21} s^2 + C_{22} s + K_{I2} n_1 d_2 + K_{I2} n_2 d_1 + K_{I1} n_1 d_2 - K_{I1} n_2 d_1)} \\ G_{\Delta Y-R-OP} = \dfrac{s(K_{P1} s + K_{I1})(C_{23} s^4 + C_{24} s^3 + C_{25} s^2 + C_{26} s + n_1 d_2 - n_2 d_1)}{C_{27} s^8 + C_{28} s^7 + C_{29} s^6 + C_{30} s^5 + C_{31} s^4 + C_{32} s^3 + C_{33} s^2 + C_{34} s + 2K_{I1} K_{I2} n_2 n_1} \end{cases} \tag{9}$$

where, $C_i$ ($i$=1,2,3,…,34) are polynomial coefficients.

There are two cases in Eq. (9):

①. $K_{I2}$ ¹ $0$ , $K_{I2} n_1 d_2 + K_{I2} n_2 d_1 + K_{I1} n_1 d_2 - K_{I1} n_2 d_1 \neq 0$

In this case, numerator constant term coefficient of $G_{Y1-R\_OP}$ is not equal to zero, constant term coefficient in denominator bracket is also not equal to zero, so $G_{Y1-R\_OP}$ is type-1 system, $G_{Y2-R\_OP}$ is also type-1 system, the system can be synchronized for step signal; if it also meets $n_1 d_2 = n_2 d_1$ , the above two systems output all exist steady-state error, but the steady-state error is equal, so the two systems can still be synchronized.

②. $K_{I2}$ ¹ $0$ , $K_{I2} n_1 d_2 + K_{I2} n_2 d_1 + K_{I1} n_1 d_2 - K_{I1} n_2 d_1 = 0$

In this case, $G_{Y1-R\_OP}$ is still type-1 system, $G_{Y2-R\_OP}$ is at least type-2 system. The two systems can be synchronized for step signal, but cannot be synchronized for ramp signal.

From the above analysis, we can see that dual-motor can be synchronized for step signal when $K_{I2}$ ¹ $0$ , dual-motor could be synchronized for ramp signal only when it meets: $K_{I2}$ ¹ $0$ , $K_{I2} n_1 d_2 + K_{I2} n_2 d_1 + K_{I1} n_1 d_2 - K_{I1} n_2 d_1 \neq 0$ and $n_1 d_2 = n_2 d_1$ .

## 4 Design of Fuzzy PI Controller

System control block diagram is shown in Fig. 2.

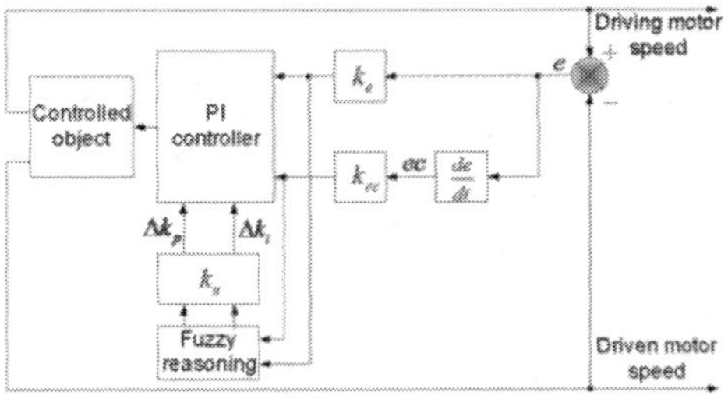


**Figure 2.** System control block diagram.

System control block diagram is shown in figure 2. The fuzzy PI controller use error $e$ and error change $ec$ as input. The control scheme is to find out the fuzzy relation between the two parameters of PI and $e, ec$ firstly, the two parameters of PI can be modified online by continuous testing of $e$ and $ec$ , so as to meet the different control parameter requirements for different $e$ and $ec$ , making the controlled object has a good dynamic and static performance.

### 4.1 Determination of Input and Output Variables

The speed error $e$ of dual-motor and the change rate $ec$ of speed error are as the input of fuzzy controller, correction parameters $\Delta k_p$ and $\Delta k_i$ are as the output of fuzzy controller, then PI regulator output parameters $k_p$ and $k_i$ are shown in Eq.(10).

$$k_p = k_p^{'} + \Delta k_p,\ k_i = k_i^{'} + \Delta k_i \qquad (10)$$

where $k_p^{'}$ and $k_i^{'}$ are setting values.

### 4.2 Determination of Membership Function

Fuzzy subsets of fuzzy controller input and output variables are $e$, $ec$, $\Delta k_p$ and $\Delta k_i$ respectively. Linguistic value of each variable is [large negative, medium negative, small negative, zero, small positive, medium positive, large positive] denoted by [NB, NM, NS, ZO, PS, PM, PB], discourse domain is integer in [-6, 6]. Taking into account the sensitivity, stability, robustness and the coverage degree of discourse domain, the membership function curve of each fuzzy subset is selected by triangle shape, shown in Fig. 3.

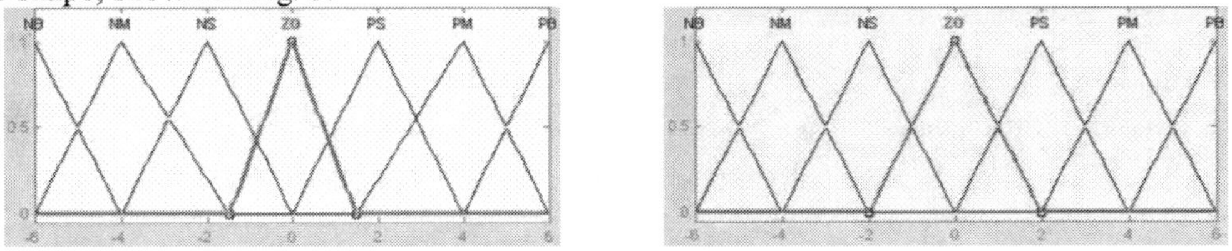


(a) Membership function of $e$ and $ec$ (b) Membership function of $\Delta k_p$ and $\Delta k_i$

**Figure 3.** Membership function curve.

### 4.3 Fuzzy Control Rules

The core of fuzzy control design is the determination of fuzzy control rules, and the selection of control rules is directly related to the performance of system. From the above parameters setting principle, through the fuzzy reasoning and experimental correction, we get the fuzzy control rules of $\Delta k_p$ and $\Delta k_i$ are shown in Table 1.

**Table 1.** Fuzzy control rules

| $e$ \ $ec$ | NB | NM | NS | ZO | PS | PM | PB |
|---|---|---|---|---|---|---|---|
| NB | PB/NB | PB/NB | PM/NM | PM/NM | PS/NS | ZO/ZO | ZO/ZO |
| NM | PB/NB | PB/NB | PM/NM | PS/NS | PS/NS | ZO/ZO | NS/ZO |
| NS | PM/NB | PM/NM | PM/NS | PS/NS | ZO/ZO | NS/PS | NS/PS |
| ZO | PM/NM | PM/NM | PS/NS | ZO/ZO | NS/PS | NM/PM | NM/PM |
| PS | PS/NM | PS/NS | ZO/ZO | NS/PS | NS/PS | NM/PM | NM/PB |
| PM | PS/ZO | ZO/ZO | NS/PS | NM/PS | NM/PM | NM/PB | NB/PB |
| PB | ZO/ZO | ZO/ZO | NM/PS | NM/PM | NM/PM | NB/PB | NB/PB |

### 4.4 Stability of Fuzzy PI Control System

In section 3, we have analyzed the stability conditions of PI control system in detail. In a conventional PI control system, if the linear PI controller is replaced by the nonlinear fuzzy PI controller, then the stability of the resulting control system remains unchanged [10]. Thus the stability conditions of fuzzy PI control system are the same as the PI control system. In the following simulation analysis, fuzzy PI control system is stable.

## 5 Simulation Experiment

Simulation block diagram is shown in Fig. 4:

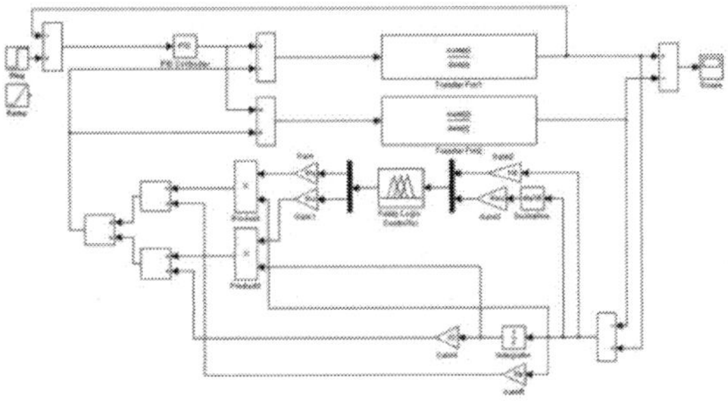

**Figure 4.** Simulation block diagram of fuzzy PI control.

Where, PID Controller is speed regulator, Transfer Fun1 is transfer function between driving motor and its current loop, Transfer Fun2 is transfer function between driven motor and its current loop, the input and output of simulation block diagram are speed given signal and dual-motor speed error respectively.

According to reference[11], we get:

$$G_I'(s)=\frac{8.7318s+6750}{1.1\times10^{-4}s^3+6.917\times10^{-2}s^2+50s+0.85} \tag{11}$$

let $G_1(s)=G_I'(s)$

In the following simulation, system parameters are chosen as follows:

$$k_e=600\,,k_{ec}=666.7\,,k_u=1.5\,,k_p=2.3\,,k_i=1.2$$

PID controller parameter is chosen as follows:kp=5.6,ki=1.6

1) In order to satisfy the equation:

$$K_{I2}n_1d_2+K_{I2}n_2d_1+K_{I1}n_1d_2-K_{I1}n_2d_1=0 \tag{12}$$

let

$$G_2(s)=\frac{8.7318s+6700}{1.1\times10^{-4}s^3+6.917\times10^{-2}s^2+50s+0.8296} \tag{13}$$

Thus, we get $n_1=6750$, $n_2=6700$, $d_1=0.85$, $d_2=0.8296$

and substituted into eq.(12), after simplification, we obtain $11295K_{I2}=95K_{I1}$.

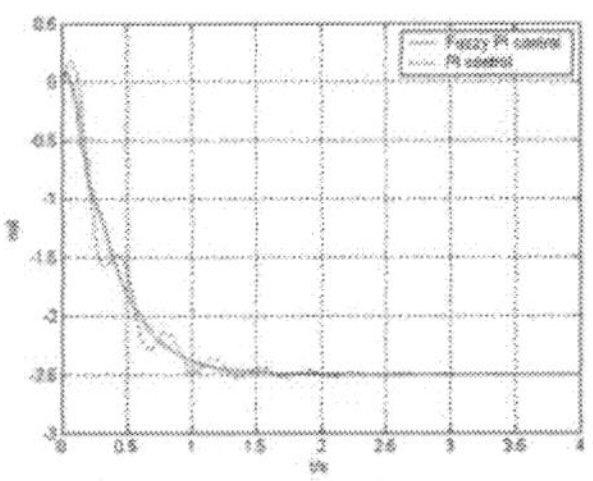

**Figure 5.** Speed error of dual-motor tracking ramp signal 1000t(mil/s).

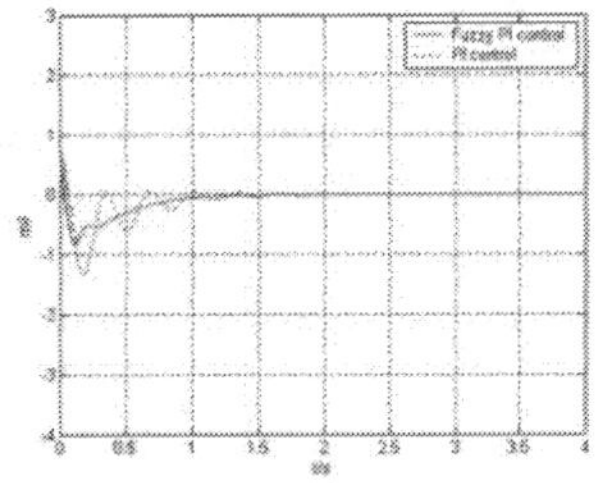

**Figure 6.** Speed error of dual-motor tracking step signal 500mil.

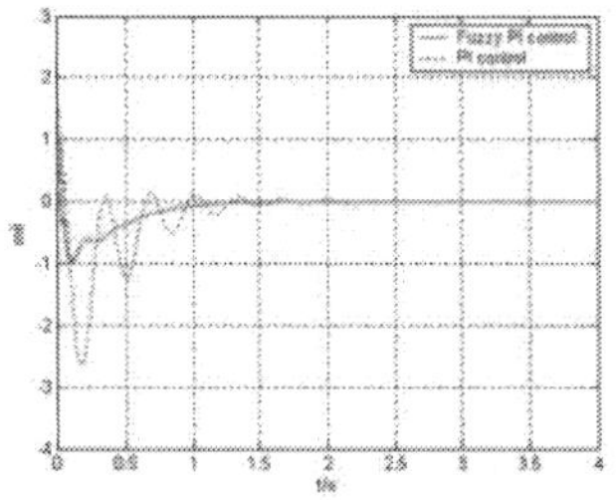

**Figure 7.** Speed error of dual-motor tracking step signal 1000mil.

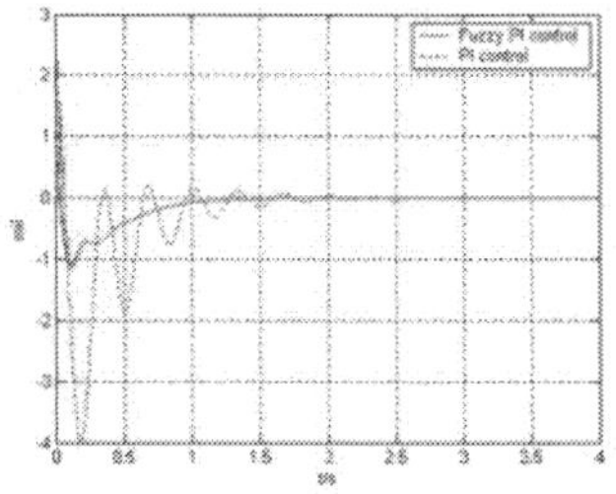

**Figure 8.** Speed error of dual-motor tracking step signal 1500mil.

2) In order to satisfy the equation:

$$K_{I2}n_1d_2 + K_{I2}n_2d_1 + K_{I1}n_1d_2 - K_{I1}n_2d_1 \neq 0 \tag{14}$$

and $n_1d_2 = n_2d_1$, let

$$G_2(s) = \frac{8.7318s + 6700}{1.1\times10^{-4}s^3 + 6.917\times10^{-2}s^2 + 50s + 0.8437} \tag{15}$$

we can get $n_1 = 6750$, $n_2 = 6700$, $d_1 = 0.85$, $d_2 = 0.8437$ and substituted into Eq.(14), after simplification, we get $K_{I2} \neq 0$.

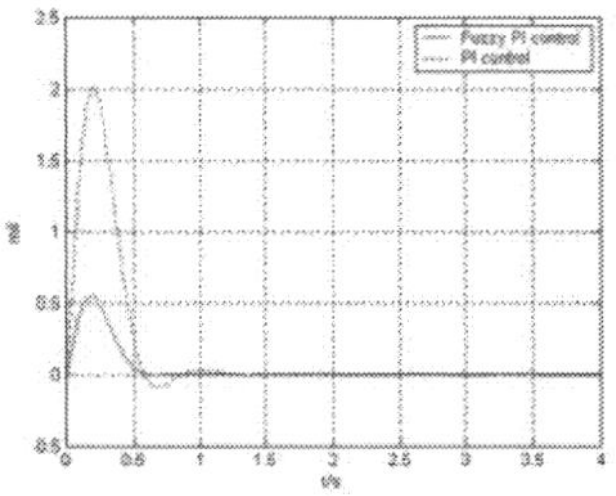

**Figure 9.** Speed error of dual-motor tracking ramp signal 1000t(mil/s).

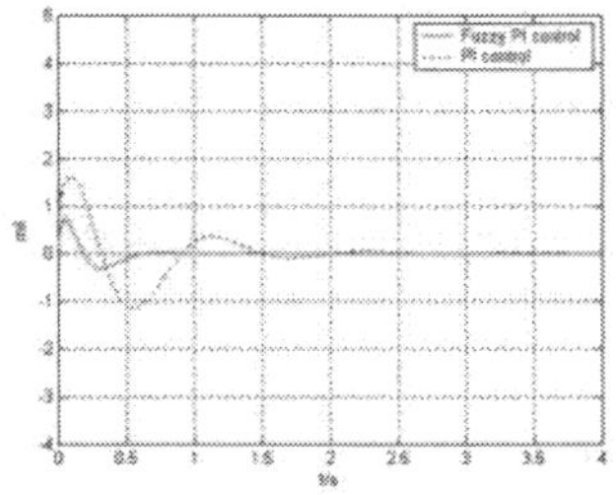

**Figure 10.** Speed error of dual-motor tracking step signal 500mil.

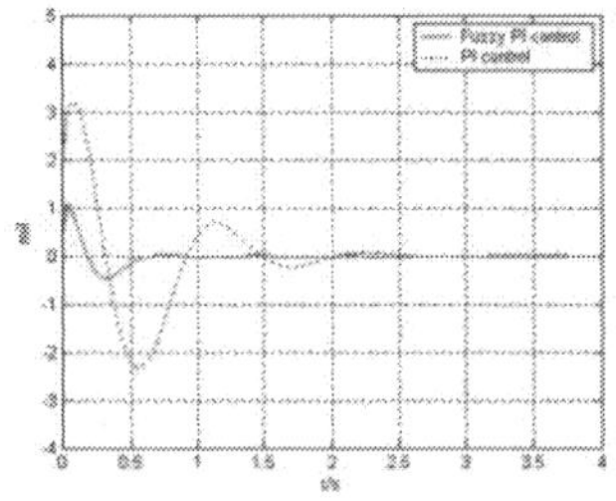

**Figure 11.** Speed error of dual-motor tracking step signal 1000mil.

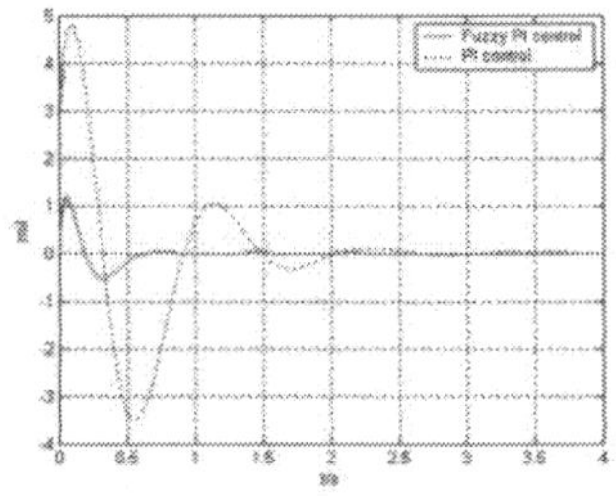

**Figure 12.** Speed error of dual-motor tracking step signal 1500mil.

From Fig. 5 to Fig. 12, we can see that when the input is a ramp signal, speed-error feedback controller using fuzzy PI compared with PI, the former start up process is better, reducing the overshoot and oscillation frequency. In the condition of input step signal amplitude changes, the speed error of dual-motor changes slightly when using fuzzy PI control, but the speed error of dual-motor changes greatly when using PI control, so the fuzzy PI controller has stronger robustness than PI controller.

## 6 Conclusion

Aiming at the dual-motor driving servo system, a simple fuzzy PI controller was presented for the first time. The conventional PI control system has a large overshoot, and the pure fuzzy control cannot eliminate the steady-state error due to its structure. In this paper, the two control methods are combined, namely fuzzy PI control, simulation results show that the fuzzy PI control not only has better dynamic and static quality, but also has better adaptability [12-15] than conventional PI control, which can significantly improve the control performance of system, validating the efficacy of the proposed control strategy. This paper provides a reference for the further research of dual-motor driving servo system.

## Acknowledgment

This work was supported by National Natural Science Foundation of China(No.61074023), Nature Science Foundation in Anhui Province of China(No.1508085MF130), Natural Science Research Key Project of Universities in Anhui Province of China (No.KJ2015A297), the Engineering Technology Research Center of Optoelectronic Appliance in Anhui Province of China, Sichuan Institute of Aerospace System Engineering in Sichuan Province of China.

The authors confirm that this article content has no conflicts of interest.

## References

1. I. X. Bogiatzidis, E. D. Mitronikas, A. N. Safacas, Investigation of backlash phenomena appearing in a twin AC cement kiln drive, 19th International Conference on Electrical Machines, 2010, 1-6.
2. I. X. Bogiatzidis, A. N. Safacas, Vibration analysis and backlash identification of a twin AC drive for a cement kiln, 5th IET International Conference on Power Electronics, Machines and Drives, 2010, 1-6.
3. L. F. Sun, Research on motion characteristic and anti-backlash control of dual-motor system, The Outstanding Master's Degree Thesis of Northeastern University, June 2012.
4. H. Z. Zhu, W. X. Song, Y. Dong, Modeling and simulation of coaxial connected dual-motor drive system, Electric Mach. Control Appl. 40(2) (2013) 23-27.
5. Q. H. Ngo, N. P. Nguyen, C. N. Nguyen, Fuzzy sliding mode control of container cranes, Int. J. Control, Autom. Syst. 13(2) (2015) 419-425.
6. C. Y. Wang, X. H. Jiao, Observer-based adaptive arbitrary switching fuzzy tracking control for a class of switched nonlinear systems, Int. J. Control, Autom. Syst. 13(4) (2015) 823-830.
7. D. W. Kim, H. J. Lee, M. H. Kim, Robust sampled-data fuzzy control of nonlinear systems with parametric uncertainties: Its application to depth control of autonomous underwater vehicles, Int. J. Control, Autom. Syst. 10(6) (2015) 1164-1172.
8. X. J. Su, X. Z. Yang, P. Shi, Fuzzy control of nonlinear electromagnetic suspension systems, Mechatron. 24(4) (2014) 328-335.
9. H. Ying, S. William, J. B. James, Fuzzy control theory: a nonlinear case, Automatica, 26(3) (1990) 513-520.
10. G. R. Chen, H. Ying, Stability analysis of nonlinear fuzzy PI control systems, The Third International Conference on Industrial Fuzzy Control and Intelligent Systems, 1993, 128-133.
11. H. B. Zhao, X. H. Zhou, Research of fuzzy PI control of dual-motors driving servo system, Journal of Chaohu College, 11(3) (2009) 71-75.
12. Y. M. Jia, H. Kokame J. Lunze, Simultaneous adaptive decoupling and model matching control of a fluidized bed combustor for sewage sludge, IEEE T. Control Syst. Technol. 11(4) (2003) 571-577.
13. S. J. Yoo, Distributed adaptive consensus tracking of a class of networked non-linear systems with parametric uncertainties, IET Control Theor. Appl. 7(7) (2013) 1049-1057.
14. B. Yao, C. X. Hu, L. Lu, Adaptive Robust Precision Motion Control of a High-Speed Industrial Gantry With Cogging Force Compensations, IEEE T. Control Syst. Technol. 19(5) (2011) 1149-115.
15. S. Y. Xu, Further results on adaptive robust control of uncertain time-delay systems, IET Control Theor. Appl. 2(5) (2008) 402-408.

# Backstepping Adaptive Fuzzy Control of Servo System With LuGre Friction

Haibo Zhao[1,2,a], Chengguang Wang[3]

[1]*Engineering Technology Research Center of Optoelectronic Appliance, Tongling University, Tongling Anhui, 244061, China*
[2]*Department of Electrical Engineering, Tongling University, Tongling Anhui, 244061, China*
[3]*Sichuan Institute of Aerospace System Engineering, Chengdu Sichuan, 610100, China*

**Abstract.** Aiming at the control problem of servo system with friction nonlinearity, we introduced the improved LuGre friction model of system. We used adaptive fuzzy system to approximate the nonlinear part and unknown parameters in system online. The purpose is to avoid the complex calculation in deducing the adaptive law of each unknown parameter. By using backstepping approach and recursively selecting the Lyapunov function, introducing the virtual control quantity, we designed an adaptive fuzzy controller with state feedback. Then we analyzed the stability of adaptive fuzzy controller. We carried out the system response analysis and system robustness analysis. We used sinusoidal signal as simulation input signal. Two cases were considered in simulation test comparing backstepping control with conventional PID control. Simulation results show that the proposed backstepping control has better position tracking performance and robustness than conventional PID control.

## 1 Introduction

In electromechanical servo system, the existence of friction nonlinearity will affect the system dynamic performance and steady-state precision [1]. Hence, it is necessary to study the control strategy of weakening the effect of friction nonlinearity to system. So far, scholars at home and abroad have carried on a large amount of researches in friction modeling. The famous models of friction nonlinearity including the Dahl model [2], the LuGre model[3], the Elasto-plastic model [4], the Leuven model [5] and the GMS model [6]. Among them, the LuGre model is widely used [7] because it is simple and accurate in describing the friction characteristics at low speed. However, when system is running in high speed, the system friction are mainly coulomb friction and viscous friction, if still using the LuGre model, it will inevitably increase the complexity of the system. In addition, due to the changes of temperature, lubrication and contact force, the parameters of friction torque will also change.

Many scholars have studied the compensation control strategies of friction nonlinearity. Y. Wang [8] proposed an effevtive observer-based compensation scheme for the effects of friction. The results show that the amplitudes of limit cycles are greatly reduced with the disturbance observer-based compensation, and the stick-slip motion at low velocities is also eliminated. S. Yang [9] proposed a new repetitive adaptive compensation scheme for the dynamic friction. Computer simulations verify

[a] Corresponding author : happyzhaohaibo@126.com

the effectiveness of the proposed scheme for the high-precision servo system. In [10], aiming at unknown dynamic friction parameters and unknown load characteristic, an adaptive friction compensation is designed for the servo system. Simulation and experiment results are shown to effectively inhibit the disadvantageous influence of friction and lay the foundation for improvement of dynamic performance of servo system. In [11], a friction torque compensation controller without distal-end feedback is designed to ensure that the output of the tendon-sheath actuation system can effectively track the expected torque trajectory, and the effectiveness of the scheme is validated by experiments. B. Feng [12] proposed a novel adaptive configuration method of friction compensation pulse characteristic parameters. The experiment results show that the proposed method has a better performance. In [13], a feedforward friction observer is proposed as formulating an explicit analytical expression for the applied observation function. The simulation results show that the proportional control combined with the observer compensates faster for frictional disturbances at suddenly changing frictional conditions than PI control. M.T.S. Aung [14] presented a new method of friction compensation that can be applied to geared actuators with high presliding stiffness. The proposed method is validated through experiments employing a harmonic drive transmission. Howerer, [8-14] have not used adaptive fuzzy control.

Adaptive fuzzy logic system proposed by Wang [15] has universal approximation property, which can effectively approximate the nonlinear function of system. In recent years, it has been widely used in the nonlinear control of various servo systems [16-20]. In this paper, an improved LuGre model [21-22] is used to describe the friction torque of system at low speed and high speed. We use an adaptive fuzzy logic system to approximate the nonlinear part of servo system with friction, design a backstepping [23-24] adaptive fuzzy controller under the condition of the unknown system parameters and load torque. Simulation and practical experiments validate the efficacy of the proposed control strategy.

## 2 System Modeling

If only considering the influence of LuGre friction nonlinear, the structure diagram of servo system is shown in Fig. 1.

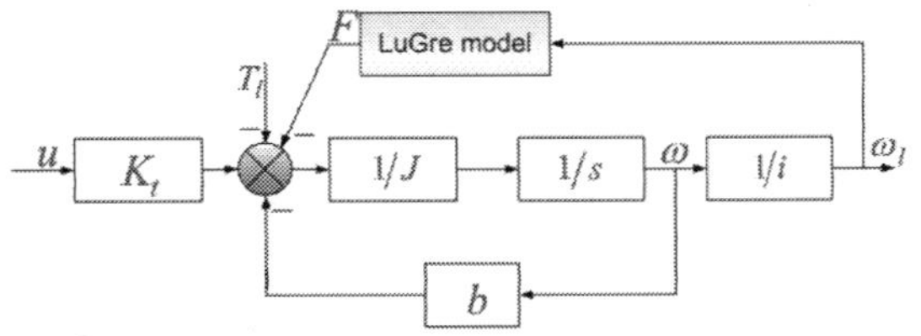


**Figure 1.** The structure diagram of servo system.

We can get the dynamic equation of servo system is as follows [1].

$$\begin{cases} \dfrac{d\theta}{dt} = \omega \\ J\dfrac{d\omega}{dt} = K_t u - b\omega - F - T_l \\ y = \theta \end{cases} \tag{1}$$

where $\theta$ and $\omega$ are the position and speed of motor output shaft respectively. $J$ and $b$ are equivalent moment of inertia and equivalent damping coefficient converted to the motor shaft respectively. $K_t$ is motor torque constant. $u$ is control variable. $F$ is load friction torque. $T_l$ is load torque converted to the motor shaft. If the LuGre model is adopted [21], then the load friction torque can be expressed as follows.

$$\frac{dz}{dt} = s(|\omega_l|)(\omega_l - \frac{|\omega_l|}{g(\omega_l)} z) \tag{2}$$

$$F = \sigma_0 s(|\omega_l|)z + \sigma_1 \dot{z} + F_c \operatorname{sgn}(\omega_l)(1 - s(|\omega_l|)) + \alpha_2 \omega_l \tag{3}$$

$$\sigma_0 g(\omega_l) = F_c + (F_s - F_c)e^{-\frac{\omega_l}{\omega_s}} = \alpha_0 + \alpha_1 e^{-\frac{\omega_l}{\omega_s}} \tag{4}$$

Where

$$s(|\omega_l|) = \begin{cases} 1 & |\omega_l| < \omega_1 \\ \dfrac{\omega_2 - |\omega_l|}{\omega_2 - \omega_1} & \omega_1 \leqslant |\omega_l| \leqslant \omega_2 \\ 0 & |\omega_l| > \omega_2 \end{cases} \tag{5}$$

where $\omega_2 > \omega_1 > 0$.

In Eq. (2)-(4), $z$ is the average deformation amount of bristles. $\omega_l$ is load speed. $\sigma_0$ is the stiffness coefficient of bristles. $\sigma_1$ is the damping coefficient of bristles. $\alpha_2$ is the viscous damping coefficient. $F_c$ is the coulomb friction coefficient. $F_s$ is the static friction coefficient. $\omega_s$ is the Stribeck velocity.

When the system is running at low speed, $s(|\omega|) = 1$, $F = \sigma_0 z + \sigma_1 \dot{z} + \alpha_2 \omega$, the friction model is equivalent to the LuGre model. When the system is running at high speed, $s(|\omega|) = 0$, $F = F_c \operatorname{sgn}(\omega) + \alpha_2 \omega$, the friction model is equivalent to a static friction model. In the actual LuGre model, parameters will change due to the influence of temperature, lubrication, material wear and other factors. So we use $\lambda$ to reflect the influence of parameter change on the friction torque. Then the load friction torque is as follows.

$$F = \lambda(\sigma_0 s(|\omega_l|)z + \sigma_1 \dot{z} + F_c \operatorname{sgn}(\omega_l)(1 - s(|\omega_l|)) + \alpha_2 \omega_l) \tag{6}$$

So the servo system model can be expressed as follows.

$$\begin{cases} \dfrac{d\theta}{dt} = \omega \\ \dfrac{d\omega}{dt} = C_0 u - C_1 \omega - \lambda C_2 \omega - \lambda \varphi z - \lambda C_5 F_c - C_6 T_l \\ y = \theta \end{cases} \tag{7}$$

where,

$C_0 = \dfrac{K_t}{J}$, $C_1 = \dfrac{b}{J}$, $C_2 = \dfrac{\sigma_1 s(|\omega_l|) + \alpha_2}{J}$, $C_3 = \dfrac{\sigma_0 s(|\omega_l|)}{J}$, $C_4 = \dfrac{\sigma_0 \sigma_1 s(|\omega_l|)}{J}$, $C_5 = \dfrac{\operatorname{sgn}(\omega_l)(1 - s(|\omega_l|))}{J}$,

$C_6 = \dfrac{1}{J}$, $\varphi = C_3 - \dfrac{C_4 |\omega|}{g(\omega_l)}$.

## 3 Backstepping Adaptive Fuzzy Control

### 3.1 Fuzzy Approximation System

In order to reduce the influence of the friction nonlinearity and the load uncertainty in servo system, we design controller employing adaptive fuzzy approximation system[19] to compensate them. The purpose is to make system can accurately track the given position signal.

We use the following fuzzy inference rules:

$R^j$ :If $x_1$ is $F_1^j$ and $x_2$ is $F_2^j$ ...and $x_n$ is $F_n^j$, then $y$ is $G^j$ $j=1,2,\cdots,M$

where $x=[x_1,x_2,\cdots,x_n]^{\mathrm{T}} \in R^n$ is the input of fuzzy system, $y \in R$ is the output of fuzzy system, $F_i^j$ and $G^j$ are fuzzy sets.

Using the following fuzzy system:

$$y=\frac{\sum_{j=1}^{M}\bar{y}^j\left(\prod_{i=1}^{n}\mu_{F_i^j}(x_i)\right)}{\sum_{j=1}^{M}\prod_{i=1}^{n}\mu_{F_i^j}(x_i)} \tag{8}$$

where, $\mu_{G^j}(\bar{y}^j)$ is fuzzy membership function, $\bar{y}^j$ is the maximum point of $\mu_{G^j}(\bar{y}^j)$.

Let

$$q_j=\frac{\prod_{i=1}^{n}\mu_{F_i^j}(x_i)}{\sum_{j=1}^{M}\left(\prod_{i=1}^{n}\mu_{F_i^j}(x_i)\right)} \tag{9}$$

define $\theta=[\bar{y}^1,\bar{y}^2,\cdots,\bar{y}^M]^{\mathrm{T}}$ as the adaptive parameter vector of fuzzy system, $q(x)=[q_1(x),q_2(x),\cdots,q_M(x)]^{\mathrm{T}}$ as fuzzy basis function vector, then Eq. (8) can be expressed as follows.

$$y(x)=\theta^{\mathrm{T}}q(x) \tag{10}$$

### 3.2 Backstepping Design

In this paper, we designed an adaptive controller so as to make system output $y$ asymptotically stable tracking the expected output $y_d$, namely $\lim_{t\to\infty}|y-y_d|=0$.

*Theorem 1:* For any given continuous function $p(x)$ on the compact set $U \in R^n$ and arbitrary $\varepsilon>0$, there exists a function $f(x)$ in the form of (8), such that:

$$\sup_{x\in U}\|f(x)-p(x)\|<\varepsilon \tag{11}$$

*Assumption 1:* System parameter $K_t/J$ is bounded. There exist $d_{\mathrm{m}}$ and $d_M$, such that

$$0<d_{\mathrm{m}}\leqslant|K_t/J|\leqslant d_M \tag{12}$$

*Assumption 2:* The first and second derivative of $y_d$ exist and are all bounded.

**Step 1** Define output error $e_1$ as

$$e_1=y-y_d=\theta-y_d \tag{13}$$

Its derivative is

$$\dot{e}_1=\dot{\theta}-\dot{y}_d=\omega-\dot{y}_d=\omega-\omega_d \tag{14}$$

Choose the following control Lyapunov function $V_1$

$$V_1=\frac{1}{2}e_1^2 \tag{15}$$

Its derivative is

$$\dot{V}_1=e_1\dot{e}_1=e_1(\omega-\dot{y}_d) \tag{16}$$

Let $\omega$ be virtual control variable, define $\alpha_1$ as feedback control input, the value of $\alpha_1$ is:

$$\alpha_1 = -k_1 e_1 + \dot{y}_d \tag{17}$$

where, $k_1 > 0$ is a design parameter, then Eq.(16) can be expressed as follows.

$$\dot{V}_1 = e_1(\omega - \alpha_1 - k_1 e_1) = -k_1 e_1^2 + e_1(\omega - \alpha_1) \tag{18}$$

**Step 2** Define the error between $\omega$ and $\alpha_1$ as

$$e_2 = \omega - \alpha_1 \tag{19}$$

Choose the following augmented control Lyapunov function $V_2$

$$V_2 = V_1 + \frac{1}{2} e_2^2 \tag{20}$$

Its derivative is

$$\dot{V}_2 = -k_1 e_1^2 + e_2(f^*(x) + g(x)u) \tag{21}$$

where,

$$f^*(x) = e_1 + (-b\omega - F - T_l)/J - \dot{\alpha}_1 \tag{22}$$

$$\dot{\alpha}_1 = \frac{\partial \alpha_1}{\partial x_1}\dot{x}_1 + \frac{\partial \alpha_1}{\partial y_d}\dot{y}_d + \frac{\partial \alpha_1}{\partial \dot{y}_d}\ddot{y}_d \tag{23}$$

We can see that $f^*(x)$ is an unknown function. For any given accuracy requirement $\varepsilon > 0$, there exist a fuzzy logic system, such that

$$f^* = \theta^{\mathrm{T}} q(x) + \delta(x),\ |\delta(x)| < \varepsilon \tag{24}$$

where $\delta(x)$ is an approximation error function.
define

$$\beta = \frac{1}{d_m}\|\theta\|^2 \tag{25}$$

Define the estimated value of $\beta$ as $\hat{\beta}$ which satisfying the following differential equation.

$$\dot{\hat{\beta}} = \frac{r}{2a^2} e_2^2 q^{\mathrm{T}} q - \sigma\hat{\beta} \tag{26}$$

where, $r > 0$, $\sigma > 0$ are design parameters.
We choose control law as follows.

$$u = -(k_2 + 0.5)e_2 - \frac{1}{2a^2} e_2 \hat{\beta} q^{\mathrm{T}} q \tag{27}$$

where, $k_2 > 0$, $a > 0$ are design parameters.
then we can get

$$e_2 f^* = e_2\theta^{\mathrm{T}} q(x) + e_2\delta(x) \leqslant \frac{1}{2a^2} e_2^2 q^{\mathrm{T}} q \|\theta\|^2 + \frac{a^2}{2}\frac{\theta^{\mathrm{T}}\theta}{\|\theta\|^2} + \frac{1}{2d_m}\varepsilon^2 \tag{28}$$

$$e_2 g(x) u \leqslant -(k_2 + 0.5) d_m e_2^2 - \frac{d_m}{2a^2} e_2^2 \hat{\beta} q^{\mathrm{T}} q \tag{29}$$

substitute Eqs.(28),(29) into Eqs.(21), we can get

$$\dot{V}_2 \leqslant -k_1 e_1^2 - k_2 d_m e_2^2 + \frac{1}{2}a^2 + \frac{1}{2d_m}\varepsilon^2 + \frac{d_m}{2a^2} e_2^2 q^T q \tilde{\beta} \tag{30}$$

where $\tilde{\beta} = \beta - \hat{\beta}$

### 3.3 Stability Analysis

We choose the augmented control Lyapunov function as follows.

$$V_3 = V_2 + \frac{d_m}{2r}\tilde{\beta}^2 \tag{31}$$

Its derivative is

$$\dot{V}_3 \leqslant -k_1e_1^2 - k_2d_me_2^2 + \frac{1}{2}a^2 + \frac{1}{2d_m}\varepsilon^2 + \frac{d_m}{2a^2}e_2^2q^{\mathrm{T}}q\tilde{\beta} - \frac{d_m}{r}\tilde{\beta}\dot{\hat{\beta}} = -k_1e_1^2 - k_2d_me_2^2 + \frac{1}{2}a^2 + \frac{1}{2d_m}\varepsilon^2$$

$$+\frac{d_m}{r}\tilde{\beta}(\frac{r}{2a^2}e_2^2q^{\mathrm{T}}q - \dot{\hat{\beta}}) = -k_1e_1^2 - k_2d_me_2^2 + \frac{1}{2}a^2 + \frac{1}{2d_m}\varepsilon^2 + \frac{\sigma d_m}{r}\tilde{\beta}\hat{\beta} \tag{32}$$

notice that

$$\tilde{\beta}\hat{\beta} = \tilde{\beta}(\beta - \tilde{\beta}) \leqslant \frac{1}{2}\beta^2 - \frac{1}{2}\tilde{\beta}^2 \tag{33}$$

So Eq. (32) can be expressed as follows.

$$\dot{V}_3 \leqslant -k_1e_1^2 - k_2d_me_2^2 + \frac{1}{2}a^2 + \frac{1}{2d_m}\varepsilon^2 + \frac{\sigma d_m}{2r}\beta^2 - \frac{\sigma d_m}{2r}\tilde{\beta}^2 \leqslant -m_0V_3 + n_0 \tag{34}$$

where, $m_0 = \min\{2k_1, 2d_mk_2, \sigma\}$, $n_0 = \frac{1}{2}a^2 + \frac{1}{2d_m}\varepsilon^2 + \frac{\sigma d_m}{2r}\beta^2$

Calculating the integral on both sides of eq.(34), we have:

$$V_3(t) \leqslant e^{-m_0t}V_3(0) + \frac{n_0}{m_0}(1 - e^{-m_0t}) \tag{35}$$

According to eqs.(31),(35), we have:

$$e_1^2 \leqslant 2V_3(t), \lim_{t\to\infty} e_1(t) \leqslant \sqrt{\frac{2n_0}{m_0}} \tag{36}$$

Because $k_1, k_2, r, \sigma, a$ are all design parameters, we can appropriately choose them to ensure that $n_0/m_0$ is small sufficiently, which can ensure that the error $e_1$ converges to a sufficient small neighborhood of the origin according to the exponential law.

## 4 Simulation Experiment

The following simulation graphics are the comparison charts of backstepping adaptive fuzzy control (BAFC) and the conventional PID control.

The parameters of servo system model are as follows: $J = 0.8\mathrm{kg\cdot m^2}$, $K_t = 1$, $b = 0.5\mathrm{N\cdot m\cdot s/rad}$, $i = 15$. The parameters of LuGre friction model are as follows: $\sigma_0 = 9500\mathrm{N\cdot m}$, $\sigma_1 = 25\mathrm{N\cdot m}$, $\sigma_2 = 0.3\mathrm{N\cdot m\cdot s/rad}$, $F_c = 0.5\mathrm{N\cdot m}$, $F_s = 0.45\mathrm{N\cdot m}$, $\omega_s = 0.01\mathrm{rad/s}$. The initial state is: $\theta = 0$, $\omega = 0$. The design parameters are: $k_1 = 90$, $k_2 = 80$, $a = 0.3$, $r = 6.2$, $\sigma = 0.6$. The fuzzy membership functions are as follows.

$\mu_{F_i^1}(x_i) = 1/(1 + \exp(-6(x_i - 6)))$, $\mu_{F_i^2}(x_i) = \exp(-0.5(x_i - 4)^2/2)$, $\mu_{F_i^3}(x_i) = \exp(-0.5(x_i - 2)^2/2)$

$\mu_{F_i^4}(x_i) = \exp(-0.5(x_i - 0)^2/2)$, $\mu_{F_i^5}(x_i) = \exp(-0.5(x_i + 2)^2/2)$, $\mu_{F_i^6}(x_i) = \exp(-0.5(x_i + 4)^2/2)$

$\mu_{F_i^7}(x_i) = 1/(1 + \exp(6(x_i + 6)))$

### 4.1 System Response Analysis

Fig. 2(a) is the system position tracking curve for sinusoidal signal input $y_d = \sin(1.2\pi t)$ by using PID control and BAFC, respectively. Fig. 2(b) is the error curve between $y$ and $y_d$ for sinusoidal signal input. Fig. 3(a) is the system velocity tracking curve for sinusoidal signal input $y_d = \sin(1.2\pi t)$ by using PID control and BAFC, respectively. Fig. 3(b) is the error curve between $\omega$ and $\omega_d$ for sinusoidal signal input. Fig. 4 is the controller output curve by using PID control and BAFC, respectively. From Fig. 2 and Fig. 3, we can see that position tracking exist flat top phenomenon and velocity tracking exists deadzone phenomenon when using PID control. But there are no flat top phenomenon and deadzone phenomenon exist when using BAFC. So the system has better tracking performance and smaller tracking error when using BAFC.

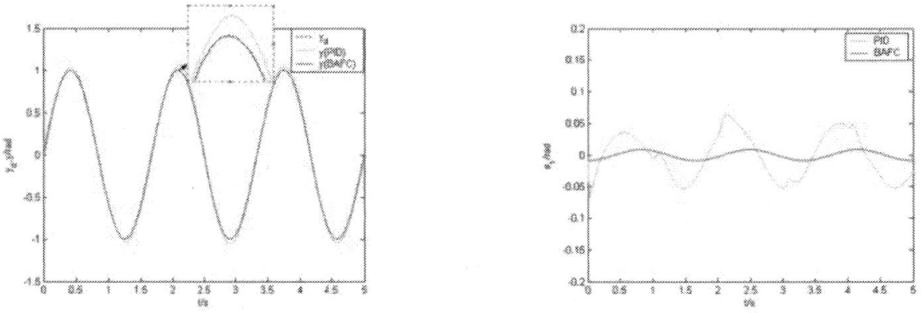


(a) System position tracking for sinusoidal signal (b) Position tracking error for sinusoidal signal

**Figure 2.** System position tracking and tracking error for sinusoidal signal input.

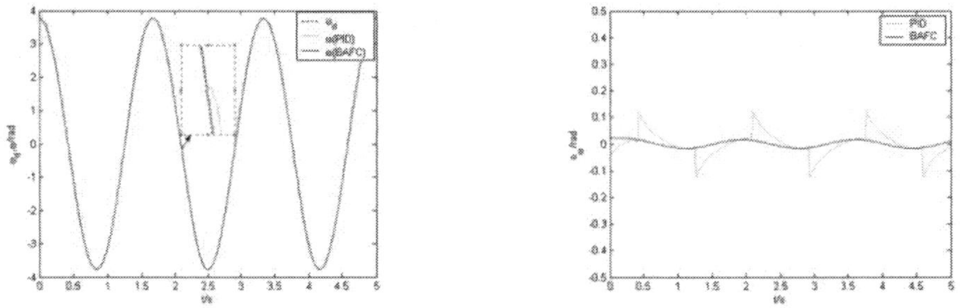


(a) System velocity tracking for sinusoidal signal (b) Velocity tracking error for sinusoidal signal

**Figure 3.** System velocity tracking and tracking error for sinusoidal signal input.

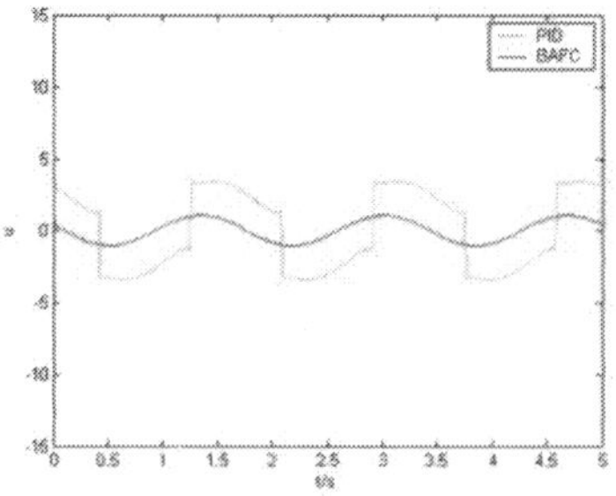


**Figure 4.** System controller output.

## 4.2 System Robustness Analysis

In Fig. 5, 10N·m disturbance signal is applied to system at 2s, the system not only has a larger fluctuation at 2s but also the tracking performance become worse after 2s when using PID control. But the system tracking performance don't become worse after 2s, only has a smaller fluctuation at 2s when using BAFC. Thus BAFC has better robustness than PID control.

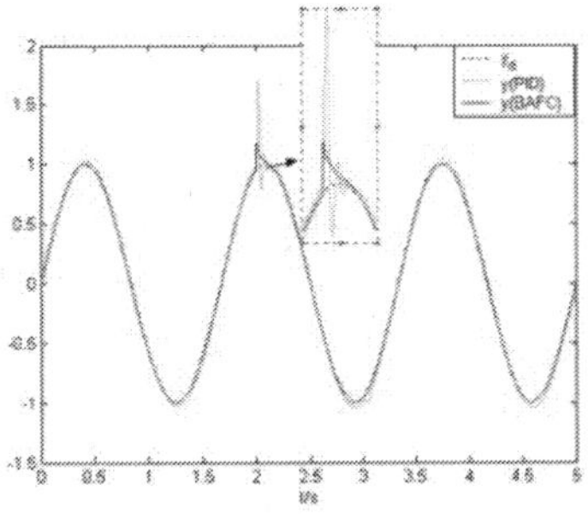

(a) System sinusoidal position tracking containing disturbance

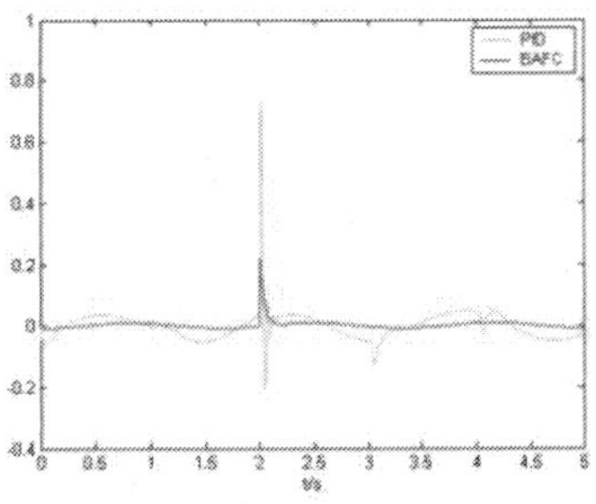

(b) Sinusoidal position tracking error containing disturbance

**Figure 5.** System position tracking and tracking error for sinusoidal signal when disturbance exists.

## 5 Conclusion

In this paper, we designed the improved LuGre friction model-based backstepping adaptive fuzzy control of servo system. The proposed control strategy has the characteristics of higher tracking accuracy, better robustness and adaptability than conventional PID control. But the system also exists backlash nonlinearity, we don't consider the coexistence of backlash and friction nonlinearlity in the system. It will be our future work considering the coexistence of them.

## Acknowledgment

This work was supported by National Natural Science Foundation of China(No.61074023), Nature Science Foundation in Anhui Province of China(No.1508085MF130), Natural Science Research Key Project of Universities in Anhui Province of China (No.KJ2015A297), the Engineering Technology Research Center of Optoelectronic Appliance in Anhui Province of China, Sichuan Institute of Aerospace System Engineering in Sichuan Province of China.

The authors confirm that this article content has no conflicts of interest.

## References

1. J. Zhou, B. Duan, J. Huang, Effect and compensation for servo systems using LuGre friction model, Control Theor. App. 25(6) (2008) 990-994.
2. S. K. Shome, M. Prakash, S. Pradhan, Robust dahl model based Sliding Mode Control for micro/nano positioning applications, 11th IEEE India Conference: Emerging Trends and Innovation in Technology, February 3, 2015.
3. H. Xie, Z. Liu, H. Yang, Advancing control for shield tunneling machine by Backstepping design with lugre friction model, Math. Probl. Eng. ten pages, 2014.
4. I. Redaelli, C. di Prisco, D. Vescovi, A visco-elasto-plastic model for granular materials under simple shear conditions, Int. J. Numer. Anal. Methods Geomech. 40(1) (2016) 80-104.
5. V. Lampaert, J. Swevers, F. Al-Bender, Modification of the Leuven integrated friction model structure, IEEE T. Autom. Control 47(4) (2002) 683-687.
6. I. Nilkhamhang, A. Sano, Robust adaptive friction compensation using linearly-parameterized GMS model, 9th European Control Conference, 2015, 1314-1319.
7. W. Tan, X. Li, H. Xiang, Research on compensation of torque ripple in servo system, J. Mech. Eng. 47(12) (2011) 1-6.
8. Y. Wang, Z. He, Friction compensation for servo systems, Electric Mach. Control 17(8) (2013) 107-112.

9. S. Yang, M. Zeng, New repetitive adaptive friction compensation scheme in high-precise servo system, J. Southeast U. 36(SUPPL) (2006) 74-48.
10. Y. Wu, D. Ma, J. Yao, Application of adaptive robust control in mechatronic servo system based on modified LuGre model, J. Mech. Eng. 50(22) (2014) 207-212.
11. Q. Wu, X. Wang, F. Du, Torque transmission characteristics and friction compensation for a tendon-sheath actuation system in pull-pull configuration, J. Mech. Eng. 51(5) (2015) 22-29.
12. B. Feng, X. S. Mei, J. Yang, Adaptive configuration method of friction compensation pulse characteristic parameters, J. Shanghai Jiaotong U. 58(5) (2014) 713-718.
13. M. Ruderman, Tracking control of motor drives using feedforward friction observer, IEEE T. Ind. Electron. 61(7) (2014) 3727-3735.
14. T. S. Aung, R. Kikuuwe, M. Yamamoto, Friction compensation of geared actuators with high presliding stiffness, J. Dyn. Syst. Meas. Control, T. ASME 137(1) (2015) 8 pages.
15. L. X. Wang, J. M. Mendel, Fuzzy basis functions, universal approximation, and orthogonal least-squares learning, IEEE T. Neural Net. 3(5) (1992) 807-814.
16. J. P. Yu, B. Chen, H. S. Yu, Adaptive fuzzy backstepping position tracking control for permanent magnet synchronous motor, Control and Decision, 25(10) (2010) 1547-1551.
17. H. Chaoui, P. Sicard, Adaptive fuzzy logic control of permanent magnet synchronous machines with nonlinear friction, IEEE T. Ind. Electron. 59(2) (2010) 1123-1133.
18. X. Guo, Q. Wang, G. Li, Adaptive fuzzy control for permanent magnet spherical motor based on friction compensation, P. Chin. Soc. Electrical Eng. 31(15) (2011) 75-81.
19. Y. F. Wang, T. Y. Chai, Adaptive fuzzy control method for dynamic friction compensation, P. CSEE, 25(2) (2005) 139-143.
20. Y. F. Wang, D. H. Wang, T. Y. Chai, Extraction and adaptation of fuzzy rules for friction modeling and control compensation, IEEE t. fuzzy syst. 19(4) (2011) 682-693.
21. L. Lu, B. Yao, Q. Wang, Adaptive robust control of linear motors with dynamic friction compensation using modified LuGre model, Automatica, 45(12) (2009) 2890-2896.
22. Q. Liang, J. Zhang, Y. Wang, Adaptive backstepping friction compensation control based on modified LuGre model, Small Special Electrical Mach. (11) (2011) 67-69.
23. D. K. Wang, Backstepping based adaptive fuzzy control and its application, J. Shengyang Normal U. 31(2) (2013) 192-195.
24. H. B. Zhao, X. H. Zhou, Backstepping adaptive control of dual-motor driving servo system, Control Theor. Appl. 28(5) (2011) 745-751.

# A mathematical model for surface roughness of fluidic channels produced by grinding aided electrochemical discharge machining (G-ECDM)

Ladeesh V.G.[1a] and Manu R.[1]

[1] *Dept. of Mechanical Engineering, National Institute of Technology Calicut, Calicut, Kerala-673 601*

**Abstract.** Grinding aided electrochemical discharge machining is a hybrid technique, which combines the grinding action of an abrasive tool and thermal effects of electrochemical discharges to remove material from the workpiece for producing complex contours. The present study focuses on developing fluidic channels on borosilicate glass using G-ECDM and attempts to develop a mathematical model for surface roughness of the machined channel. Preliminary experiments are conducted to study the effect of machining parameters on surface roughness. Voltage, duty factor, frequency and tool feed rate are identified as the significant factors for controlling surface roughness of the channels produced by G-ECDM. A mathematical model was developed for surface roughness by considering the grinding action and thermal effects of electrochemical discharges in material removal. Experiments are conducted to validate the model and the results obtained are in good agreement with that predicted by the model.

## 1 Introduction

Machining of ceramics and composites are gaining more importance in the modern industries due to its wide spread application in various fields. Development of new energy efficient and accurate techniques for ceramic machining is a challenging task for scientific community. Grinding aided electrochemical discharge machining is a promising technology, which can be effectively utilized for machining brittle materials like glass and advanced ceramics. A G-ECDM machine consist of a diamond coated solid engraving tool maintained as cathode and an auxiliary electrode kept at a distance of 4-6cm from the tool as anode. Both the electrodes will be dipped in an electrolyte. When a direct current (DC) is applied across the electrodes, hydrogen bubbles will be liberated due to electrochemical reaction. These bubbles act like a series of switches connected to the circuit. The current flows from the tool electrode to the electrolyte through the interface between bubbles called bubble bridges. Due to sudden ohmic heating, some of the bubble bridges blow off, thereby opening the circuit quickly. This switching-off action [1] causes a sudden fluctuation in current thereby inducing an e.m.f. This will cause ionization of the gas inside the bubble and a discharge will be produced. The cycle repeats several thousands of times per second. When a workpiece is placed in the vicinity of these discharges, material removal takes place by melting. Thus the material beneath the tool will be softened by the discharges and the rotating diamond coated tool will remove the material

[a] Corresponding author : ladeesh.vg@gmail.com

from the machining zone by grinding action. A portion of the material will be removed by chemical etching action in which the molten workpiece material reacts with the electrolyte to form complex compounds. These combined mechanisms enhance the material removal rate and the grinding action of diamond grits remove the recast layer produced by solidification of molten material.

Basak and Ghosh [1] experimentally found that the inductance of the circuit plays a predominant role in controlling material removal rate (MRR). An enhancement in MRR was observed with an extra inductance in the circuit. They developed a mathematical model to predict MRR in electrochemical discharge machining of glass. Chak and Rao [2] performed trepanning of alumina by ECDM using an abrasive tool. He observed that the use of pulsed DC reduced the cracking at high voltage and machining ability got improved with the abrasive tool. Jain et al. [3] performed electrochemical discharge machining of alumina and glass using conventional cutting tool (CCT) and abrasive cutting tool (ACT). The material removed for different voltages and temperatures was found to be high for ACT when compared with CCT. Ladeesh and Manu [4] performed drilling of borosilicate glass using G-ECDM and observed that thermal melting, grinding action and chemical etching are the different mechanisms contributing material removal. Liu et al. [5] used grinding aided ECDM for the machining of composites (alumina particles reinforced aluminium alloy) for the first time. In most of the studies mentioned above, researchers tried to improve the process of drilling using ECDM with the help of abrassive tools. No previous work has been reported on using abrasive tool for machining channels or slots on ceramics with the aid of ECDM.

In the present study an attempt has been made to develop fluidic channels on glass using the technique of G-ECDM. The significant machining parameters are investigated with the help of systematic experiments. A mathematical model is developed to predict the surface roughness of the channels produced using G-ECDM.

## 2 Experimental methodology

### 2.1 Experimental setup

A CNC router machine was used to facilitate the motion of the tool in X, Y, Z direction which is shown in Fig. 1(a). The tool used is a diamond coated solid engraving tool of diameter 3mm, which is shown in Fig. 1(b). A stainless steel plate of dimension 200mm x 10mm x 0.7mm was used as the auxiliary electrode maintained as anode. The electrolyte used was potassium hydroxide. A customized DC power supply with provision for varying frequency and duty ratio was used for the experiments. Workpiece used was a borosilicate glass of dimension 200mm x 50mm x 5mm

(a) (b)

**Figure 1.** (a) Machining setup (b) diamond core engraving drill

### 2.2 Design of experiments

An experiment was conducted using Plackett-Burman design. This design is widely accepted for conducting preliminary experiments. The design is more efficient and economical when compared to two level full factorial designs. In the present study, six factors are selected which includes tool rotational speed, feed rate, voltage, duty factor, frequency and electrolyte concentration. Each factor has two levels, which makes a total of 12 experimental runs. Two replications are taken. The selected factors and their levels are given in Table 1. The Placket-Burman design layout used for developing experimental runs are shown in Table 2. Surface roughness of the machined channels are measured using a contact type surface roughness tester (Mitutoyo- Surftest SJ-210). From the main effects plot shown in Fig. 2., it is evident that the Ra value reduces with increase in feed, increase in frequency, reduction of voltage and reduction of duty factor. Analysis of variance (ANOVA) was conducted using a software MINITAB-17 to identify the significant factors. Results of ANOVA are given in Table 3. From the F-test, factors with high F-values like voltage, duty factor, frequency and feed rate are found to be significant and the effect of concentration and tool rotation speed are found to be insignificant.

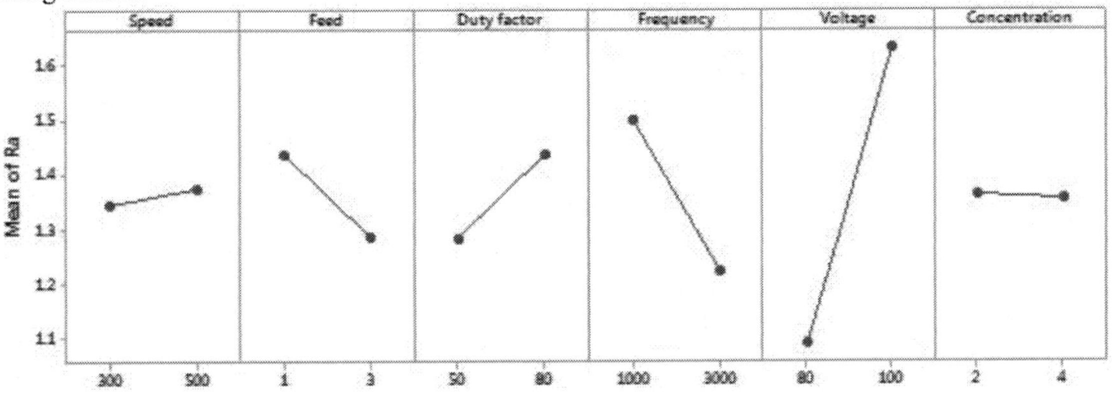


**Figure 2.** Main effects plot for Surface Roughness (Ra)

**Table 1.** Factors and levels

| Factors & Notation | Unit | Levels Low (-1) | High (+1) |
|---|---|---|---|
| Speed (A) | rpm | 300 | 500 |
| Feed (B) | mm/min | 1 | 3 |
| Duty factor (C) | % | 50 | 80 |
| Frequency (D) | Hz | 1000 | 3000 |
| Voltage (E) | V | 80 | 100 |
| Concentration (F) | M | 2 | 4 |

**Table 2.** Plackett-Burman Design

| Runs | A | B | C | D | E | F |
|---|---|---|---|---|---|---|
| 1 | 1 | -1 | 1 | -1 | -1 | -1 |
| 2 | 1 | 1 | -1 | 1 | -1 | -1 |
| 3 | -1 | 1 | 1 | -1 | 1 | -1 |
| 4 | 1 | -1 | 1 | 1 | -1 | 1 |
| 5 | 1 | 1 | -1 | 1 | 1 | -1 |
| 6 | 1 | 1 | 1 | -1 | 1 | 1 |
| 7 | -1 | 1 | 1 | 1 | -1 | 1 |
| 8 | -1 | -1 | 1 | 1 | 1 | -1 |
| 9 | -1 | -1 | -1 | 1 | 1 | 1 |
| 10 | 1 | -1 | -1 | -1 | 1 | 1 |
| 11 | -1 | 1 | -1 | -1 | -1 | 1 |
| 12 | -1 | -1 | -1 | -1 | -1 | -1 |

**Table 3.** ANOVA for Ra

| Source | DoF | Sum of squares | Mean squares | F-Value |
|---|---|---|---|---|
| A | 1 | 0.0053 | 0.0053 | 0.39 |
| B | 1 | 0.1406 | 0.1406 | 10.48 |
| C | 1 | 0.1437 | 0.1437 | 10.71 |
| D | 1 | 0.4551 | 0.4551 | 33.91 |
| E | 1 | 1.7642 | 1.7642 | 131.46 |
| F | 1 | 0.0004 | 0.0004 | 0.03 |
| Error | 17 | 0.2281 | 0.0134 | |
| Total | 23 | 2.7373 | | |

**DoF – Degrees of Freedom*

## 3 Mathematical Modelling

Three stages can be identified in an ECDM wave form. In stage-I, a gradual rise of voltage happens till the critical voltage ($V_c$) is reached and suddenly the voltage drops to an arc maintaining voltage. In stage II the arc maintaining voltage continues till the pulse-off time. In the third stage a gradual drop in the voltage to zero can be observed. Stage I can be called as the ignition delay in which the chemical action causing hydrogen liberation and gas film formation happens. Hence the major energy available for the spark is given by arc maintaining voltage ($V_a$). For ECDM using voltage greater than the critical voltage, ignition delay is almost absent as shown in Fig. 3(b) when compared to normal ECDM wave form shown in Fig. 3(a)and the duration for arcing is more, which means that more energy can be supplied by the discharge. This is because of the drastic rise in the bubble generation rate at very high voltage above critical voltage. This high bubble concentration permits the discharge to initiate immediately soon after the pule is switched on after a discharge cycle thereby reducing the ignition delay period significantly. Thus, the voltage pulses can be simplified and assumed as rectangular pulses with the arc duration $t_a$.

### 3.1 The model for material removal rate (MRR)

The following assumptions are made for developing a MRR model for G-ECDM.

- The workpiece is an electrically non conductive material and the material removal takes place mainly by melting due to electrochemical discharges.
- ECDM happens by a series of sparks and each spark contributes for the formation of a crater.
- The crater produced will be in the form of a portion of sphere (spherical cap).
- Applied voltage is well above critical voltage and the Ignition delay is absent in all discharge cycles
- The tool feed rate is very low, so that the grinding action will be only used to flush the material removed by thermal effects of electrochemical discharges.

From Fig. 4, the total cycle time can be expressed as a sum of pulse-on time or arcing period ($t_a$) and pulse-off time or idle time ($t_i$) which is given by (1).

$$T_c = t_a + t_i \tag{1}$$

Accordingly, the arc energy can be determined using (2).

$$E_a = \int_0^{t_a} I_a(t) V_a(t) dt \tag{2}$$

The energy transferred to the workpiece will be much less than the energy supplied by arcing. The percentage of energy transferred to the workpiece is given by a term called coefficient of energy transfer ($C_T$) whose value is around 0.2 [6]in the case of ECDM. Hence the total energy transferred to the work can be expressed as (3). The transferred energy will cause the material removal in the form of micro craters.

$$E_T = \int_0^{t_p} C_T I_a(t) V_a(t) dt \tag{3}$$

The volume of the crater produced by a single spark is proportional to the transferred energy and the volume of the crater can be determined by (4) and (5)

$$W_c = \mu_1 \int_0^{t_a} C_T V_a(t) I_a(t) dt \tag{4}$$

$$W_c = \mu \int_0^{t_a} V_a(t) I_a(t) dt \tag{5}$$

where $\mu_1$ is the proportionality constant and $\mu = \mu_1 . C_T$.

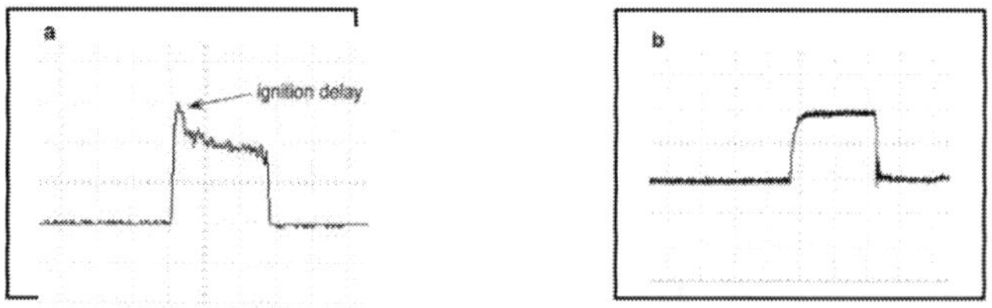


**Figure 3.** Typical voltage waveform of (a) Normal ECDM voltage wave form (b) ECDM above critical voltage

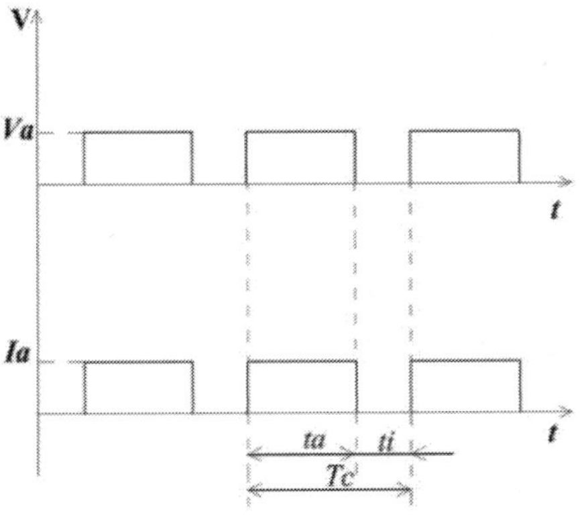


**Figure 4.** Voltage and current waveforms of G-ECDD

Basak et al. [1] found experimentally a relation connecting volume of the material removed per spark and the energy supplied per spark for soda lime glass, which is given by (6)

$$W_c = 0.7x10^{-5} E_a^{1.5} \tag{6}$$

If $f$ is the frequency of discharge, the material removal rate can be approximated using the expression (7)

$$MRR = \delta . W_c . f \tag{7}$$

Where $\delta$ is the density of the workpiece material. Frequency of discharge $f_s$ can be calculated as [7]:

$$f_s = \frac{R_T}{L} \ln\left[\frac{V_0 - V_w}{V_0 - V_c}\right] \tag{8}$$

where $R_T$ is the total resistance, which is a sum of electrolyte bulk resistance ($R_b$) and resistance at the interface of tool and electrolyte at the time of discharge ($R_c$). $L$ is the circuit inductance, $V_0$ is the applied voltage, $V_w$ is the water decomposition potential and $V_c$ is the critical voltage. In G-ECDM, the major portion of material removal happens by the discharges at the lateral face of the tool. Hence, the spark frequency at the lateral face of the tool ( $f_L$ ) can be determined by (9)

$$f_L = f_s A_L / A_s \tag{9}$$

where $A_s$ and $A_L$ are area of the tool in the electrolyte and area of lateral face of the electrode dipped in the electrolyte. Thus, the equ. (7) for MRR can be modified as:

$$MRR = 0.7x10^{-5} E_a^{1.5} x\delta \left[\frac{A_L}{A_s} x \frac{R_T}{L} \ln\left(\frac{V_0 - V_w}{V_0 - V_c}\right)\right] \tag{10}$$

The above equation for MRR is applicable for glass only. A generalized model for MRR is given by (11)

$$MRR = \mu E_a^{\lambda} x\delta \left[\frac{A_L}{A_s} x \frac{R_T}{L} \ln\left(\frac{V_0 - V_w}{V_0 - V_c}\right)\right] \tag{11}$$

Where $\mu$ and $\lambda$ are constants, which depends mainly on workpiece material.

### 3.2 Model for surface roughness

In ECDM, the material removal initially happens by melting due to the thermal energy of the spark and small craters will be produced. The removed molten material forms a build-up- edge (BUE) around the crater as shown in Fig. 5(a). The random craters finally overlap and removes material layer by layer. The grinding action of the grits will remove the remaining build-up-edges and finally the peak of the crusts of overlapped craters will be ground by the rotating diamond grits. The final finished surface of G-ECDD will appear as shown in Fig. 5(b). The average peak to valley value of surface roughness will be given by the final average peak height of the crusts ($b$) on finished surface.

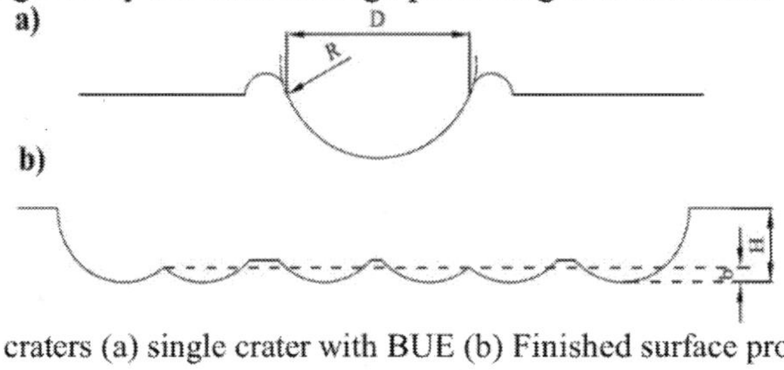


**Figure 5.** Illustration of the craters (a) single crater with BUE (b) Finished surface produced by G-ECDM.

Volume of the crater produced by a single spark is given by (12)

$$W_c = \frac{1}{3}\pi H^2(3R - H) \tag{12}$$

Substitute the Eq. (6) on (12), we will get:

$$0.7x10^{-5}xE_a^{1.5} = \frac{1}{3}\pi H^2(3R - H) \tag{13}$$

The ratio of the crater depth ($H$) to the crater diameter ($D$) for ECDM is in the range 0.2- 0.3 [8]. Assuming that, the average diameter of the crust is approximately equal to the radius of curvature ($R$) of the spherical crater. Taking H/D ratio as 0.25 and simplifying Eq. (13), we will get the following expression (14) for $H$.

$$H = 8.467x10^{-3}E_a^{0.5} \tag{14}$$

The average peak to valley height $R_t$ is given by the average crater height $b$ of the overlapped craters. From microscope observation, the value of $b$ is found to be $1/k^{th}$ of H. $k$ depends on the diamond grit size, whose value can be chosen from Table 4. *Hence, Rt* can be written as:

$$R_t = \frac{1}{k}(8.5x10^{-3}E_a^{0.5}) \tag{15}$$

Substituting (2) in (15) and simplifying, we get:

$$R_t = \frac{1}{k}8.5x10^{-3}(V_aI_at_p)^{0.5} \tag{16}$$

Thus, surface roughness during G-ECDM can be expressed as a function of arc maintaining voltage, current and pulse-on period. The surface roughness measurements using contact type method revealed that the centre line average surface roughness value ($Ra$) is 1/9 th of $R_t$ value for G-ECDM.

**Table 4.** Value of $k$ for different grit size

| U. S Mesh | In microns | $k$ |
|---|---|---|
| 60-80 | 250-177 | 5 |
| 100-140 | 149-105 | 7 |
| 170-230 | 88-63 | 9 |

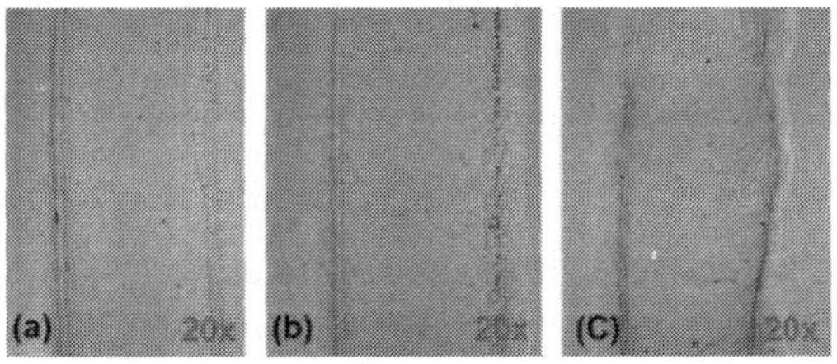


**Figure 6.** Fluidic channels machined at: a) 80V, 50%, 500Hz, 3M b) 90V, 60%, 2kHz, 3M c) 100V, 70%, 3 kHz, 3M

## 4 Experimental Validation of the Model

Experiments are conducted to validate the model for surface roughness. The microscopic images of machined channels are shown in Fig. 6. The same experimental setup used for conducting G-ECDM was used for performing validation experiments. Fig. 7 and Fig. 8 shows the Ra value obtained for channels machined using different voltages and different pulse-on time respectively. A low feed rate of 10µm/s was used so that the effect of feed rate is insignificant.

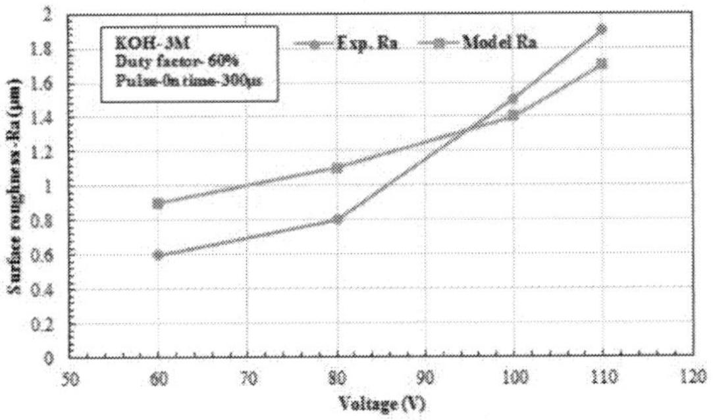


**Figure. 7** Effect of voltage on surface roughness

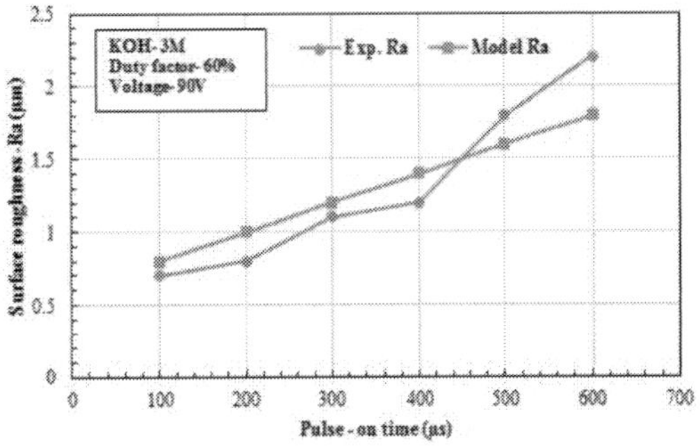


**Figure. 8** Effect of pulse-on time on surface roughness

It can be observed from Fig. 7 and Fig. 8, that Ra increases with an increase in voltage and pulse- on time. Discharges at higher voltage produce craters of more depth, which will increase the Ra value. Similarly, a pulse with more time period will produce an arc, which will allow to dissipate the discharge energy for long time. This will increase the depth of the craters and thereby increases the *Ra* value. The obtained experimental values for Ra are in close agreement with the mathematical model for the channels machined at different voltages. The effect of feed rate will become significant in controlling surface roughness at higher feed values.

## 5 Conclusions

- From the experiment, the significant factors that contributes surface roughness of the channel machined using G-ECDM are identified as voltage, frequency, duty factor and feed rate. The effect of tool rotation speed and electrolyte concentration was found to be insignificant.
- Surface roughness decreases with decrease in voltage, duty factor, feed rate and increase in frequency
- A mathematical model is developed, which can fairly predict the surface roughness of channels machined using G-ECDM. The model correlates arc maintaining voltage, current and pulse-on period with surface roughness.

## References

1. Indrajit Basak, Amitabha Ghosh, J. of Mat. Proc. Tech. **71**, 350-359 (1997)
2. S. K. Chak, P. V. Rao, Int. J. of Mach. Tools & Manuf. **47**, 2061–2070 (2007).
3. V. K. Jain, S. K. Choudhury, K. M. Ramesh, Int. J. of Mach. Tools & Manuf. **42**, 1269-1276 (2002).
4. V. G. Ladeesh, R. Manu, Procedia Tech. **25**, 1122 – 1128 ( 2016)
5. J. W. Liu, T. M. Yue, Z. N. Guo, Int. J. of Mach. Tools and Manuf. **50**, 86-96 (2010).
6. Chenjun Wei, Kaizhou Xu, Jun Ni, Adam John Brzezinski, Dejin Hu, Int. J Adv. Manuf. Tech. **54**, 987-995 (2011).
7. Rolf Wuthrich, Philippe Mandin, Electrochimica Acta **54**, 4031-4035 (2009).
8. Baoyang Jiang, Shuhuai Lan, Jun Ni, Zhaouyang Zhang, J. of Mat. Proc. Tech. **214**, 892-898 (2014)

# Stress analysis and deformation prediction of sheet metal workpieces based on finite element simulation

Ren Penghao[1], Wang Aimin[1,a], Wang Xiaolong[1] and Zhang Yanlin[2]

[1]*School of Mechanical Engineering, Beijing Institute of Technology, 100081 Beijing, China*
[2]*Beijing Xin Li Machinery Co., Ltd, 100039 Beijing, China*

**Abstract:** After aluminum alloy sheet metal parts machining, the residual stress release will cause a large deformation. To solve this problem, this paper takes a aluminum alloy sheet aerospace workpiece as an example, establishes the theoretical model of elastic deformation and the finite element model, and places quantitative initial stress in each element of machining area, analyses stress release simulation and deformation. Through different initial stress release simulative analysis of deformation of the workpiece, a linear relationship between initial stress and deformation is found; Through simulative analysis of coupling direction-stress release, the superposing relationship between the deformation caused by coupling direction-stress and the deformation caused by single direction stress is found. The research results provide important theoretical support for the stress threshold setting and deformation controlling of the workpieces in the production practice.

## 1 Introduction

Aluminum alloy sheet workpieces has a wide application in aerospace industry because of its light weight and enough strength. However, the rigi[a]dity of the parts is poor, and the residual stress release of the parts in the machining process causes deformation, which often makes the parts do not meet the accuracy requirements.

In order to solve this problem, the researchers have studied the distribution characteristics and evolution of stress after pre-stretching plate parts cutting specimen[1]; As for test and analysis of stress, they develop and improve the effective stress testing and controlling technology[2-3]; researchers also study on cutting engineering, such as a machining process and tool geometry or clamping method producing influence on distribution of residual stress and deformation of parts[4-6];Taking mechanism on residual stress as the starting point, the cutting force, machining stress, thermal and mechanical force are comprehensive analysed and calculated[7]. These studis on the residual stress concentrate on deformation caused by the release of stress during specific processing conditions, results of which provide important reference for the production practice, but there are also problems such as limited applicable conditions, and poor versatility. For stress conditions under other conditions, it is difficult to get a reasonable solution.

In terms of the nature of the problem, the workpiece surface and internal stress balance is broken during or after the machining process. A new equilibrium condition is established through the

[a] Corresponding author : wangam@bit.edu.cn

deformation. Therefore, it is necessary to research how stress direction and numerical value do effect on deformation of the workpiece. And on this basis, we can predict the deformation according to the stress condition of the workpiece. Then provide important theoretical support to controlling the numerical value and direction of residual stress in sensitive area by adjusting the process parameters.

## 2 Theoretical model of stress and deformation of sheet metal workpieces

According to the theory of elastic-plastic mechanics, the metal material can be regarded as an ideal elastic-plastic model, and the stress-strain behavior of metal materials is shown in Figure 1.

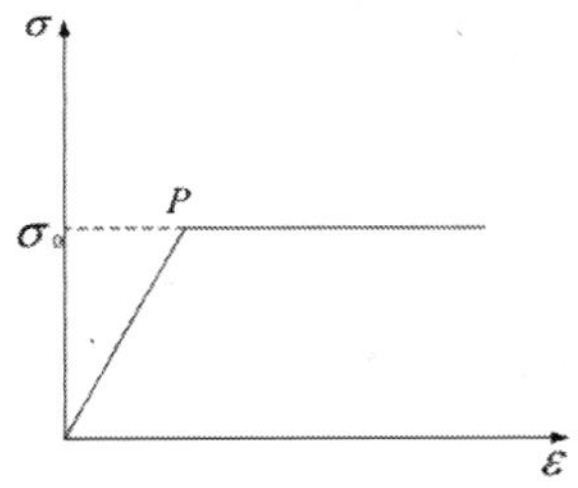


**Figure 1.** Stress-strain behavior of an ideal elastic-plastic model

In the figure, the stress which point P corresponds to is the limit yield stress $\sigma_0$. While the stress $\sigma<\sigma_0$, the material will produce elastic strain. The strain will change with the release of stress, and the workpiece, as a continuum, will produce corresponding deformation; while $\sigma>\sigma_0$, the material accumulates permanent strain and the deformation of the workpiece will not change even if the stress changes. Therefore, the deformation of the workpiece caused by the stress release is elastic deformation[8].

The arbitrary orientation of grain metal material is isotropic elastic material. When the machining process is completed, the residual stress will be released. The stress redistribution of the hexahedral element stress meets the following equilibrium condition:

$$\begin{cases} \dfrac{\partial \sigma_x}{\partial x} + \dfrac{\partial \tau_{yx}}{\partial y} + \dfrac{\partial \tau_{zx}}{\partial z} = 0 \\ \dfrac{\partial \sigma_y}{\partial y} + \dfrac{\partial \tau_{zy}}{\partial z} + \dfrac{\partial \tau_{xy}}{\partial x} = 0 \\ \dfrac{\partial \sigma_z}{\partial z} + \dfrac{\partial \tau_{xz}}{\partial x} + \dfrac{\partial \tau_{yz}}{\partial y} = 0 \end{cases} \tag{1}$$

According to the generalized Hooke's law[9], the strain caused by the stress release is :

$$\varepsilon_{ij} = \frac{1+\mu}{E}\sigma_{ij} - \frac{\mu}{E}\sigma_{kk}\delta_{ij} \tag{2}$$

where

$$\varepsilon_{ij} = \begin{bmatrix} \varepsilon_x & \varepsilon_{xy} & \varepsilon_{xz} \\ \varepsilon_{yx} & \varepsilon_y & \varepsilon_{yz} \\ \varepsilon_{zx} & \varepsilon_{zy} & \varepsilon_z \end{bmatrix}$$

DOI: 10.1051/matecconf/201710402009

$$\sigma_{ij} = \begin{bmatrix} \sigma_x & \tau_{xy} & \tau_{xz} \\ \tau_{yx} & \sigma_y & \tau_{yz} \\ \tau_{zx} & \tau_{zy} & \sigma_z \end{bmatrix}$$

$$\delta_{ij} = \begin{bmatrix} 1 & 0 & 0 \\ 0 & 1 & 0 \\ 0 & 0 & 1 \end{bmatrix}$$

$$\sigma_{kk} = \sigma_{11} + \sigma_{22} + \sigma_{33}$$

and ($E$) is the elastic modulus and the ($\mu$) is the Poisson ratio.

According to the relationship between the strain and deformation described by Lagrange displacement, the deformation of continuum can be expressed as :

$$\varepsilon_{ij} = \frac{1}{2}(u_{i,j} + u_{j,i} + u_{r,i}u_{r,j}) \tag{3}$$

The typical item expressed by engineering symbols is:

$$\varepsilon_x = \frac{\partial u_x}{\partial x} + \frac{1}{2}\left[\left(\frac{\partial u_x}{\partial x}\right)^2 + \left(\frac{\partial u_y}{\partial x}\right)^2 + \left(\frac{\partial u_z}{\partial x}\right)^2\right] \tag{4}$$

## 3 Finite element analysis modelling

Stress release simulation is realized by the finite element analysis, in which the ABAQUS software is used for modelling. Taking a sheet aerospace workpiece (as in Figure 2) as an example, the material is forged aluminum 2A12, whose density is 2270Kg/m$^3$, elastic modulus is 71.7Gpa, and Poisson's ratio is 0.3. The accuracy requirement of the deformation is the edge AB, which is vertical to the plane on the left side of the workpiece. In order to ensure a higher simulation accuracy, and avoid the tedious operation, the workpiece is divided into 47611 hexahedron swept meshes. In order to study the influence of stress magnitude and orientation on the deformation of the workpiece, the stress condition of the workpiece is used as initial condition with quantitative input. When the process is finished, the stress is released free, so the boundary condition is not set in the simulation process.

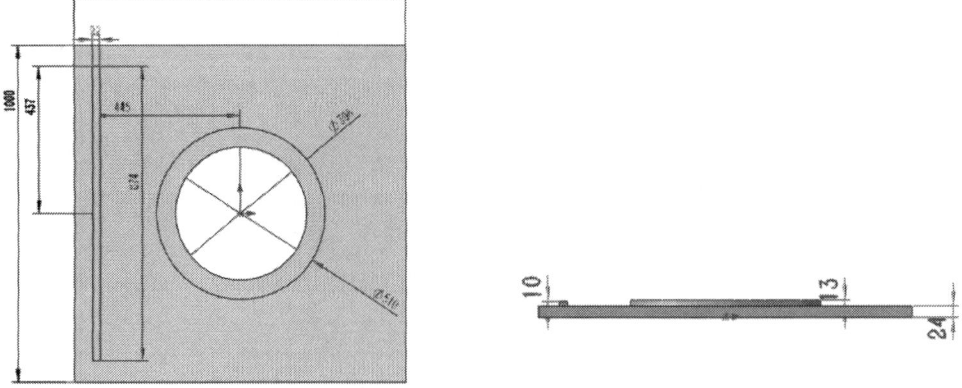


**Figure 2.** A sheet aerospace workpiece

DOI: 10.1051/matecconf/201710402009

# 4 Stress release simulation and deformation analysis

## 4.1. Analysis of deformation of workpiece under different stress numerical value

The initial condition of the workpiece is shown in Figure 3. In order to explore the relationship between the stress and deformation of the workpieces, the initial stress is put into the red area in the figure manually. According to the literature and practice, the milling process will cause 100Mpa-300Mpa residual stress on the surface. So place initial stress in the elements which are on the surface of the workpiece, whose numerical values are 100Mpa. The precision of the sheet is required to be perpendicular to the plane of the sheet, so after the release of the stress free (Figure.4), the deformation of the edge AB in the vertical direction of the plane is observed and analysed, and the deformation of each node of the AB side is shown in Figure 5.

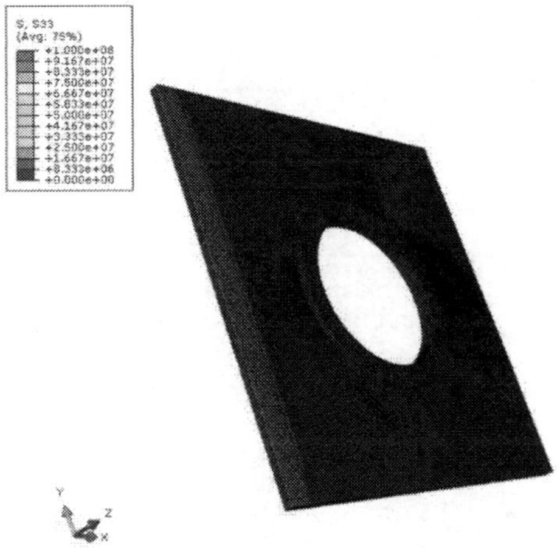


**Figure 3.** Initial condition of workpiece

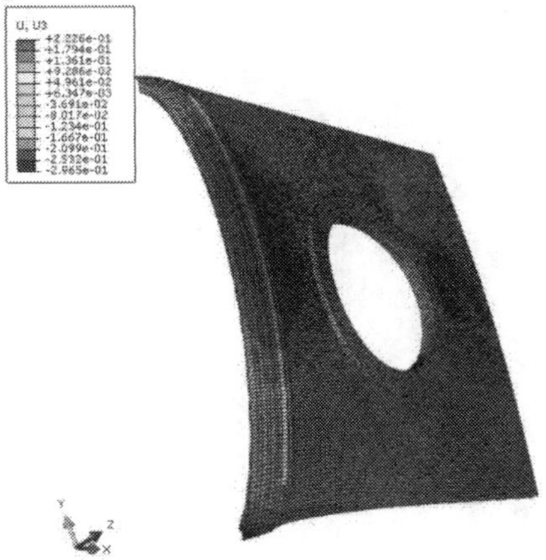


**Figure 4.** Workpiece after stress release free

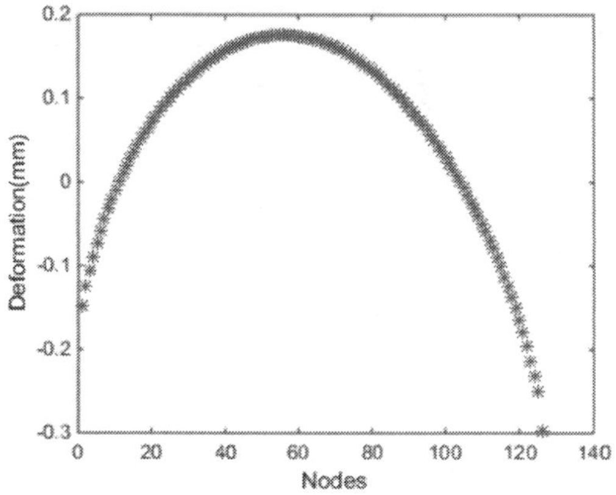


**Figure 5.** Deformation of edge AB in Z direction

To explore the effects of different stress numerical values on deformation of the workpiece, we do multiple analysis of different initial stress conditions and the deformation after the release of stress, the figures are similar with Figure 6. The values of the maximum deformation are shown in table 1.

**Table 1.** Deformation of different stress numerical values

| Initial stress/Mpa | The deformation of edge AB /mm | | |
|---|---|---|---|
| | Increasing direction | Minus direction | Deflection |
| 50 | 0.0877 | -0.1448 | 0.2325 |
| 100 | 0.1754 | -0.2965 | 0.4719 |
| 150 | 0.2639 | -0.4430 | 0.7069 |
| 200 | 0.3508 | -0.5929 | 0.9437 |
| 250 | 0.4371 | -0.7449 | 1.1820 |
| 300 | 0.5278 | -0.8860 | 1.4138 |

For the visual display of the maximum deformation of the workpiece under different stress numerical values, the stress-deformation figure of the workpiece is shown in Figure 6.

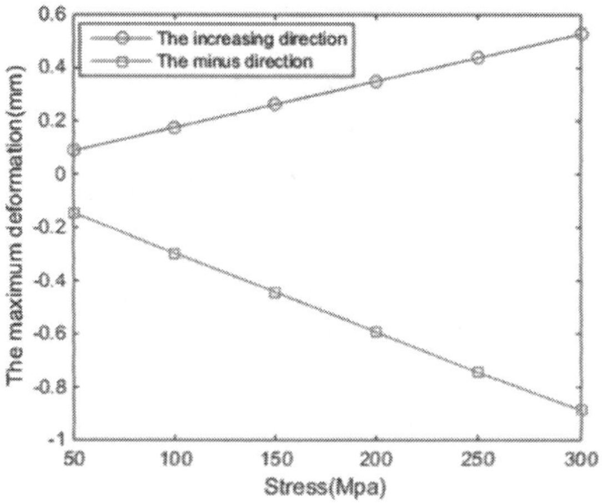


**Figure 6.** Deformation of edge AB

By the figure and data, it is known that the deformation of the edge of the workpiece is linearly related to the initial stress. The relationship between maximum deformation ($u_{max}$, mm) and the stress in Z direction ($\sigma$, Mpa) can basically fit for:

$$u^{+}_{max} = 1.756\times10^{-3}\sigma$$

$$u^{-}_{max} = -2.961\times10^{-3}\sigma \quad (5)$$

### 4.2 Analysis of deformation of parts under the condition of multi-direction stress

According to the theory of elastic mechanics, the deformation caused by the stress release in different directions is also different. In order to explore the influence of the stress direction on the deformation of the workpiece, two more simulation experiments are carried out:

(1) Place 100Mpa initial stress in the Y direction into each element and then release free. The deformation figure is shown in Figure 7.

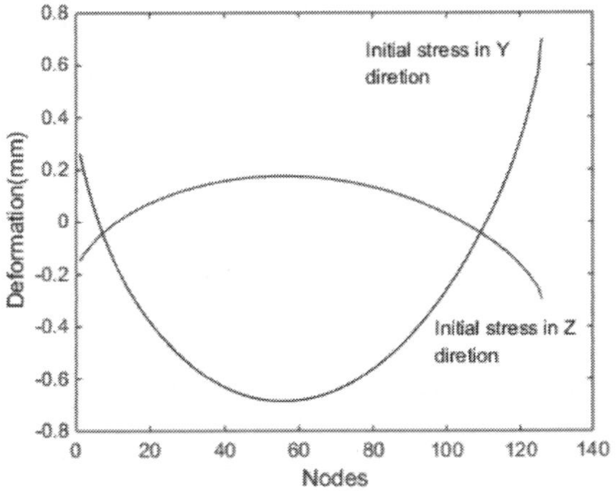


**Figure 7.** Deformation of edge AB under different stress

(2)Place 100Mpa initial stress both in the Y direction and Z direction into each element and then release free. And then observe and analyse the deformation.

In the finite element model, we select a sampling point every 15 nodes along edge AB, and there are 8 sampling points in total, the simulation results obtained at various that the deformation data of sampling points whose initial stress in different direction and in multi-directions are shown in Table 2. And deformation of each sampling point is plotted as curves which are shown in Figure 8.

As the table and the figure shown, deformation caused by the stress of deferent directions is deferent. After the initial stress of Y direction the release ,the workpiece bend in the minus direction, and the deformation is large; However, after the initial stress of Z direction release, the deformation of workpiece is in normal direction and the value is small; When the initial stress is in Y-Z coupling directions, the deformation of the edge is vector sum of which are caused by the stress in single direction.

**Table 2.** Deformation of each sampling point

| Sampling points | Deformation of edge AB in Z direction | | |
|---|---|---|---|
| | Initial stress in Y direction | Initial stress in Z direction | Initial stress in Y-Z coupling direction |
| 1 | -0.2772 | 0.0363 | -0.2406 |
| 2 | -0.5356 | 0.1242 | -0.4113 |
| 3 | -0.6614 | 0.1669 | -0.4943 |
| 4 | -0.6823 | 0.1741 | -0.5082 |
| 5 | -0.6081 | 0.1488 | -0.4593 |
| 6 | -0.4358 | 0.0902 | -0.3458 |
| 7 | -0.1494 | -0.0073 | -0.1569 |
| 8 | 0.3137 | -0.1648 | 0.1486 |

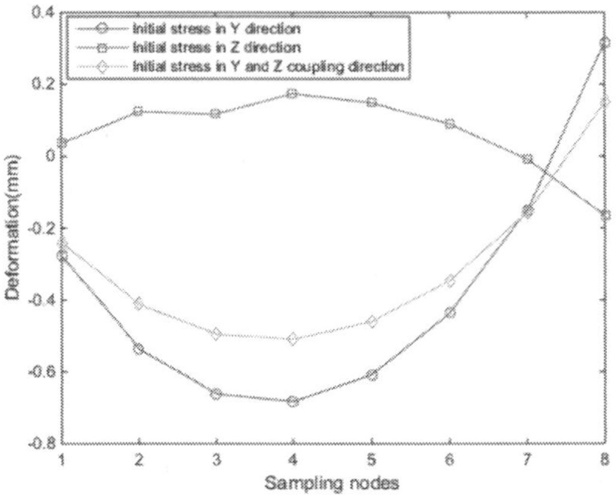


**Figure 8.** Deformation of sampling nodes

## 5 Conclusions

In this paper, the theoretical model of the stress and deformation of the workpiece is established based on the knowledge of elastic mechanics, and the finite element model is established by ABAQUS software. The stress release simulation experiment is carried out, and the following results are obtained:

(1) When stress is in one direction, the deformation and the numerical value of stress get into a linear relationship. The deformation increases (decreases) with initial stress increasing (decreasing). To guarantee the precision of the workpiece, we should control the stress reasonably

(2) The initial stress in different directions cause different sizes and directions of deformation of the workpiece. For sheet metal workpieces, the stress release which is parallel to the plane direction of the sheet will cause large deformation .And this kind of stress should be controlled rigorously.

(3) The deformation caused by multi-direction stress release is vector sum of which are caused by the single direction stress release. According to this, in order to guarantee the precision of the workpiece, if the milling process produce larger stress in Y direction, stress in Z direction can be applied into the workpiece to get deformation of vertical offset in the direction of the plate surface to ensure the straightness of the edge or the flatness of the plate plane.

## References

1. Wu Yunxin, Liao Kai, JMT, 1, 42-46(2010)
2. Yu Wei. *Analysis of machining deformation of Aeronautical thin wall parts based on residual stress* (2004)
3. Wu Yunxin, Gong Hai, Liao Kai. Journal of South China University of Technology (NATURAL SCIENCE EDITION), **39**, 90-94(2011)
4. Wang Zhaojun, Chen Wuyi, AAAS, **17**, 175-179(2005)
5. Omar Fergani .Xiaohui Jiang . Yamin Shao . Int J Adv Manuf Technol **83**, 1153-1160. (2016)
6. Li Chun, Cheng Hui Ming, Yang Feng Jun New Technology & New Process **11**, 70-73 (2012)
7. Jiang Xiaohui. *The formation mechanism of residual stress and the control method of machining accuracy of complex thin wall parts.*(2014)
8. Xu Zhilun. *Elastic mechanics.* (Higher Education Press, 2013)
9. Chen Huifa, A.F. Salipu. *The elastic and plastic mechanics.* (Chinese Building Industry Press, 2013)

# Complex Method Mixed with PSO Applying to Optimization Design of Bridge Crane Girder

He Yan[a] and Wang Zongyan

[1]*School of Mechanical & Power Engineering, North University of China, Shanxi, Taiyuan, China*

**Abstract.** In engineer design, basic complex method has not enough global search ability for the nonlinear optimization problem, so it mixed with particle swarm optimization (PSO) has been presented in the paper,that is the optimal particle evaluated from fitness function of particle swarm displacement complex vertex in order to realize optimal principle of the largest complex central distance.This method is applied to optimization design problems of box girder of bridge crane with constraint conditions.At first a mathematical model of the girder optimization has been set up,in which box girder cross section area of bridge crane is taken as the objective function, and its four sizes parameters as design variables, girder mechanics performance, manufacturing process, border sizes and so on requirements as constraint conditions. Then complex method mixed with PSO is used to solve optimization design problem of cane box girder from constrained optimization studying approach, and its optimal results have achieved the goal of lightweight design and reducing the crane manufacturing cost . The method is reliable, practical and efficient by the practical engineer calculation and comparative analysis with basic complex method.

## 1 Introduction

Bridge crane is widely used in steel, metallurgy, chemical industry, building materials and other industries, because its forces in various working conditions are very complicated, the traditional design methods are often difficult to obtain accurate solutions. Designers have to use overlarge safety factor to ensure crane safety case, resulting the increase of steel material waste and construction cost[1].Tian Deyu[2]using ANSYS optimization module ,Fan Xiaoning[3] using genetic algorithm, and Jin Chenguang[4] applying the chaos genetic algorithm for optimization design of crane girder, their optimization results than that of the traditional design are the weight reduction of girder . But the genetic algorithm usually uses binary coding, and also need to carry out complex replication, genetic, variation and other operations. Considering the optimization constraints of main girder in the optimization mathematical model including not only the requirements of strength, stiffness but also stability, so its optimization is a complex multidimensional optimization problem involving many constrains and nonlinear discrete variables. Many years ago, many researchers used the traditional constrained nonlinear algorithm, but there are shortcomings and limitations, such as such as large grid calculated amount and random direction method can only obtain the local optimal solution, Lagrange

[a]Corresponding author : Heyan@zbhy91@163.com

multiplier method is suitable for optimization problem of solving equality constraints, and complex method needs the continuity and differentiability of objective function, so far the intelligent algorithm is the best way to solve this kind of problem.

Particle swarm optimization (PSO) is a global optimization algorithm simulating biological swarm intelligence theory. Particles usually carry on searching by guiding the optimization through individual particle information sharing and competition between groups, and by using iterative method to find the optimal solution. This method has more simple encoding rules than the genetic algorithm (GA), has shown its superiority in solving practical problems in the engineering field [5-6].

Aiming at optimization design problems of box girder of bridge crane with constraint conditions, the complex method mixed with PSO algorithm, in which the best particle evaluated from PSO will replace complex vertexes to obtain maximum central distance of complex. From the constraint processing technology compounding swarm intelligence algorithm, optimization design problem of bridge crane box girder section size will be solved and the direction of biological swarm intelligence applying to mechanical optimization design will be expanded.

## 2 Complex method mixed with particle swarm optimization algorithm

### 2.1 Complex method

Complex method is an important direct method for solving constrained optimization problems. The idea is to construct an initial complex shape with K vertices in the feasible region. The objective function values of each vertex of the composite shape are compared to find the vertex of maximum objective function (called the worst point), then according to certain rules find the practical new vertex which objective function value decline, and using this point instead of the dead, a new complex will be constituted, when its shape every changing once, it is moving a step to the most solution, until to the best approximation.

The calculation steps of the complex method are as follows [7]:

1) Select the vertex numbers $k$ of the complex shape, in general, $k$ is set as $n+1 \le k \le 2n$, the initial complex consisted of $k$ vertexes is formed in the feasible region .

2) Calculate the objective function value of complex each vertex and compare them, then find the best point $X_L$, the worst point $X_H$ and the next to the worst point $X_G$.

3) Calculate the center $X_C$ of other ($k$-1) vertex aside from the worst point $X_H$. Determine whether $X_C$ is feasible, if it is feasible point, then go to step 4);otherwise, determine the lower limit and the upper limit value of design variables , even let a=$X_L$, b=$X_H$, and then go to step 1), reconstructing initial complex.

4) Calculate the reflection point $X_R$ according to the equation (1), when necessary, change the value of the reflection coefficient until the reflection is successful, meeting the following equation (2). Then $X_R$ replaces $X_H$, and the new complex is constituted.

$$X_R = X_C + \alpha(X_c - X_H) \tag{1}$$

where, $\alpha$ is reflection coefficient, generally taking as equation (2).

$$\alpha = 1.3 \begin{cases} g_j(x_R) \le 0 \\ f(x_R) < f(x_H) \end{cases} j = 1,2, \mathrm{L}, m) \tag{2}$$

5) If the convergence condition which the root-mean-square of difference of the function value between each vertex and the good point is less than the error limit, namely equation (3) is satisfied, the calculation is terminated. Finally get the optimal solution, namely X*=$X_L$, f(X*)=f($X_L$).Otherwise, turn to step 2).

$$\left\{ \frac{1}{k-1} \sum_{j=1}^{k} \left[ f(x_j) - f(x_L) \right]^2 \right\}^{\frac{1}{2}} \le \varepsilon \tag{3}$$

### 2.2 Particle swarm optimization algorithm

In the PSO algorithm, each particle moves according to its velocity in the search space. The whole optimization process is that particle adjusts to its position and speed based on its optimal location and that of other individuals in its neighbour, then to accelerate moving toward global optimal value.

The standard particle swarm algorithm is shown in equation (4) (5):

$$v_{id}(t+1) = wv_{id}(t) + c_1r_1(p_{id} - x_{id}(t)) + c_2r_2(p_{gd} - x_{id}(t)) \tag{4}$$

$$x_{id}(t+1) = x_{id}(t) + v_{id}(t+1) \tag{5}$$

Where, $x_{id}(t)$ is the current position of the particle $i$, $p_{id}$ is the superior position of the particle $i$, which searches till now, $p_{gd}$ is the entire particle swarm's superior position. $c_1$ and $c_2$ are acceleration constants with positive values, usually set as $0 \sim 2$; $r_1$ and $r_2$ are random numbers in range[ 0 1]. $v_{id}$ is the current flying velocity of the particle $i$ in range $[-v_{\max}, v_{\max}]$, $v_{max}$ is the maximum particle velocity, being non-negative number; $w$ is positive constant, called inertia; weight, which is linearly decreasing reducing with iteration , namely:

$$w = w_{\max} - \frac{w_{\max} - w_{\min}}{T_{\max}} t \tag{6}$$

Where $w_{\max}$ and $w_{\min}$ are the maximum and minimum inertia weight values. Typically $w_{\max} = 0.9 \sim 1.4$, $w_{\min} = 0.4$; $t$ is current iteration number, $T_{\max}$ is the maximum iteration number.

### 2.3 The composite shape vertex is replaced with the best particle of PSO to achieve the maximum central distance

At present, the reason resulting in the global search ability of complex method is that the domain surrounded by a new complex not large enough, which is randomly generated in a particular search domain or complex calculation process , Namely Hamming distances between the complex central points and each complex vertex are not large enough. Now the sum of Hamming distance is defined complex central distance represented as equation (7).

$$D_c = \sum_{i=1}^{k} D_i \tag{7}$$

If the optimal solution obtained by PS0 solution can replace a vertex of the complex to make the central distance of complex the largest, so that the search capability of the modified complex will be improved, and the size of the search has been expanded. The replacement criterion based on maximum complex central distance is as follow:

1) In PSO calculation, the initial population of each group of the complex shape method is corresponding to that of the PSO algorithm.
2) Replace $k$ vertexes of the current complex one by one with the better solution solved by the PSO, at the same time keeping the ( $k-1$)vertexes no changed to form $k$ new complex;
3) Calculate the central distance of $k$ new complexes, and set the complex having maximum central distance as that in next iteration.

## 3 The establishment of optimization mathematical model of the main girder of bridge crane

### 3.1 Composition and working principle of the bridge crane

The bridge crane is mainly composed of three parts, including car, bridge and electrical equipment. The three sets of motions constitute a bridge crane work three-dimensional space. Thus the bridge girder consisted of main beam and end beam is important component part of the bridge. Therefore, the main beam is the key point in the design of the crane, and its structure is a box type structure [8].

### 3.2 Selection of design variables

The optimization design goal of the crane girder is to reduce its weight on the basis of satisfying the performance of the whole machine. In the span being certain, the key factor affecting the weight of the main beam is its section area . It is mainly determined with the dimensions of upper and lower flange, main side webs and the reinforced rib plates whose weights are relate to the span[9,10]. Therefore, there are 4 design variables of main girder simplified in all, namely web height, flange width, web thickness, flange thickness, respectively, set as $x_1$, $x_2$,$x_3$,$x_4$ shown in Figure 1.

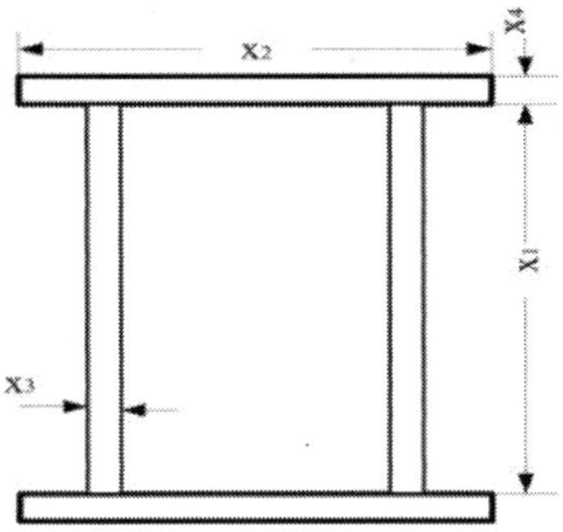


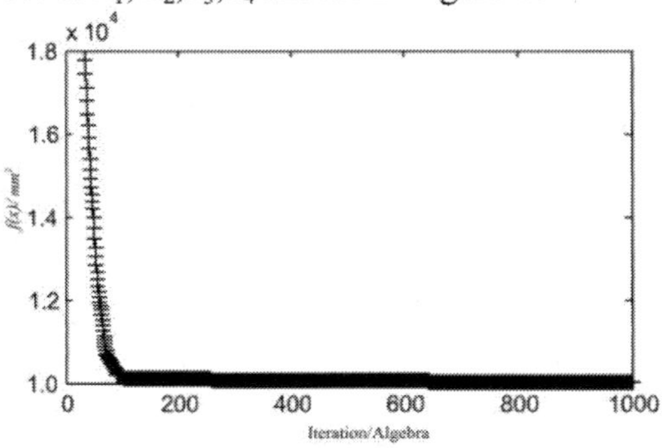


**Figure 1.** Box girder section of bridge crane **Figure2.**Evolution curve of PSO algorithm

### 3.3 Objective function

When the main beam span and steel density are certain, the weight of the girder is proportional to its cross-sectional area, so the section area of the main beam is taken as the objective function. It is established for the quality of single girder in double girder bridge crane to be the lightest, which can be expressed as equation (8).

$$f(x) = 2(x_1x_3 + x_2x_4) \tag{8}$$

### 3.4 Constraint conditions

(1) Intensity constraint

Corner point stress of the flange in cross section is the largest, so the bending stress constraint can be represented as equation (9).

$$\sigma = \frac{T_V x_1}{2I_{xx}} + \frac{T_H x_2}{2I_{yy}} \le [\sigma] \tag{9}$$

Where, $T_v$ denotes the sum of the bending moment in the vertical direction, which is caused by the fixed load and moving load on the beam section; $T_H$ denotes the sum of the bending moment generated by the horizontal inertia load when the cart is braking; $I_{xx}$,$I_{yy}$ denote inertia moment of beam cross section to the X axis and the Y axis, respectively; $[\sigma]$ is allowable stress of main beam design.

$$g_1(x) = \frac{3L}{4}\left[\frac{P_1 + \rho(x_1x_3 + x_2x_4)L}{3x_1x_2x_4 + x_1^2x_3} + \frac{P_2}{3x_1x_2x_3 + x_2^2x_4}\right] - [\sigma] \le 0 \tag{10}$$

Where, $P_1$ and $P_2$ denote the maximum vertical static load and the maximum horizontal dynamic load, respectively. $\rho$ is the density of main beam steel material , $L$ is span of the main beam.

(2) Stiffness constraint

$$\frac{P_1 L^3}{48EI_{xx}} \leq [f_v] \tag{11}$$

Where E is the elasticity modulus of steel material, in general, E=2.06×105MPa;$[f_v]$ is main girder allowable deflection, $[f_v]$=L/700;After finishing the simplification, the stiffness constraint is expressed as equation (12).

$$g_2(x) = \frac{P_1 L^3}{3x_1^2 x_2 x_4 + x_1^3 x_3 \times 9.9\times10^6} - [f_V] \leq 0 \tag{12}$$

(3) Process constraint

$$g_3(x) = \frac{x_2}{x_4} - 60 \leq 0 \tag{13}$$

$$g_4(x) = \frac{x_1}{x_3} - 160 \leq 0 \tag{14}$$

(4) Boundary constraint

$$g_5(x) = 5 - x_3 \leq 0 \tag{15}$$

$$g_6(x) = 5 - x_4 \leq 0 \tag{16}$$

# 4 Realization of optimization design

## 4.1 Parameters Setting

(1) Parameters setting in the particle swarm optimization algorithm

Set the various parameters of PSO algorithm, including particle numbers $m$=20, the inertia weight $w_{\max} = 1.2$, $w_{\min} = 0.4$,acceleration constants $c_1$=$c_2$=2, the maximum iterations number $T_{max}$=1000, the maximum particle velocity $v_{max}$=1, the fitness goal $\varepsilon = 0.1$.

(2) The crane structure parameters setting

Taking the double girder crane in common use as an example, the basic parameters are set as follows:

$P_1 = 1.2\times10^5 N$, $P_2 = 1.2\times10^4 N$, $L = 10.5m$, $[\sigma] = 140MP_a$, $[f_v] = L/700$, $x_1 \in [660,900]$, $x_2 \in [240,400]$, $x_3 \in [5,32]$, $x_4 \in [5,32]$,and four variables of each dimension are mm as the unit. Generally the variation intervals of web height and flange width of the main beam are set to an integer multiple of 5*mm*;when flange thickness and web thickness are less than 30*mm*, the variation intervals are set to 0.5mm;when they are more than 30*mm*, the change intervals are set to 1*mm*[10].

## 4.2 The Optimization Results

The optimization problem involves one objective function and six constraints. Complex method mixed with PSO (CM-PSO) is programmed by MATLAB for optimization. The *f(x)* is taken as fitness in PSO, and particle swarm evolutionary iteration process has been shown in Figure 2, the optimization results with the complex mixed with PSO method and simple complex method (CM) are compared .They are shown in Table 1.

## 4.3 Results analysis

The optimal results in Table 1 show that the area of the main beam section is 14819 $mm^2$ by the complex method, and that is 11890$mm^2$ by complex method mixed with PSO, which has reduced 19.77%. Seen from Figure 2, the PSO algorithm running in groups quickly converges to the optimal solution. After the optimal particle replacing complex vertex, enlarging the maximum complex central distance, and expanding the search space of feasible solutions, meanwhile increasing the diversity of population in optimal process, so the global convergence of algorithm has greatly improved.

**Table 1.**Optimization results of main girder structure size

| Method / Variable (unit) | CM | CM-PSO |
|---|---|---|
| web height $x_1$ (mm) | 784 | 775 |
| flange width $x_2$ (mm) | 356 | 345 |
| web thickness $x_3$ (mm) | 6 | 5 |
| flange thickness $x_4$(mm) | 7.6 | 6 |
| Section area f(x) ($mm^2$) | 14819 | 11890 |
| Iterative algebra (g) | 690 | 270 |
| Time (s) | 14 | 3 |

## 5 Conclusions

In the complex method mixed with PSO, the optimal particle in PSO replaces complex vertex to realize the maximum complex central distance. Starting from method of constraint handling techniques and swarm intelligence, the problems of optimization design are solved, which improves this modified complex search ability, and makes the search scale enlargement, so that the optimal solution speed and accuracy are enhanced. Through new algorithm the girder cross-sectional sizes of the bridge crane are optimized, the section area of the main beam has greatly decreased, and we achieve the goal of both the compact lightweight structure and mechanical properties. This method can be applied to the similar operating mechanism.

## References

1. L. Riqi. Analysis and Optimization Research on the structure of box girder of overhead traveling crane[D].Harbin: Northeast Forestry University, **4** (2012)
2. T. Deyu, S. Dawen, S. Tingting etc. Optimization design of main girder of bridge crane [J]. *New Technology & New Process*, 11 (2013)
3. F. Xiaoning, C. Hui, X. Gening. Optimization design software for overhead traveling crane based on annealing genetic algorithm [J]. *Hoisting and conveying machinery*, **5** (2013)
4. J. Chenguang. Optimization Design of Bridge Crane Box Girder Based on Chaotic Genetic Algorithm [D].North University of China. **6** (2014)
5. H. Yourui. Intelligent optimization algorithm and its application [M]. Beijing: National Defense Industry Press, **1** (2008)
6. Y. Yafei. The Method of Particle Swarm Optimization Solving Constrained Problems [D]. Beijing: China University of Geosciences, (2012)
7. L. Zhifeng. Mechanical optimization design [M]. Beijing: Higher Education Press, 2011.3
8. X. Gening. The metal structure of mechanical equipment[M].Beijing: Mechanical Industry Press, (2009)
9. T. Fang, S. Juan, Z. Yu. Technique of constrain processing based on punishment and reparation strategy [J]. *Computer engineering & design*, **12**(2006)
10. Y. Qinglin, T. Yuanfang, S. Nanxi etc. Particle swarm optimization algorithm combined with penalty function is used to optimize the main beam of bridge crane [J]. *Hoisting and conveying machinery*, **4**(2015)

# Dynamic Analysis of the Tilted Furuta Pendulum

Ali Wadi[1,a], Jin-Hyuk Lee[1] and Lotfi Romdhane[1,a]
[1]MCE-College of Engineering, American University of Sharjah, Sharjah, UAE

**Abstract**— This paper deals with the dynamic analysis of a tilted passive Furuta pendulum. The pendulum is tilted to ensure the existence of a stable equilibrium position. An analytical and ADAMS models are developed. These models take into account the full inertia of the pendulum and the viscous and dry friction in the joints. The only external force applied to the model is the weight. An experimental setup is used to validate the obtained models. The simulated time response of the system is compared to the experimental one. A good agreement between the two results is obtained. A frequency analysis of the response is then performed, which showed the existence of different harmonics depending on the initial angular position. Some differences are noticed, however, between the experimental results and those obtained through the simulation. These differences could be attributed to the chosen values of the viscous and dry friction parameters. A future work is underway, which uses this model to identify the friction parameters of the joints.

## 1 Introduction

The Furuta pendulum (also known as a rotary inverted pendulum) is an under-actuated system, which is formed by two bodies in series. The first one is rotating around a vertical axis followed by a pendulum rotating around a horizontal axis. It was first proposed by Furuta [1] mainly to test different control laws[2]. This mechanism is a popular one that can be found in most control laboratories. The Furuta pendulum is mainly used as a testbed of nonlinear control strategies or as an educational apparatus. Most of the works in the literature used the Furuta pendulum to illustrate their proposed control law [1]. However, most of these works did not elaborate on the dynamic model of the Furuta pendulum and they limited their work to a simplified model [3].

Few papers dealt with the modeling of the Furuta pendulum. These works highlighted the complexity of the dynamic model and the necessity to simplify based on several assumptions. The authors of [1, 4] proposed a linearized model based on a small angle assumption. Others neglected one or more of the cross coupling terms relating the two rotations [1]. The authors of [2,4] showed, through simulations, that this assumption is not acceptable and it could have an important effect on the dynamic response. To the best knowledge of the authors, there is no work in the literature that carries time-extended experimental tests and validates the results against the deduced model. The authors of derived a significantly simplified system and carried a brief 3-second test. All the experimental works on the Furuta pendulum are directed at testing various control strategies [5]. The dynamics of the system is often overlooked or linearized. However, the knowledge of the dynamic behavior of a

[a] Corresponding Author: ali.wadi@alumni.aus.edu

system is highly desirable to design an advanced controller for the given nonlinear system, no matter how robust the controller is. This work presents an analytical model, which is validated through experimentation.

When the Furuta pendulum is used in a passive mode, the gravity, which is the only external load, actuates only the second joint. Therefore, the dynamic modeling of the system without an input torque, is incomplete and could not be validated through experimentation. This fact could explain the reason of the nonexistence in the literature of dynamic models of passively actuated Furuta pendulum. However, a dynamic model of the passive system could capture the physics of the system without the perturbation of the input torque. In this paper, to actuate both joints in the passive mode, we propose to tilt Furuta pendulum. In this case, the Furuta pendulum has a stable equilibrium position, defined by the gravity, and the oscillations of the pendulum around this equilibrium could be studied, analytically and experimentally. Indeed, the model shows that there is an equilibrium position around which the system will oscillate when one of the joints is moved from this position. The developed model includes the effect of the viscous and dry friction. An analytical model along with a model under ADAMS are derived. These models are then compared to the experimental results. The experimental setup uses a commercial Furuta pendulum manufactured by Quanser Inc [6]. A set of experiments were conduct with different initial positions of the pendulum. The simulation results are then compared to the experimental ones, in order to validate the model.

The rest of the paper is organized as follows: after the introduction, Section II introduces the developed models under MATLAB and ADAMS. The comparison of the results, along with a discussion, is presented in Section III. Some concluding remarks and future work are presented in Section IV.

## 2 Modeling of the system

### 2.1 The Furuta Pendulum

Figure 1 shows the Furuta pendulum in its tilted position. The angle $\phi$ is taken around the fixed y-axis. The obtained tilted reference frame is $x_0 - y_0 - z_0$, where $z_0$ is now the axis of the first joint between the ground and the arm (1). The joint is disconnected from the existing motor and the pendulum is subjected only to its weight. Figure 2 shows the schematic of the Furuta pendulum in its tilted position. The arm is rotating around the fixed axis $z_0$ by an angle $\theta_0$ and the pendulum is rotating around the moving axis $x_1$ by an angle $\theta_1$. The different parameters of the two links are defined in Table 1, whereas those defining the joints are listed in Table 2. All these values are obtained from the manufacturer of the system [6], except for the dry friction coefficient. Several values were tested for the friction and the ones listed in Table 2 are the ones that gave the best results. It is worth noting that the ADAMS model is limited to the dry friction, whereas the analytical model uses both viscous and dry friction.

 DOI: 10.1051/matecconf/201710402011

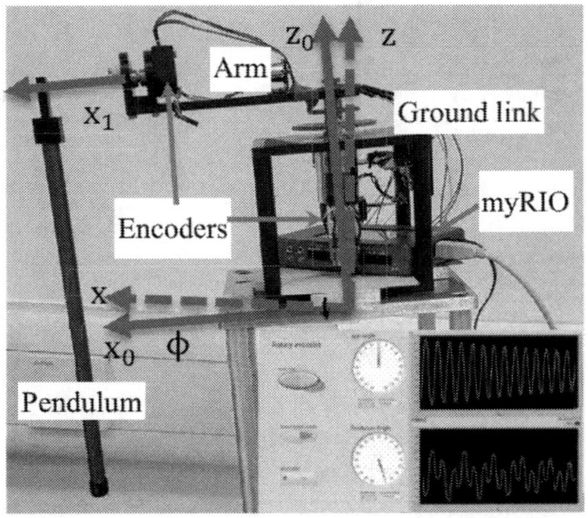


**Figure 1.** The tilted Furuta pendulum

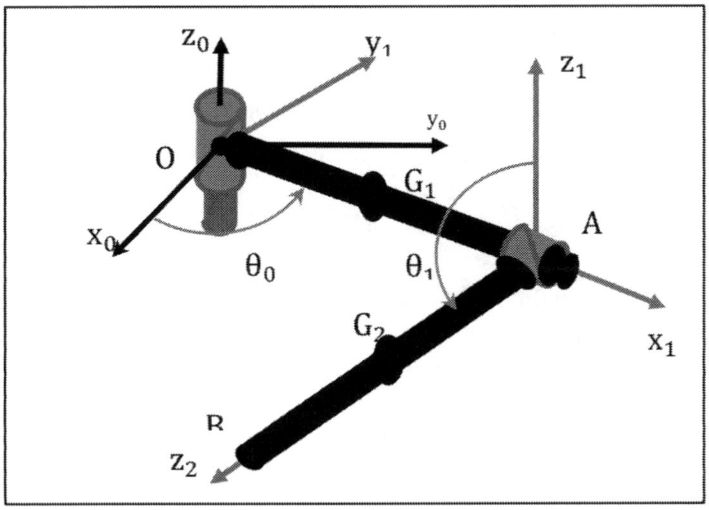


**Figure 2.** Schematic of the tilted Furuta pendulum.

**Table 1.** Parameters of the two Links

| Parameter | Arm (1) | Pendulum (2) |
|---|---|---|
| Mass $(Kg)$ | $m_1 = 0.370$ | $m_2 = 0.128$ |
| Moment of Inertia matrix $\boldsymbol{J}_i$ $(Kg.m^2)$ | $j_{1z} = 1.1 \times 10^{-3}$ | $j_{2xx} = j_{2yy} = 5.43 \times 10^{-3}$ $j_{2zz} = 2.33 \times 10^{-6}$ |
| Distance from pivot point to center of gravity $(m)$ | $l_1 = OG_1 = 0.0062$ | $l_2 = AG_2 = 0.185$ |
| Total length $(m)$ | $L_1 = OA = 0.216$ | $L_2 = AB = 0.316$ |

**Table 2.** Parameters of the two Joints

| Parameter | Joint 1 | Joint 2 |
|---|---|---|
| Radius of the shaft | $r_1 = 4$ | $r_2 = 4$ |

| | | |
|---|---|---|
| Dry coefficient of friction | $\mu_1 = 3 \times 10^{-4}$ | $\mu_2 = 3 \times 10^{-4}$ |
| Viscous coefficient of friction $(N.m/\frac{Rad}{s})$ | $\eta_1 = 1.3 \times 10^{-4}$ | $\eta_2 = 1.3 \times 10^{-4}$ |

## 2.2 Dynamic Model of the Furuta Pendulum

### 2.2.1 Analytical Model

Based on the schematic of the Furuta pendulum shown in Figure 2, the dynamic model is elaborated. The Euler-Lagrange formalism is used to derive the equations of the motion of the system. Two generalized coordinates are required to fully describe the dynamics of the two-degrees-of-freedom system, and they are the angular displacements of the Arm and Pendulum links given by

$$\mathbf{q} = \begin{bmatrix} \theta_0 \\ \theta_1 \end{bmatrix} \tag{1}$$

The Euler-Lagrange equation, for the i<sup>th</sup> generalized coordinate of the system can be written as follows:

$$\frac{d}{dt}\left(\frac{\partial T}{\partial \dot{q}_i}\right) - \frac{\partial T}{\partial q_i} + \frac{\partial V}{\partial q_i} = Q_i, \ (i = 0,1) \tag{2}$$

The kinetic energy, T, is written as the sum of linear and rotational kinetic energies of the two links, given by

$$\mathrm{T} = \sum_{i=1}^{2} \frac{1}{2} \mathrm{m_i v_{Gi}^2} + \sum_{i=1}^{2} \frac{1}{2} \omega_{i-1}^{T} (\mathrm{J_i} \omega_{i-1}) \tag{3}$$

where

$\boldsymbol{\omega}_0 = \dot{\theta}_0 \mathbf{z}_0$, $\boldsymbol{\omega}_1 = \dot{\theta}_0 \mathbf{z}_0 + \dot{\theta}_1 \mathbf{x}_1$ and $\mathbf{v}_{Gi}$ is the velocity of $G_i$.

The potential energy of the gravity is written as the result following scalar product between the global vertical axis and the center of gravity position vector in the local tilted frame of reference.

$$V = -m_1\, g\, (\mathbf{z} \,.\, \mathbf{OG_1}) - m_2\, g\, (\mathbf{z} \,.\, \mathbf{OG_2}) \tag{4}$$

The generalized forces, representing the viscous and dry friction in the joints, are written as the sum of frictional torques acting on the links.

$$Q_j = -\eta_j\, \dot{\theta}_j - \mu_j\, r_j\, F_{nj}\, \mathrm{sign}(\theta_j) \tag{5}$$

where $\eta_j$ and $\mu_j$ are, respectively, the viscous and dry friction of joint j. $F_{nj}$ is the normal force in the joint j.

The final model is given as follows:

$$\mathrm{H}\, \ddot{q} + \mathrm{B}(q, \dot{q}) + \mathrm{G}(q) = \mathrm{Q} \tag{6}$$

where:

The generalized inertia 2x2 matrix is given by:

$$\mathbf{H} = \begin{bmatrix} \left(j_{1z} + m_1 l_1^2 + m_2 L_1^2 + \left(m_2 l_2^2 + j_{2y}\right)\sin(\theta_1)^2 + j_{2z}\cos(\theta_2)^2\right) & -m_2 l_2 L_1 \cos(\theta_1) \\ -(m_2 l_2 L_1 \cos(\theta_1)) & j_{2x} + m_2 l_2^2 \end{bmatrix} \tag{7}$$

The centrifugal and Coriolis forces are given by the following vector:

$$\mathbf{B}(\mathbf{q}, \dot{\mathbf{q}}) = \begin{bmatrix} -\left(\left(m_2 l_2^2 - j_{2y} + j_{2x}\right)\sin(2\theta_1)\right)\dot{\theta}_0\dot{\theta}_1 + (m_2 l_2 L_1 \sin(\theta_1))\,\dot{\theta}_1^2 \\ \left(\left(-m_2 l_2^2 - j_{2y} + j_{2z}\right)\sin(\theta_1)\cos(\theta_1)\right)\dot{\theta}_0^2 \end{bmatrix} \tag{8}$$

The effect of gravity is given by the following vector

$$\mathbf{G}(\mathbf{q}) = \begin{bmatrix} (m_1 g l_1 \sin(\theta_0)\sin(\phi) + m_2 g (L_1 \sin(\theta_0) - l_2 \sin(\theta_2)\cos(\theta_0))\sin(\phi)) \\ (-m_2\, g\, l_2\, (\sin(\theta_0)\cos(\theta_1)\sin(\phi) - \sin(\theta_1)\cos(\phi))) \end{bmatrix} \tag{9}$$

The generalized forces due to friction are given by the following vector:

$$\mathbf{Q} = \begin{bmatrix} -\eta_0\, \dot{\theta}_0 - \mu_0\, r_0\, F_{n0}\, \mathrm{sign}(\theta_0) \\ -\eta_1\, \dot{\theta}_1 - \mu_1\, r_1\, F_{n1}\, \mathrm{sign}(\theta_1) \end{bmatrix} \tag{10}$$

### *2.2.2 ADAMS Model*

Figure 3 shows the model under the software ADAMS. The gravity vector is shown downwards. It is worth mentioning that in the ADAMS model, the gravity is tilted by the angle $\phi$ with respect to the axis of the first joint. The friction is modeled in ADAMS using its building model. The parameters used are given by the menu shown on Figure 3. Only the dry friction parameter values are changed, all the other parameters are kept as default values given by ADAMS. It is worth mentioning that the viscous friction is not included in this model.

The output of the model is the motion of the first and second joints, for different initial positions of the second joint.

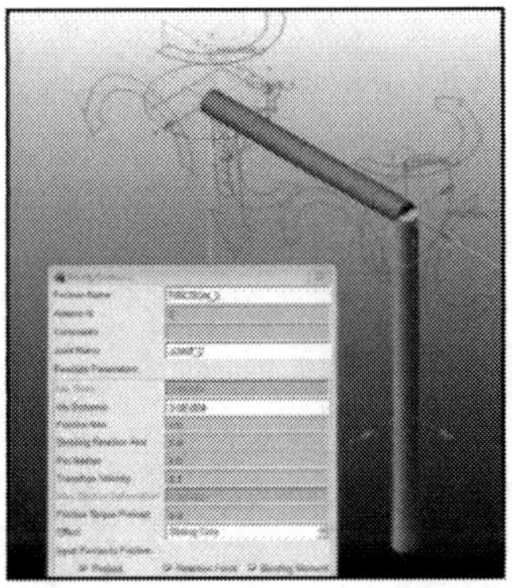

**Figure 3.** The ADAMS model of the tilted Furuta pendulum

### *2.2.3 Experimental Setup*

The experimental setup is made of an instrumented Furuta pendulum, an acquisition system (myRIO Embedded System, National Instruments), and a computer with the LabVIEW software (Figure 1). The system is equipped by two rotary encoders (E2 Optical Kit Encoder by US Digital) to record the angular positions of the two joints as a function of time.

The sampling period is chosen to be 10 ms and the acquisition time covers the full motion until the pendulum stops. This time ranged from 12 to 45 s, depending on the initial angular position of the motion.

## 3 Results and Discussion

The simulation results (analytical and ADAMS) and the experimental results are obtained for two different initial positions of the pendulum (2). These positions are chosen to be 15 degrees and 55 degrees, and they are all taken with respect to the equilibrium position, which is along the negative z-axis. These values are chosen to have a small angle (15 degrees) and a large angle (55 degrees).

In all the experiments the pendulum starts with a zero initial angular velocity. The motion in the two joints is monitored until the system returns to the equilibrium position. The obtained results from the three cases (Analytical, ADAMS, Experimental) are the time responses of the two joint angles. For the experimental setup, the recording was repeated several times and no differences were observed between different runs having the same initial angular position. Based on this time response, the decay of the amplitude as a function of time and the frequency analysis were performed.

### 3.1 Time response

The time response for the two cases of initial angular positions are shown in Figure 4. These results show that the pendulum has a motion similar to a simple pendulum, however the arm is animated by highly complex motion. This complexity is predicted by the analytical model, due to the coupling between the two motions and the nonlinear terms. None of the two models (analytical and ADAMS) fitted the experimental results. For the pendulum motion, the analytical model was closer to the experimental result despite the slight difference in its natural frequency, which led to a shift between the two curves after a few periods of motion. The ADAMS model had a similar problem with the frequency but also the amplitude was not predicted correctly. This difference in the two cases could not only be attributed to the high nonlinearity in the model but also to the models used in describing the viscous and dry frictions. Indeed, the analytical model used Eq. 4 along with Table 2 to describe the model of the joints. In ADAMS, the Coulomb friction parameters are shown on Figure 3. However, the model of the joints is not taking into account the viscous friction.

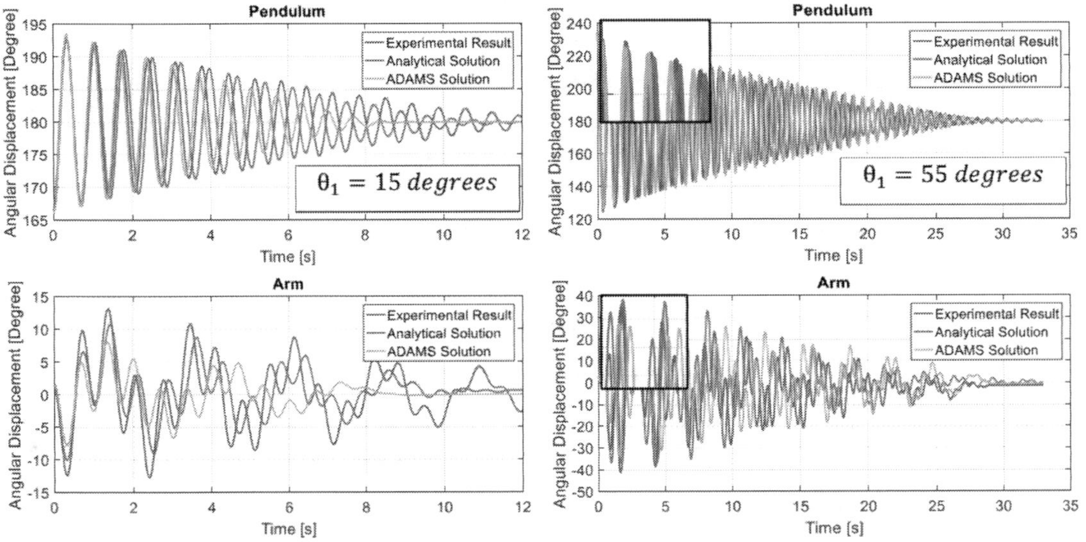


**Figure 4.** Time response of the arm and pendulum

### 3.2 Frequency Response

The frequency analyses of the responses are shown in Figure 5 for the arm and the pendulum. These figures show the frequency responses for three initial angles, i.e., 15, 55 and 90 degrees. One can notice that the analytical model predicted the results better than the ADAMS model in all cases. Another important result from these plots is the number of harmonics, in the response of the arm, that changes from one, in the case of $\theta_1 = 15$ degrees to five in the case of $\theta_1 = 90$ degrees. This is result shows again the high nonlinearity of the model. At $\theta_1 = 15$ degrees, there is only one dominant frequency at around 1.5 Hz. Lower frequencies cannot be detected due to the high signal to noise ratio. For the case of $\theta_1 = 55$ degrees, four harmonics, 0.4-1.2-3.6-5.8 Hz, are detected for the arm and five harmonics at $\theta_1 = 90$ degrees, i.e., 0.4-1-3-5-7 Hz. For the pendulum, the response is similar in all cases, except at $\theta_1 = 90$ degrees, where a second harmonic at 3 Hz starts to appear. This type of behavior shows the high nonlinearity of the system. Despite this complexity, the proposed model succeeded in capturing the physics of the system in all three cases. Some differences persist which could be attributed to the values used for the friction parameters.

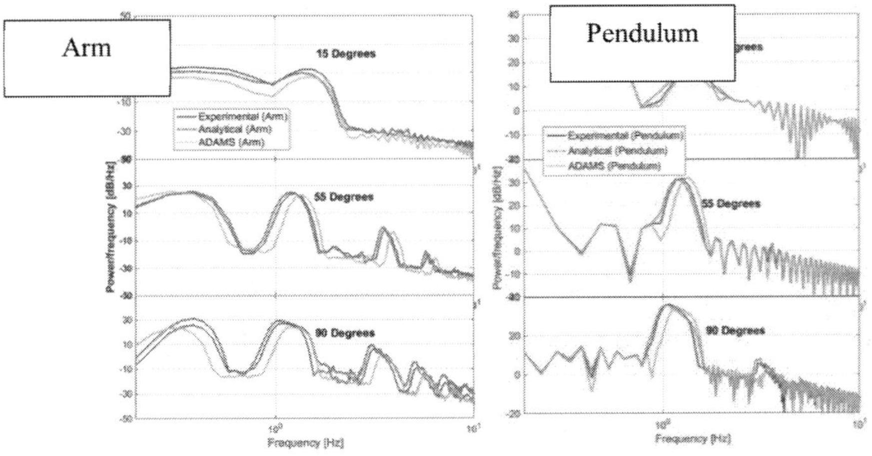


**Figure 5.** Frequency response for the arm and pendulum at different initial angles

## 4 Conclusion

This paper presented a dynamic analysis of a tilted Furuta pendulum. The pendulum was tilted to ensure the existence of a stable equilibrium position. A complete analytical model was derived where all the full inertia of the pendulum was taken into account. ADAMS was also used to simulate the system. The obtained results showed that the developed analytical model gave results that are closer to the experimental ones and hence predicted better the real dynamic behavior of the system.

A frequency analysis was performed and it showed that the system is highly nonlinear. Different harmonics were detected depending on the initial angular position. The difference between the simulation results and the experimental ones could be attributed to the values of the viscous and dry friction parameters and the model chosen for these frictions.

## Acknowledgment

This work was financed by the American University of Sharjah under the grant FRG15-40.

## References

1. K. Furuta, M. Yamakita and S. Kobayashi, "Swing-up control of inverted pendulum using pseudo-state feedback", Proceedings of the Institution of Mechanical Engineers, Part I: Journal of Systems and Control Engineering 1991-1996 (vols 205-210), **206**, no. 49, pp. 263-269, 1992.
2. I. S. Diniz, D. Colón, R. Carrion, R. Aguera, L. Góes, "The Rotary Inverted Pendulum: Modeling, Simulation and Control Aided by Computer: CAD-CAE", 22nd International Congress of Mechanical Engineering, Ribeirão Preto, Brazil, 2013.
3. S. Jadlovský and J. Sarnovský, "Modelling of Classical and Rotary Inverted Pendulum Systems – A Generalized Approach," Journal of Electrical Engineering, **64**, no. 1, Jan. 2013.
4. B. S. Cazzolato and Z. Prime, "On the Dynamics of the Furuta Pendulum," Journal of Control Science and Engineering, **2011**, pp. 1–8, 2011.
5. L. T. Aguilar, "Identification based generation of self-excited oscillations for underactuated mechanical systems via two-relay algorithm," 2008 International Workshop on Variable Structure Systems, Turkey, doi:10.1109/VSS.2008.4570680.
6. "Quanser - Rotary Inverted Pendulum", Quanser.com, 2016. [Online]. Available: http://www.quanser.com/products/rotary_pendulum. [Accessed: 15- Jul- 2016].

# Modeling and Simulation on Errors of Feed Unit by Considering Change of Force Bearing Point

Wei Zhang [1, 2, a], Min Wang [1] and Bing-bing Le[1]

[1]*Advanced Manufacturing technology of the Key Laboratory of Beijing Municipality College of Mechanical Engineering and Applied Electronics Technology, Beijing University of Technology, Beijing 100124, China*

[2]*Engineering Training Center, Inner Mongolia University of Science and Technology, Baotou 014010, China*

**Abstract:** The linear feed unit is a type of precision linear motion component that is widely used in computer numerical control (CNC) machine tools. The contact stiffness and error influences the performance of the feed unit directly. Thus, investigating contact stiffness and error is important in optimizing the design and improving the performance of the linear feed unit. In this study, the contact mechanics and the deformation of the roller between the roller and rail are analyzed. Calculation model of the contact stiffness and error based on the Hertz theory and multi-body kinematics are established and the change of the contact angle is also considered. Errors and the contact stiffness curve of five directions and the changes of the slope of the stiffness curve after the load increases to a certain size are obtained. The Motion precision errors of The roller linear feed unit are analyzed. The effectiveness of the proposed models on contact stiffness and error is verified through simulation on a specialized test system of the linear feed unit.

## 1 Introduction

The roller linear feed unit has widely used as a guide and load-bearing member in machine tools and mechanical devices. The feed unit has high positioning accuracy, low coefficient of friction, good precision retaining ability and good maintainability among others; thus, roller linear feed unit is a key feature of CNC machining equipment. The roller linear feed unit is mostly used for small or midsized precision machine tools, whereas the roller linear feed nnit has higher load capacity and higher

[a] Corresponding author: zhangwei2009035@163.com

rigidity and is applicable to large CNC machine tools and machining centers. In recent years, the growing trend of high speed, high precision, and high durability in CNC machine tools has propelled high performance and reliability to be the most important index for linear feed unit. Thus, in-depth researches on the dynamic characteristics of linear feed unit are greatly needed. This research can lay a good foundation for enhancing the performance of linear feed unit.

Existing researches on the dynamic characteristics of linear feed unit were focused on contact stiffness, vibration during operation, and friction characteristics of roller linear guides. Wu [1, 2] and Liu [3] studied the contact stiffness between the roller and raceway and its related influence factors were analyzed and calculated based on the Hertz theory. In addition, vibration synthesis and prediction by the method of employing the finite element were performed. After Shimizu's study [4,5] on the characterization of frictional hysteresis in roller-bearing guide ways, In article Jui Pin Hung[6], the linear rolling guide under different loading finite element model was established based on Hertz contact theory, the influence of external load to guide stiffness was studied and the finite element model was modified based on test results. In Wei Jun T [7], the relationship of the force between roller and raceway was analyzed and the stiffness model of horizontal, vertical and deflection angle around the spindle rotational was built based on Palmgren formula. The hertz contact model of roller among nails of linear roller guide is established firstly. Then the coordinate transformation matrix between the two coordinate systems is obtained with multi-body kinematic theory. Due to the analysis the stress of the roller linear motion unit, five mechanical equilibrium equations are obtained, then the amount of deformation of its five directions is determined. Thereafter, the analysis of roller linear feed units is simulated based on contact stiffness model proposed and the motion precision errors of the roller linear feed unit feed unit are analyzed. Finally, the proposed models are simulated by computer.

## 2 Contact stiffness of linear feed unit

The roller linear feed unit (hereinafter referred to as the table) contains four sliders; each slider also contains a number of rollers. The feed unit produces three rotation errors and two straightness errors with the role of the force of three directions, which are represented respectively as: $\delta y$、$\delta z$、$\varepsilon x$、$\varepsilon y$、$\varepsilon z$. $\delta y$、$\delta z$ are the deformation along *Y* and *Z* axis respectively, and $\varepsilon x$、$\varepsilon y$、$\varepsilon z$ are the deflection angle around the *X, Y, Z* axis respectively. The coordinate system $O_0$ in the center of the working table is set up. The three axes are parallel to the movement direction, horizontal direction and vertical direction of the work table, and the center of the table and the slider are high in the vertical direction. The coordinate system $O_1$ (0,δy,δz) is established in the center of the working table after the deformation, as shown in Figure. 1, In Figure 2, the black graphic slider with external loading produces linear roller guide way side position coordinates; the red graphic slider with external loading generates the position coordinates of elastic deformation.

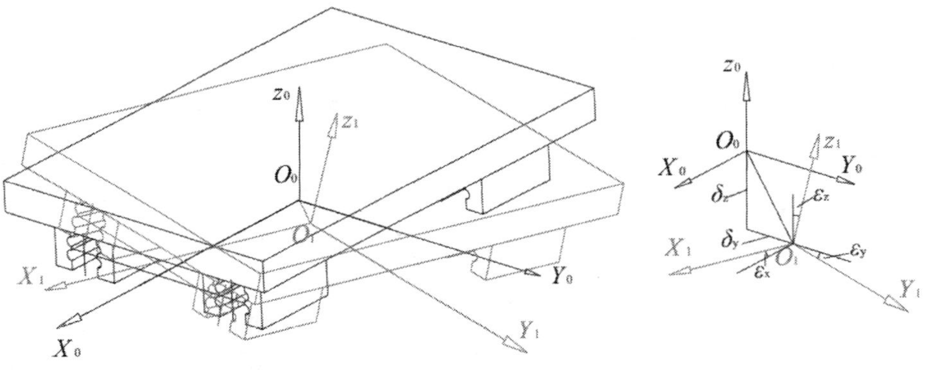


**Figure.1** Coordinate system of Linear feed unit

The transformation matrix ${}^{0}\boldsymbol{T}_{1}$from the coordinate system of $O_0$ to the coordinate system of $O_1$ is:

$$
{}^{0}T_{1}=\begin{bmatrix} 1 & -\varepsilon_{z} & \varepsilon_{y} & 0 \\ \varepsilon_{z} & 1 & -\varepsilon_{x} & \delta_{y} \\ -\varepsilon_{y} & \varepsilon_{x} & 1 & \delta_{z} \\ 0 & 0 & 0 & 1 \end{bmatrix} \qquad (1)
$$

In the rolling linear guide, the roller, the slider and the track have contact surface. When the slider is subjected to external load, the contact stress and the elastic deformation are generated in the contact area at the same time. Under the condition of applying the pre pressure, the cross section of the slider 1 is analyzed. In Figure 2, the position of each column roller in slider and force are showed below.

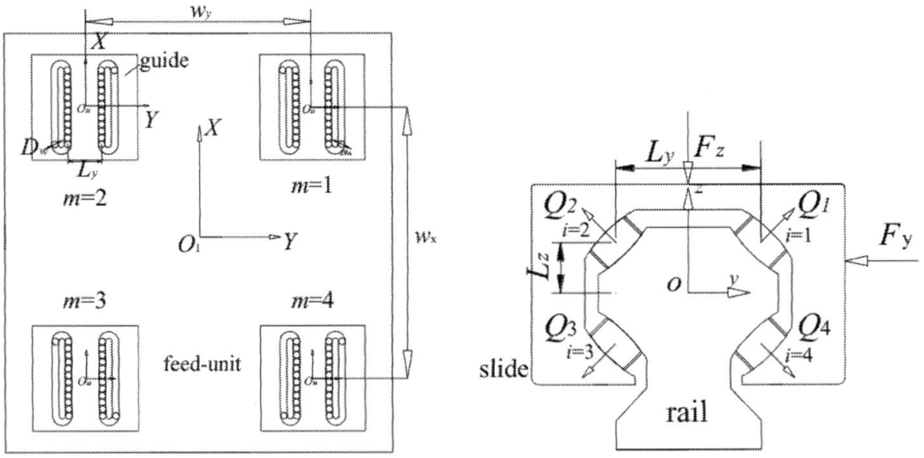


**Figure.2** Locations and force of every roller in feed unit

As is shown in Figure.3, when there is no external load, according the geometry relationship of roller and rail the distance between the curvature center of the slider raceway and the guide raceway, is obtained $\boldsymbol{O}_{s}\boldsymbol{O}_{g}$: equation (2)

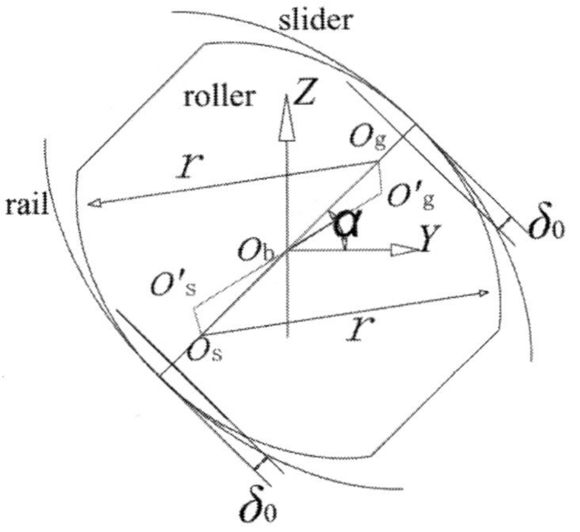


**Figure.3** The geometry relationship of roller and rail

$$O_sO_g = 2r - D_w = (2f-1)D_w \tag{2}$$

Where, $D_w$ is the diameter of the roller, $O_s$ means the curvature center the slider rail. $O_g$ means the curvature center the guide rail. $r=fD_w$, and $f$ is the curvature ratio. $r$ is the curvature radius of roller paths in the slider and guide, Because of the elastic deformation of the sliding block and the guide rail is very small. The model will be considered as a rigid body, and only the deformation of the roller is considered. After the external loading, the curvature center coordinate of the j-th roller in the j-th column in the coordinate system of $O_1$ is $X_{ij(1)}$, $Y_{ij(1)}$, $Z_{ij(1)}$.

$$\begin{aligned} X_{ij(1)} &= W_x - (0.5 \cdot n \cdot D_w) + D_w \cdot j \\ Y_{ij(1)} &= \begin{cases} W_y + L_y - 0.5 \cdot (2 \cdot f - 1)\cos\alpha, & i = 1,2 \\ W_y - L_y + 0.5 \cdot (2 \cdot f - 1)\cos\alpha, & i = 3,4 \end{cases} \\ Z_{ij(1)} &= \begin{cases} L_z - 0.5 \cdot (2 \cdot f - 1)\sin\alpha, & i = 1,4 \\ -L_z + 0.5 \cdot (2 \cdot f - 1)\sin\alpha, & i = 2,3 \end{cases} \end{aligned} \tag{3}$$

Where, $n$ is the number of every column of guide. By the coordinate conversion, the coordinates obtained in the coordinate system can be expressed as equation (4):

$$\left[X_{ij(0)}\ Y_{ij(0)}\ Z_{ij(0)}\ 1\right]^T = {}^0T_1\left[X_{ij(1)}\ Y_{ij(1)}\ Z_{ij(1)}\ 1\right]^T \tag{4}$$

Also the curvature center coordinate of the $j$-th roller in the $i$-th column in the coordinate system of $O_0$ is $X_{ij(0)}$, $Y_{ij(0)}$, $X_{ij(0)}$.

$$\begin{aligned} X'_{ij(0)} &= X_{ij(1)} - \varepsilon_z Y_{ij(1)} + \varepsilon_y Z_{ij(1)} \\ Y'_{ij(0)} &= \varepsilon_z X_{ij(1)} + Y_{ij(1)} - \varepsilon_x Z_{ij(1)} + \delta_x \\ Z'_{ij(0)} &= -\varepsilon_y X_{ij(1)} + \varepsilon_x Y_{ij(1)} + Z_{ij(1)} + \delta \end{aligned} \tag{5}$$

After receiving the loading, the distance between the curvature center of the slider and the curvature center of the guide is expressed below equation (6):

$$O_s^{'}O_g^{'} = \left(Y_{ij(0)} - Y'_{ij(0)}\right)^2 + \left(Z_{ij(0)} - Z'_{ij(0)}\right)^2 \quad (6)$$

The elastic deformation of each roller is equation (7):

$$\delta_{ij} = \frac{O_s^{'}O_g^{'} - O_sO_g}{2} + \delta_0 \quad (7)$$

The model of contact force and roller elastic deformation base on hertz theory are showed below:

$$\delta = KQ^{2/3} \quad (8)$$

Where, K is parameters related to the curvature. The contact angle of the roller after deformation is:

$$\alpha_{sij} = \arctan\frac{\left|Z_{ij(0)}' - Z_{ij(0)}\right|}{\left|Y_{ij(0)}' - Y_{ij(0)}\right|} \quad (9)$$

Therefore, the resultant forces of four columns rollers in slider 1 in *Y*, *Z* axis directions are shown as equation (10):

$$F_{1z} = \sum_{j=1}^{n}(Q_{1j} + Q_{2j} - Q_{3j} - Q_{4j})\cdot\sin\alpha_{sij}; F_{1y} = \sum_{j=1}^{n}(Q_{1j} + Q_{2j} - Q_{3j} - Q_{4j})\cdot\cos\alpha_{sij} \quad (10)$$

In the same method of analysis, the resultant forces of all four columns rollers in slider 2, 3, 4 are obtained respectively in *Y* axis, and *Z* axis directions, which are represented as: $F_{2y}$, $F_{3y}$, $F_{4y}$, $F_{2z}$, $F_{3z}$, $F_{4z}$. Figure.4 shows the situation of force and torque of the worktable installed four sliders. Force and torque analysis of the entire table lists the mechanical balance equations, as follow equation (11):

$$\begin{cases} F_{1z} + F_{2z} + F_{3z} + F_{4z} - F_z = 0 \\ F_{1y} + F_{2y} + F_{3y} + F_{4y} - F_y = 0 \\ W_y\left(F_{2z} + F_{3z} - F_{1z} - F_{4z}\right) + 2M_x + F_zd_y = 0 \\ W_x\left(F_{1z} + F_{2z} - F_{3z} - F_{4z}\right) + 2M_y + F_zd_x = 0 \\ W_x\left(F_{3y} + F_{4y} - F_{1y} - F_{2y}\right) + 2M_z + F_yd_x = 0 \end{cases} \quad (11)$$

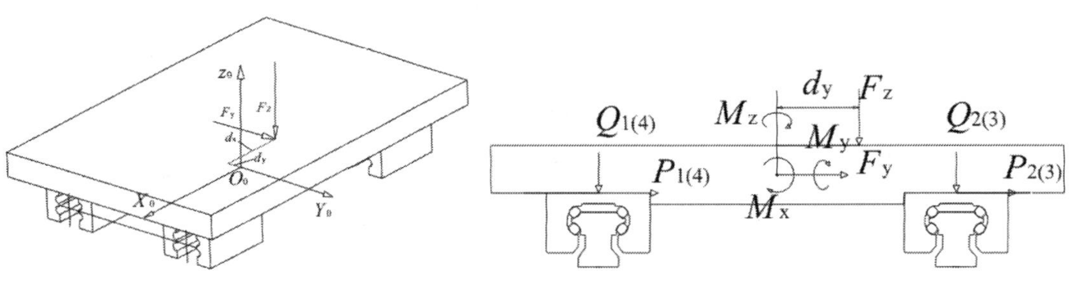


**Figure.4** Stress analysis of linear feed unit

## 3 Calculating and Simulation

By use computer, the vertical deformation and the lateral deformation of the linear feed unit are calculated. As shown In Figure.5, the thin line represent that it does not consider the change of contact angle. And the star line represent that it takes into the change of contact angle.

The tensional stiffness of the linear feed unit around $X_0$、$Y_0$ and $Z_0$ axis are calculated. As shown in Figure.6, the star line represent that it does not consider the change of contact angle. And the triangular line represent that it takes into the change of contact angle.

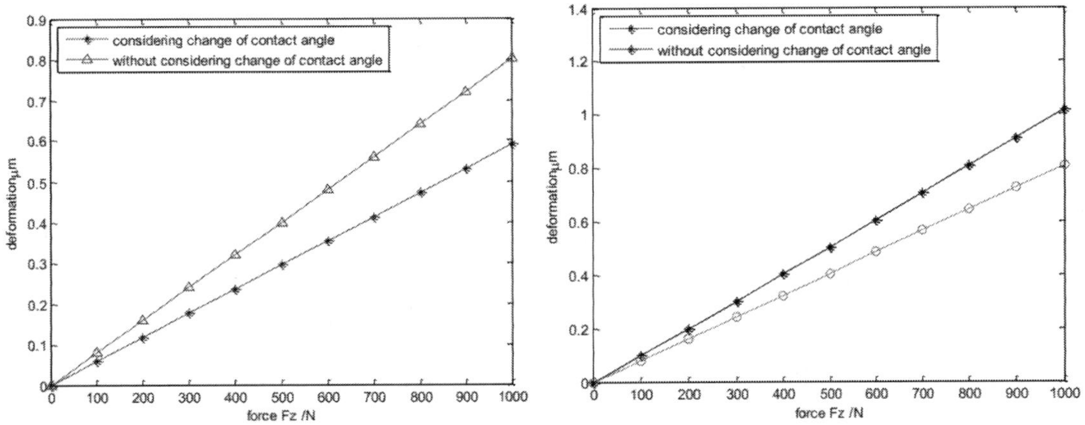


**Figure.5** Deformation analysis of linear feed unit

The movement precision error of linear feed unit is showed as follow. In Figure.7, the curves show the three rotational errors respectively. The curves in Figure.8 present two straightness errors and relationship between the value of the errors of three directions *X, Y, Z* and the distant from force point to geometry center of feed unit.

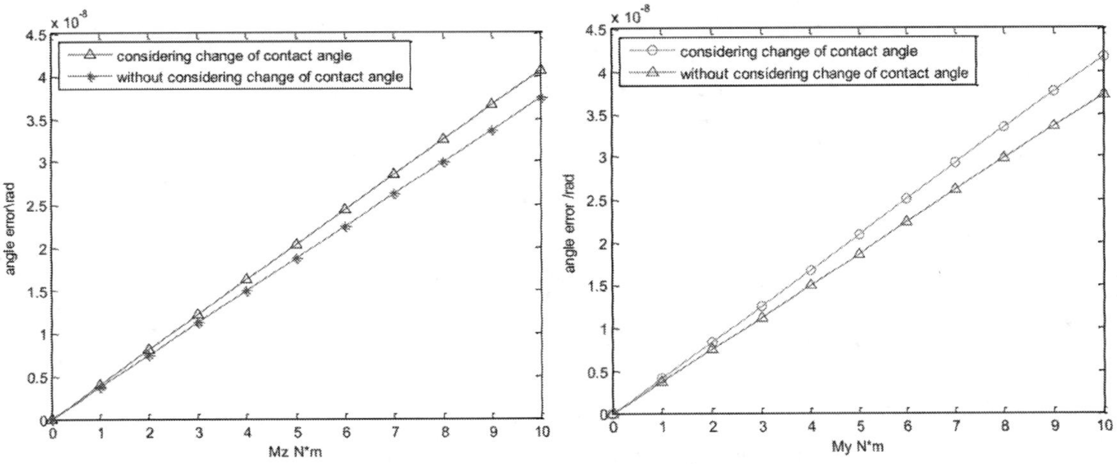


**Figure.6** Deformation analysis of linear feed unit

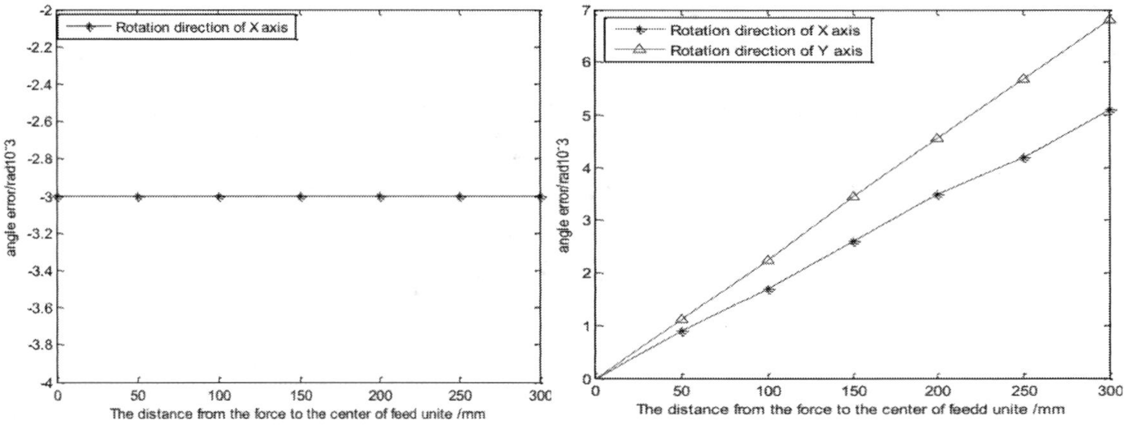


**Figure.7** Angle errors and force acting point position

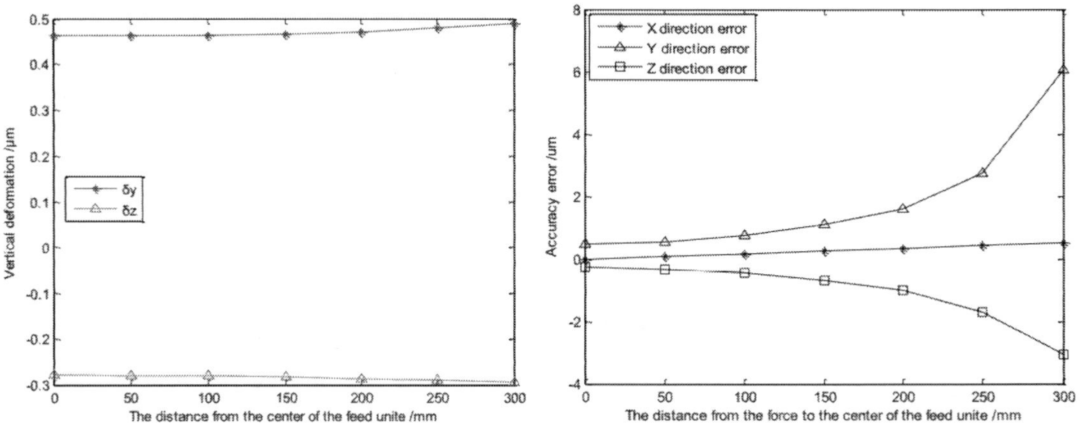


**Figure.8** Errors and force acting point position

## 4 Conclusions

1. By comparing the analysis results, it shows that the vertical error and lateral error of the linear feed unit is greater than the error of in *X* direction, because the movement of the linear feed unit is free in *X* direction.

2. After considering the change of the contact angle, we found that, both vertical stiffness values and horizontal stiffness of linear feed unit are decrease. The larger the value of the contact angle change, the curve of the vertical stiffness and the curve of the horizontal stiffness is not consistent. the horizontal stiffness is falling even faster.

3. With the distant from force point to geometry center of feed unit become bigger, the angle error around *X*, *Y*, and *Z* axis become bigger respectively. As shown in above section, the there angle error of the linear feed unit are very different. The angle error around *Y* axis and the angle error around *Z* axis are bigger the angle error around *X* axis. The angle error around *Y* axis is smaller the angle error

around *Z* axis.

4. When the load moves far from the geometry center of the linear feed unit, three rotational errors and two straightness accuracy errors all become bigger in different degrees.

## Acknowledgment

This work is supported by National Natural Science Foundation of China, Grant No. 51575014; Natural Science Foundation of Beijing, Grant No. 3154029 and KZ201410005010; Important National Science & Technology Specific Projects of China, Grant No. 2012ZX04010021-001-004; China Postdoctoral Science Foundation Funded Project, Grant No. 2016M591033; Beijing Postdoctoral Research Foundation, Grant No. 2015XX-13.

## References

1. J.S.S. Wu, J.C. Chang, J.P. Hung, The effect of contact interface on dynamic characteristics of composite structures. J. Mathematics and Computers in Simulation **74** , 6 (2007), 454–467.
2. J.S.S. Wu, J.C. Chang, G.A. Tsai, C.Y. Lin, F.M. Ou, The effect of bending loads on the dynamic behaviors of a rolling guide. J. Journal of Mechanical Science and Technology **26** , 3 (2012), 671–680.
3. J.S. Liu, S.X. Xiong, Z. Li, J.L. Sun, The rigidity model of linear motion guide in the situation of LMGs standing under several loads. J. Proceedings of the International Conference on Mechanical Transmissions, (2001), **pp**. 664–666.
4. S. Shimizu, E. Saito, H. Uchida, Tribological studies of linear motion roller guide systems. J. Tribology Transactions **41**, 1 (1998), 49–59.
5. S. Shimizu. Load distribution and accuracy/rigidity of linear motion roller guides system. J. Journal of the Japan Society for Precision Engineering **27**, 2(1993), 119–125.
6. J. P Hung. Load effect on the vibration characteristics of a stage with rolling guides. J. Journal of Mechanical Science and Technology, **23,** 1(2009), 9-99.
7. J.T Wei, Y Zhong. Model for wear prediction of roller guides. J. Wear (2013), 305:260-266

# Comparison of a 10 DOF quarter vehicle model with the FE analysis of tire impact against a road obstacle

Nenad Kranjcevic[1, a], Kristian Gruicic[1] and Marko Jokic[1]

[1]*Faculty of Mechanical Engineering and Naval Architecture, University of Zagreb, Zagreb, 10002, Croatia*

**Abstract.** In order to investigate the dynamic behaviour of tires on extreme irregularities on the road, a 10 degrees of freedom lumped 2D quarter vehicle model, comprising the two point rigid ring follower tire model, is presented and implemented in Matlab Simulink. The simulation results of tire impact against a step on the road, at various speeds and step heights, are compared to the results achieved by the FE analysis of the same problem.

## 1 Introduction

This paper deals with a relatively fast, yet still fairly accurate simulation of the dynamic behavior of a tire impacting an obstacle. Among various papers [1,2] which gave different approaches to the problem, [3] studied a quarter vehicle with different contact, i.e. follower models: point contact, constant footprint, rigid ring, two point and five point rigid ring followers. Compared to the industrial Flexible ring tire model (Ftire) in Simpack, the two point rigid ring follower model was found to be the most accurate. The idea of the two point follower model originates from [4], where the enveloping model with tandem elliptical cams was introduced and implemented as part of a 3D semi-empirical dynamic model of a tire rolling on uneven, but gradually changing road paths. The approach to the problem presented in [4] was adopted in this work, despite the fact that higher steps are under scope. The equations of motion of the 2D quarter vehicle model with 10 degrees of freedom (DOF) are presented in section 2, while the time consuming, but more accurate, finite element (FE) quarter vehicle model is presented in section 3. Sections 4 and 5 give a comparison of these two models and some conclusions, respectively.

## 2 The 2D mechanical model for Matlab Simulink implementation

The mechanical model for the simulation of dynamic behavior of a tire passing over a curb-like obstacle or pothole on the road consists of two sub-models:

- enveloping model and
- quarter vehicle model with rigid ring tire model.

The enveloping model produces from the real road profile the effective profile which is then used as input to the quarter vehicle model to obtain the dynamic response.

[a] Corresponding author : nenad.kranjcevic@fsb.hr

## 2.1 The quarter vehicle model with the rigid ring tire model

The basis for the quarter vehicle model is the tire rigid ring model which comprises three inertial elements: rim $m_{\text{rim}}$, ring $m_{\text{b}}$ and contact mass $m_{\text{c}}$. The fourth inertial element of the system is the mass of the quarter of the vehicle $m_{\text{car}}$. These inertial elements are connected with lumped (discrete) springs and dampers (Fig. 1). For the purpose of controlling the equations of motion, when the wheel is lifted above the road, an additional spring in the vertical direction is added between the contact patch and the road.

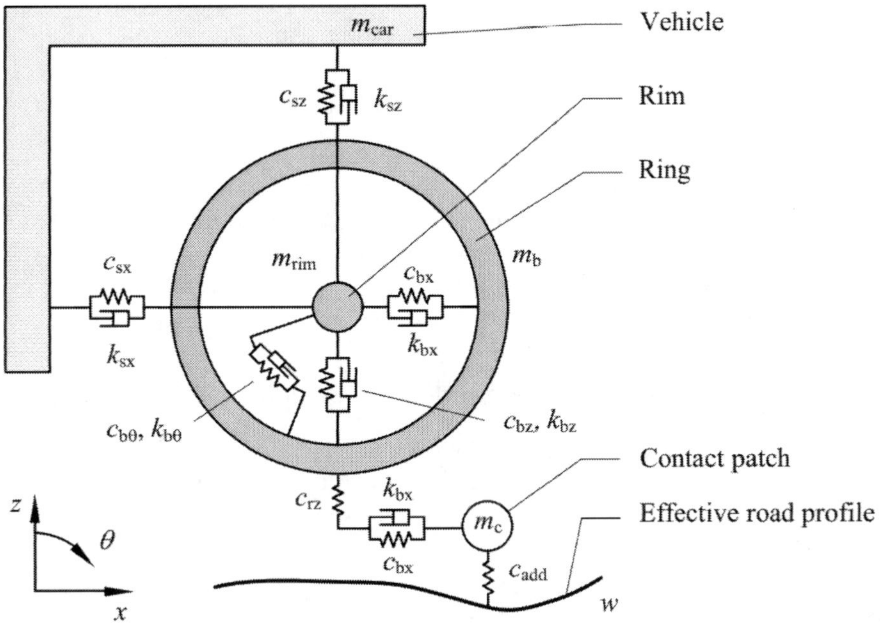


**Figure 1.** Quarter vehicle model with rigid ring tire model.

The equations of motion of the contact patch can be easily derived:

$$m_{\text{c}}\ddot{x}_{\text{c}} = -k_{\text{rx}}(\dot{x}_{\text{c}} - \dot{x}_{\text{b}}) - c_{\text{rx}}(x_{\text{c}} - x_{\text{b}}) + C_{\text{c}\kappa}\kappa_{\text{c}} \tag{1}$$

$$m_{\text{c}}\ddot{z}_{\text{c}} = -g - c_{\text{rz}}(z_{\text{c}} - z_{\text{b}}) - c_{\text{add}}(z_{\text{c}} - w) \tag{2}$$

The last term in Eq. (1) presents the sliding force [4]. Similarly, the equations of motion of the ring are:

$$m_{\text{b}}\ddot{x}_{\text{b}} = -k_{\text{bx}}(\dot{x}_{\text{b}} - \dot{x}_{\text{rim}}) - c_{\text{bx}}(x_{\text{b}} - x_{\text{rim}}) + k_{\text{bz}}\dot{\theta}_{\text{rim}}(z_{\text{b}} - z_{\text{rim}}) - k_{\text{rx}}(\dot{x}_{\text{b}} - \dot{x}_{\text{c}}) - c_{\text{rx}}(x_{\text{b}} - x_{\text{c}}) + \beta_{\text{y}}F_{\text{cN0}}\,, \tag{3}$$

$$m_{\text{b}}\ddot{z}_{\text{b}} = -g - k_{\text{bz}}(\dot{z}_{\text{b}} - \dot{z}_{\text{rim}}) - c_{\text{bz}}(z_{\text{b}} - z_{\text{rim}}) - k_{\text{bx}}\dot{\theta}_{\text{rim}}(x_{\text{b}} - x_{\text{rim}}) - c_{\text{rz}}(z_{\text{b}} - z_{\text{c}}) \tag{4}$$

and

$$I_{\text{b}}\ddot{\theta}_{\text{b}} = -k_{\text{b}\theta}(\dot{\theta}_{\text{b}} - \dot{\theta}_{\text{rim}}) - c_{\text{b}\theta}(\theta_{\text{b}} - \theta_{\text{rim}}) + r_{\text{e0}}k_{\text{rx}}(\dot{x}_{\text{b}} - \dot{x}_{\text{c}}) + r_{\text{e0}}c_{\text{rx}}(x_{\text{b}} - x_{\text{c}}) - r_{\text{e0}}f_{\text{r}}c_{\text{rz}}(z_{\text{b}} - z_{\text{c}}). \tag{5}$$

The last terms in Eq. (3) and (5) are the component of the normal force acting on the ring in horizontal direction and the rolling resistance, respectively.

The equations of motion of the rim are:

$$m_{\rm rim}\ddot{x}_{\rm rim} = -k_{\rm bx}(\dot{x}_{\rm rim}-\dot{x}_{\rm b})-c_{\rm bx}(x_{\rm rim}-x_{\rm b})+k_{\rm bz}\dot{\theta}_{\rm rim}(z_{\rm rim}-z_{\rm b})-k_{\rm sx}(\dot{x}_{\rm rim}-\dot{x}_{\rm car})-c_{\rm sx}(x_{\rm rim}-x_{\rm car}), \quad (6)$$

$$m_{\rm rim}\ddot{z}_{\rm rim} = -g-k_{\rm bz}(\dot{z}_{\rm rim}-\dot{z}_{\rm b})-c_{\rm bz}(z_{\rm rim}-z_{\rm b})-k_{\rm bx}\dot{\theta}_{\rm rim}(x_{\rm rim}-x_{\rm b})-k_{\rm sz}(\dot{z}_{\rm rim}-\dot{z}_{\rm car})-c_{\rm sz}(z_{\rm rim}-z_{\rm car}) \quad (7)$$

and

$$I_{\rm rim}\ddot{\theta}_{\rm rim} = -k_{\rm b\theta}(\dot{\theta}_{\rm rim}-\dot{\theta}_{\rm b})-c_{\rm b\theta}(\theta_{\rm rim}-\theta_{\rm b})+M_{\rm rim}, \quad (8)$$

where $M_{\rm rim}$ is the driving torque. The equations of the vehicle are:

$$m_{\rm car}\ddot{x}_{\rm car} = -k_{\rm sx}(\dot{x}_{\rm car}-\dot{x}_{\rm rim})-c_{\rm sx}(x_{\rm car}-x_{\rm rim}), \quad (9)$$

$$m_{\rm rim}\ddot{z}_{\rm car} = -g-k_{\rm sz}(\dot{z}_{\rm car}-\dot{z}_{\rm rim})-c_{\rm sz}(z_{\rm car}-z_{\rm rim}). \quad (10)$$

Whenever the wheel is lifted above the road surface, there is no action exerted from the road on the wheel, so the vertical contact force in the additional spring, together with the horizontal component of the normal force and the rolling resistance, i.e. the last terms in Eq. (2), (3) and (5), respectively, must be set to zero.

## 2.2 The enveloping tire model – the tandem model with elliptical cams

The enveloping model and the initial simulation parameters are adopted from [4], together with the additional necessary equations: rolling radius polynomial approximation, approximation of velocity dependent tire sidewall stiffness and sliding force equation, where the $C_{\rm c\kappa}$ parameter value is adopted from [5].

With non-dimensional parameters $p_{\rm ae}$, $p_{\rm be}$, $p_{\rm ce}$ and $p_{\rm ls}$, which depend on the tire geometry, and the outer radius of an unloaded and un-deformed tire $r_0$ according to [4], the ellipse length, height and power, as well as the tandem base length (Fig. 2) are given by:

$$a_{\rm e} = p_{\rm ae}r_0\,,\ b_{\rm e} = p_{\rm be}r_0\,,\ c = p_{\rm ce} \text{ and } l_{\rm s} = p_{\rm ls}2a\,, \quad (11\text{-}14)$$

respectively.

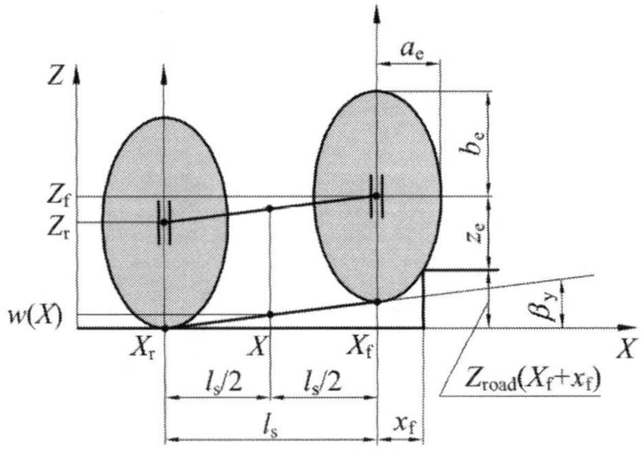


**Figure 2.** The tandem model with elliptical cams [4].

For a given tire and the normal force to the effective road surface

$$F_N = F_{az} \cos \beta_r / \cos(\beta_y + \beta_r), \quad (15)$$

half of the effective contact patch length $a$ is given by the empirical equation:

$$a = q_{a2} \sqrt{F_N^2} + q_{a1} \sqrt{F_N}, \quad (16)$$

where the coefficients $q_{a1}$ and $q_{a2}$ depend on the tire dimensions. Thus, the effective road height is:

$$w = (Z_f + Z_r) / 2 - b_e \quad (17)$$

and the effective forward slope of the road:

$$\beta_y = \text{atan}[(Z_r - Z_f) / l_s] \quad (18)$$

where

$$Z_i = \max\left[Z_{road}(X_i + x_i) + z_e(x_i)\right], \quad i = \text{f, r} \quad (19)$$

are center heights of the front and rear elliptical cams, while $z_e$ is the vertical distance from the contact point to the cam center, given by:

$$z_e = \left| b_e \left(1 - \left(|x_f| / a_e\right)^c\right)^{1/c} \right|. \quad (20)$$

Thus, in the simulation process of rolling the wheel over the road obstacle, the actual road is replaced with the effective road profile, where the enveloping characterics depend on the horizontal position of the wheel center. An example of the enveloping characteristics for a particular tire dimension and road profile is presented in Fig. 3.

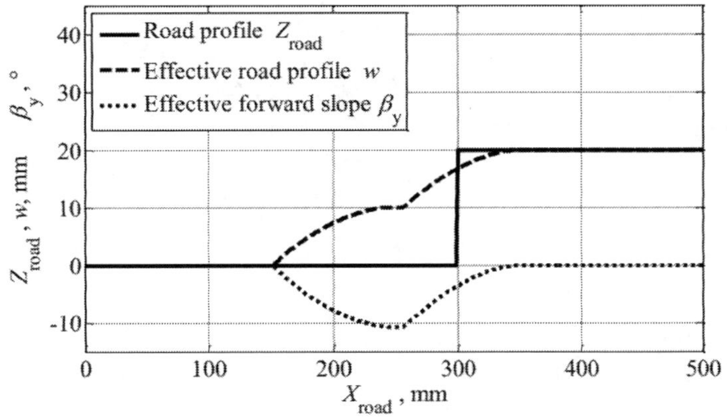


**Figure 3.** An example of the enveloping characteristics of a 205/60 R15 tire and a road profile with a 20 mm high step.

## 3 The finite element model of the quarter vehicle

The purpose of the designed FE quarter vehicle model (Fig. 4) is to accurately simulate the tire and obstacle edge impact and establish the resulting quarter vehicle model response.

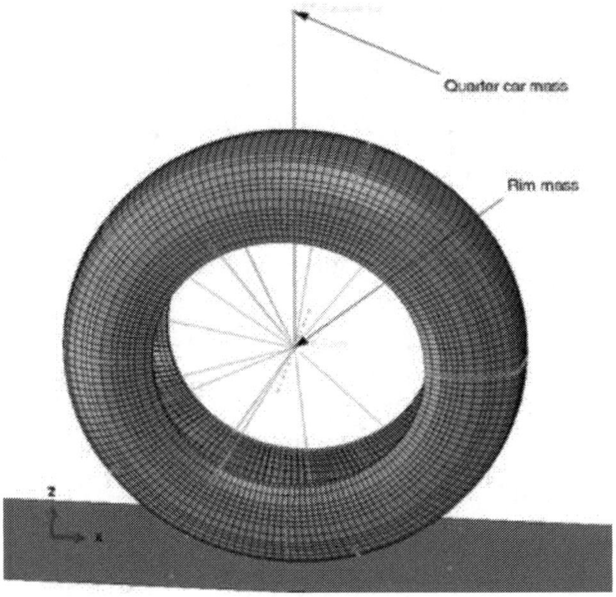


**Figure 4.** Finite element model.

It comprises the following elements:
• quarter car mass: discrete inertia with mass 240 kg,
• tire (FE model),
• rim mass: discrete inertia located at the rim center, adjusted such that the sum of the total mass of the tire and the rim equals 36 kg,
• spring/dashpot element that connects quarter car mass and rim mass: spring stiffness 16000 N/m, viscous damping coefficient 980 Ns/m,
• road surface: rigid surface.

The FE model of the tire has been adopted from [6], and the tire dimensions have been adjusted to correspond to the standard 175/65 R14 tire. The tire comprises rubber components (tread, inner liner, sidewall and apex) and reinforcements (carcass and two belts).

The rubber material has been modeled using the 8 node linear continuum elements and the rubber reinforcements are modeled as the embedded rebar elements (i.e. unidirectional fibers). The reinforcements are assumed to be linearly elastic with the modulus of elasticity 172.2e+09 N/m$^2$. The material properties for the rubber material (hyperelastic and viscoelastic properties) are adopted from [4]. The rubber density is 1100 kg/m$^3$.

## 4 Comparison of results

The simulation was carried out for different initial vehicle speeds (30, 40 and 50 km/h) and step heights from 25 to 150 mm. For a particular initial vehicle speed of 40 km/h and 150 mm step height, the rim center displacements for both methods, and for the Simulink model with initial parameter values as in [4], are presented in Fig. 5a. The initial parameter values of the residual stiffness $c_{rz}$, sidewall stiffness and damping coefficients $c_{bx}$, $c_{bz}$, $k_{bx}$ and $k_{bz}$ in Simulink are then slightly modified to match the FE results in the whole range of initial velocities and step heights. The response obtained using these modified parameters, for the chosen velocity and step height, is presented in Fig. 5b.

DOI: 10.1051/matecconf/201710402013

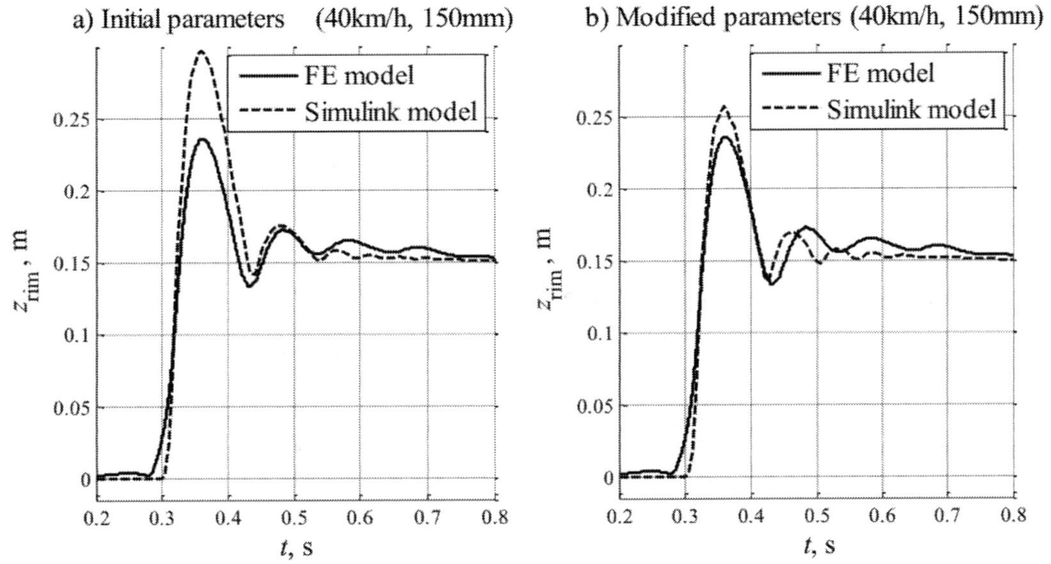


**Figure 5.** Responses of Simulink model compared to FE model: a) with initial and b) modified Simulink model parameters. Initial vehicle speed is 40 km/h and step height 150 mm.

## 5 Conclusion

The presented lumped quarter vehicle with rigid ring tire model gives a significant advantage in calculation speed over the FE simulation (minutes over hours) and fairly accurate results for the amplitude and frequency response during the short period of impact. It could be worth to consider additional optimization of system parameters, which are normally experimentally achieved, to bring the dynamic response of the lumped system even closer to the FE simulation. Both methods remain to be experimentally verified.

## Acknowledgments

The authors gratefully acknowledge the support from Ford Motor Company.

## References

1. A.V.Pesterev, L.A. Bergman, C.A. Tan, J. Sound Vib. **275** 127-149, (2004)
2. G.Verros, S. Natsiavas, C. Papadimitriou, J. Vib. Control **11**, 581-606, (2005)
3. N.W. Frey, *Development of a Rigid Ring Tire Model and Comparison Among Various Tire Models for Ride Comfort Simulations*, (Master's thesis, Clemson Univ., 2009).
4. A.J.C. Schmeitz, *A Semi-Empirical Three-Dimensional Model of the Pneumatic Tyre Rolling over Arbitrarily Uneven Road Surfaces*, (PhD thesis, TU Delft, 2004).
5. M. Schmid, *Tire Modeling for Multibody Dynamics Applications*, (Thesis, UW-Madison, 2011).
6. C. Wai, O. A. Olatunbosun, J. Terramech. **56**, 1—16, (2014).

# Calculation of displacements in a half-space from a linear array based on analytical solutions of transient wave

Xuecong Zhang[1], Cunfu He [1], Yan Lu [1,a] and Bin Wu[1]

[1] *College of Mechanical Engineering and Applied Electronics Technology, Beijing University of Technology, Beijing, China*

**Abstract.** In this paper, we propose a theoretical calculation method for the displacements of transient wave excited by a linear array coupled to an elastic solid with a longitudinal wave couplant. Firstly, the vertical and horizontal displacements excited by a vertical line source are solved based on exact and analytical solutions of transient elastic waves induced by a line source normally acting on an elastic half-space. Secondly, in this case, the displacements from a linear array can be determined by the superposition of the solutions of a number of discrete line forces mentioned above. All the theoretical results are compared to the finite element simulation results. The theoretical results are consistent with finite element simulation data. Therefore, the calculation method can save a lot of computing time. This paper provides a fast and accurate calculation method of transient wave displacements in a half-space generated by a linear array. The calculation method can be used to explore the sound field characteristics of a linear array transducer in the future.

## 1 Introduction

The ultrasonic array includes many active transducer elements which can be excited individually or simultaneously. Ultrasonic array has the flexibility to perform inspections along different directions with controllable beam patterns. These advantages accelerated the development of ultrasonic linear arrays for engineering and industry applications in the past several decades [1].

For the scanning over a larger area, a linear array transducer can be coupled to a sample with a longitudinal wave couplant, as shown in Fig. 1. Because the transient wave displacements in the detected object excited by the linear array can indicate characteristics of sound field, so it's very important to the transducer's design. As shown in Fig. 1, transducer elements are coupled to a sample with a longitudinal wave couplant. The inter-element spacing is $p$; element width is $l$; total element number is $N$; array size is $L$ $(=N\cdot p)$. In the previous studies, many great contributions to ultrasonic transducer sound field calculation have been made[2-8]. Wooh et al. regarded each transducer element as many discrete line sources loading on the surface of a elastic half-space, to predict the ultrasound field radiated by the linear phased array [4-6]. It assumed that the sound wave from each line source could be modeled as a cylindrical wave generated from an infinitely long line source pulsating radially and harmonically in a unbounded acoustic medium.

[a] Corresponding author: lvyan@bjut.edu.cn

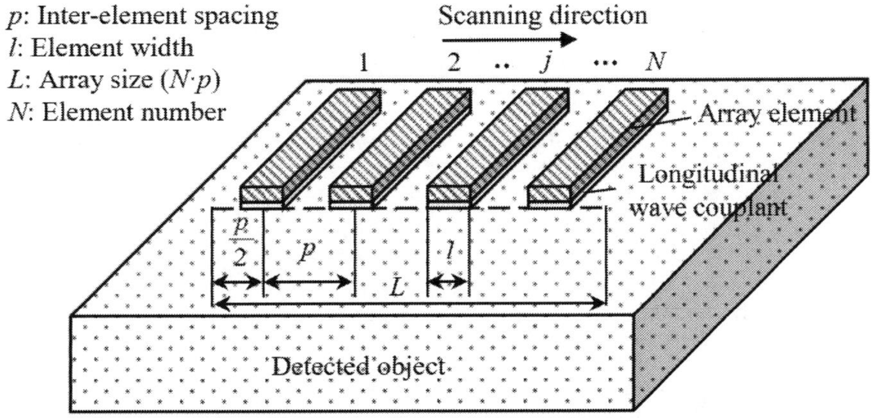


**Figure 1.** Geometry of a linear phased array transducer scanning over the surface of an object using longitudinal wave couplant.

In previous studies, theoretical calculation of ultrasound fields was largely simplified: First of all, for a vertical line force acting on a solid half-space, both longitudinal and shear waves are generated inside the solid and none of them can be characterized by a simple cylindrical wave with equal amplitude along all directions.Secondly, in real applications, transducer elements are not actuated by harmonic excitation but by pulsed waveforms with finite duration.

Consequently, the sound field based on analytical solutions of transient wave could provide precise characteristics of a linear array transducer. In this paper, we propose a theoretical calculation method for the displacements of transient wave excited by a linear array coupled to an elastic solid with a longitudinal wave couplant. Firstly, the vertical and horizontal displacements excited by a line source normally loading on an elastic half-space are solved by analytical solutions. Secondly, in this case, the displacements from a linear array can be determined by the superposition of the transient elastic wave displacements induced by a number of discrete line sources mentioned above. All the theoretical results are compared to the finite element (FE) simulation results and the consistent comparison results are obtained. This paper provides a fast and accurate calculation method for the transient displacements in a half-space generated by a linear array. The calculation method will be used to guide the design of a linear array transducer in the future.

## 2 Transient wave displacements in an elastic solid half-space from a vertical line source loading on the surface

For an ultrasonic linear array coupled to the surface of an elastic half-space with a longitudinal wave couplant, elastic waves are excited by the piezoelectric elements and transmitted to the elastic solid through normal stresses acting on the surface. Consequently, we should focus on the analytical solutions of elastic waves generated by a line vertical force loading on an elastic half-space, and derive the displacements of this transient elastic wave. As shown in Figure 2, a concentrated force is normally applied at a point of an elastic half-space defined in a Cartesian coordinate $(x, y)$, where $Q$ is a constant amplitude and $f(t)$ is an arbitrary function of time $t$.

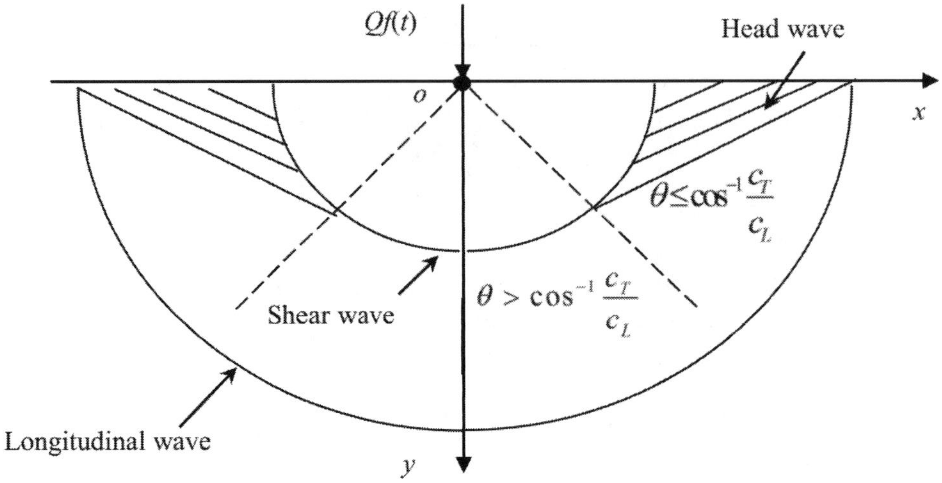


**Figure 2.** Wave fronts pattern from a vertical line force loading on an elastic half-space.

The problem has been studied thoroughly before[9]. The displacement components analytical solutions are shown as

$$u(x,y,t) = -\frac{Q}{\mu}\left[(u)_L - (u)_{TL} - (u)_T\right], \tag{1}$$

$$v(x,y,t) = \frac{Q}{\mu}\left[(v)_L + (v)_{TL} + (v)_T\right], \tag{2}$$

where $u(x,y,t)$ is the horizontal displacement, $v(x,y,t)$ is the vertical displacement. $(u)_L$ and $(v)_L$ are longitudinal wave terms, $(u)_T$ and $(v)_T$ are shear wave terms, $(u)_{TL}$ and $(v)_{TL}$ are head wave terms, $\mu$ is the shear modulus of the elastic solid. The analytical solutions of $(u)_L$, $(u)_{TL}$, $(u)_T$, $(v)_L$, $(v)_{TL}$ and $(v)_T$ could be obtained form Ref. [9]. The pattern of waveforms is shown in Figure 2. As shown in the figure, three modes exist: longitudinal wave, shear wave and head wave. Head wave only exists in the area where $\theta \leq \cos^{-1}\frac{s_L}{s_T}$.

Figure 3 and 4 show the theoretical $u$ and $v$ displacements at $x$=26 mm, $y$=147 mm (where $\theta > \cos^{-1}\frac{s_L}{s_T}$) and $x$=141 mm, $y$=51 mm (where $\theta < \cos^{-1}\frac{s_L}{s_T}$) generated by a vertical line source which is excited by 5 cycles of sinusoidal function at a frequency of 400 kHz and modulated by a Hanning window. Structural steel is used for the example with the following parameters: $\rho$=7850 kg/m$^3$; $E$=2 GPa; $\sigma$=0.33; $c_L$= 6182 m/s; $c_T$= 3091 m/s; $\mu$= 75 GPa; $t$= 0~0.07 ms. The analytical solutions will be compared with numerical solution calculated from a finite element (FE) simulation software COMSOL Multiphysics® (COMSOL Inc., Burlington, USA). The blue lines indicate theoretical results and the red dotted lines show FE simulation results. The two kinds of results are consistent with each other. This allows us to calculate the displacements at any point in a half-space excited by linear array.

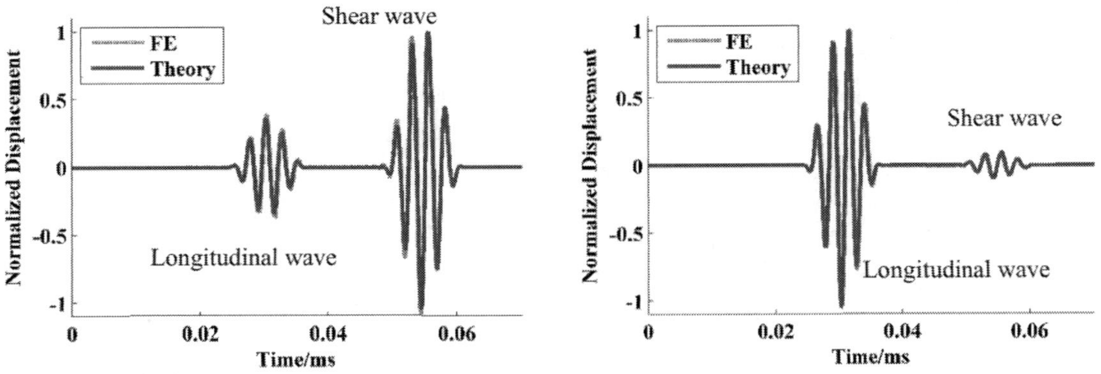


(a) $u$ displacement (b) $v$ displacement

**Figure 3.** Theoretical and simulated $u$ and $v$ displacements at $x$=26 mm, $y$=147 mm.

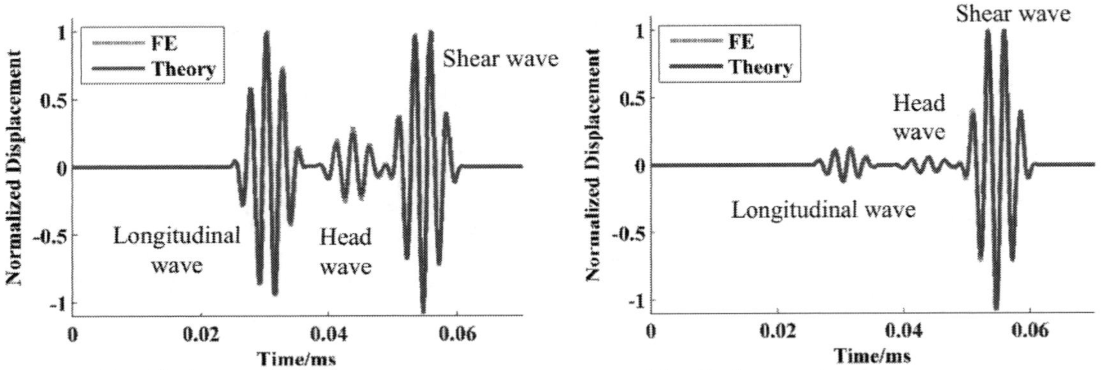


(a) $u$ displacement (b) $v$ displacement

**Figure 4.** Theoretical and simulated $u$ and $v$ displacements at $x$=141 mm, $y$=51 mm.

## 3 Transient displacements generated by a linear array based on analytical solutions

The calculation of displacements in a half-space from a linear phased array based on the analytical solutions is discussed below. As shown in Figure 5, each transducer element is modeled by $n$ discrete vertical line sources ($i$ =1, 2, ..., $n$), which are uniformly distributed within the transducer element width ($l$). The $N$ transducer elements are labelled as $j$, $j$=1, 2, ..., $N$. The line sources within one transducer element are excited simultaneously, whereas line sources of different transducer elements will have a given time delay.

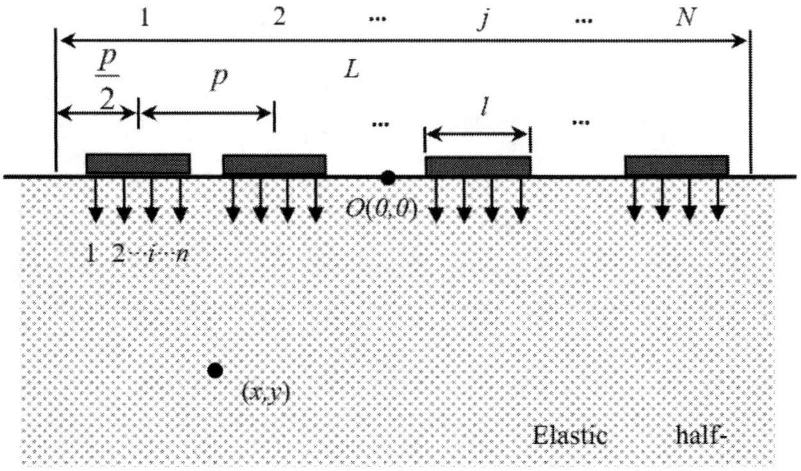


**Figure 5.** The model of vertical forces loading on an elastic half-space by a linear phased array transducer.

Figure 5 shows the discrete line forces of a linear phased array with an element number $N$, inter-element spacing $p$, element width $l$, and array size $L$, and the array is coupled on an elastic half-space with a thin longitudinal wave couplant. As shown in Fig. 5, the coordinate $(x, y)$ is originated from the center of the linear phased array. Let $u_0(x,y,t)$ and $v_0(x,y,t)$ respectively be the displacement components along $x$- and $y$-axis of the transient elastic wave radiated by a concentrated line force vertically loaded at the origin point $(0, 0)_{(x, y)}$ with a given time domain function of $f(t)$. Based on the superposition principle, the displacements from $j$th transducer element are

$$u_j(x,y,t)=\sum_{i=1}^{n}u_0\left[x-(i-\frac{n+1}{2})\frac{l}{n}-\left(j-\frac{N+1}{2}\right)p,\ y,t-\Delta t_j\right], \tag{3}$$

$$v_j(x,y,t)=\sum_{i=1}^{n}v_0\left[x-(i-\frac{n+1}{2})\frac{l}{n}-\left(j-\frac{N+1}{2}\right)p,\ y,t-\Delta t_j\right], \tag{4}$$

where $\Delta t_j$ is the time delay assigned to the $j$'th transducer element. The displacements from the linear array can be derived,

$$U(x,y,t)=\sum_{j=1}^{N}u_j(x,y,t), \tag{5}$$

$$V(x,y,t)=\sum_{j=1}^{N}v_j(x,y,t), \tag{6}$$

where $U(x,y,t)$ and $V(x,y,t)$ are respectively the displacement components along $x$- and $y$-axis.

Figure 6 shows the theoretical and simulated $u$ and $v$ displacements generated by a linear array at $x$=195, $y$= 160. The excited signal are 6 cycles of sinusoidal function and modulated by a Hanning window.An example with the following parameters are used: $f$=200 kHz, $n$= 20; $p$= 20 mm; $N$= 5; $l$= 10 mm; $t$=0~0.14 ms. The parameters of the steel used here are the same with those of the last example. The blue lines indicate theoretical results and the red dotted lines indicate simulated results. The theoretical results of displacements are consistent with simulated results of displacements. Consequently, this method can provide the accurate transient wave displacements in a half-space from a linear array and replace numerical simulation method, thus saving a lot of computing time. This allows us to calculate the sound field of the linear array, which usually plays a key role in ultrasound inspection.

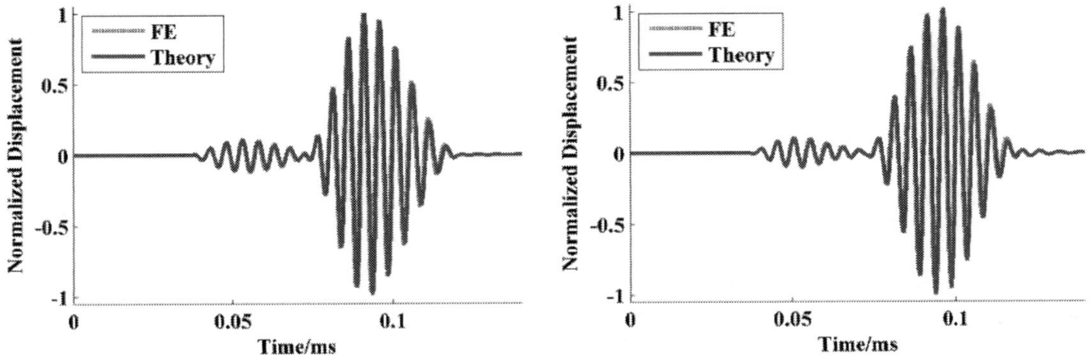


(a) *u* displacement (b) *v* displacement

**Figure 6.** Theoretical and simulated u and v displacements at $x$=195 mm, $y$=160 mm

## 4 Conclusion

This paper proposes a theoretical calculation method for the displacements of transient wave excited by a linear array coupled to an elastic solid with a longitudinal wave couplant. The displacements generated by the linear array mentioned above, are treated as the superposition of the displacements of a number of discrete vertical line forces acting on the surface of a half-space. The displacement amplitude can be easily and accurately calculated at any point in a half-space with minimum assumption and simplification. The theoretical results are consistent with finite element simulation data. Therefore, the method can save a lot of computing time.

This work made improvements for the displacements calculation of transient wave in a half-space from a linear array. It provides precise results for linear array applications. This paper provides a fast and accurate calculation method for the transient displacements in a half-space generated by a linear array. The method can be used to explore the characteristics of the sound field of a linear array transducer.

## Acknowledgements

This work was supported by National Natural Science Foundation of China (grant number 51235001).

## References

1. B.W. Drinkwater, P.D. Wilcox, NDT&E Int. **39**, 525 (2006)
2. M. Spies, J. Nondestruct. Eval. **13**, 85 (1994)
3. M. Spies, Ultrasonics **42**, 213 (2004)
4. S. C. Wooh, Y. Shi, Wave Motion **29**, 245 (1999)
5. S. C. Wooh, Y. Shi, J. Nondestruct. Eval. **18**, 39 (1999).
6. S. C. Wooh, Y. Shi, Ultrasonics **36**, 737 (1998)
7. L. Azar, Y. Shi, S.C. Wooh, NDT&E Int. **33**, 189 (2000)
8. J. H. Lee, S. W. Choi, IEEE T. Ultrason. Ferr. **47**, 644 (2000).
9. J. D. Achenbach, *Wave Propagation in Elastic Solids* (North-Holland Pub. Co., Amsterdam, 1973)

# Modelling guided waves in anisotropic plates using the Legendre polynomial method

Mingfang Zheng [1] , Cunfu He [1] , Yan Lu[1,a] and Bin Wu [1]

[1] *College of Mechanical Engineering and Applied Electronics Technology, Beijing University of Technology, Beijing, 100124, China*

**Abstract.** A numerical method to compute phase dispersion curve in unidirectional laminate is described. The basic feature of the proposed method is the expansion of fields quantities in single layer on different polynomial bases. The Legendre polynomial method avoid to solve the transcendental dispersion equation of guided wave. Guided waves that have very close propagation constants are calculated with great accuracy. Numerical solution of dispersion relation are calculated for guided waves propagation in orthotropic unidirectional fiber composites. The validation of the polynomial approach is depicted by a comparison between the associated solution and those obtained using Transfer matrix method.

## 1 Introduction

Composite materials play an important role in many structural components, such as auto parts, boat hulls, and aircraft structures [1].In recent years, much attention has been paid to nondestructive testing method for composite. Ultrasonic guided waves is effective way to detect the presence of structure defect such as damage, disbands and delaminations. For any analysis of guided waves propagation in structures, the velocity of the guided wave mode are essential for further study [2]. In many cases, accurate calculation of wave's speeds are very benefit to conduct experimental studies.

It is also true that the multi-mode characters, dispersion of guided waves, and the anisotropic behaviour of composite materials, guided waves are exceedingly complex and not easy to analyse. There are many methods to investigate the propagation characteristics of guided waves in composite laminates. These method include the Transfer matrix method (TMM)[3, 4], the Global matrix method (GMM)[5], the scattering matrix method(SMM)[6], the stiffness transfer matrix method(STMM)[1], reverberation-ray matrix method(RMMM)[7, 8], the finite element method (FEM)[9-11], and the semi-analysis finite element method (SAFE) [12]. Guided waves propagation can then be found by imposing appropriate boundary. However, the search of root still remains a rather difficult task, and part of them may be missed.

In order to avoid these drawbacks, an expansion polynomial method has been presented to obtain wave propagation without solving any transcendental equation. Lefebvre et al. [13] studied wave propagation in piezoelectric plates using the Legendre polynomial method since then this approach is used extensively to investigate the guided waves propagation in various structure [14-18]. The key aspect of Legendre polynomial method is the expansion of the unknown displacement quantities on a

[a] Corresponding author : lvyan@bjut.edu.cn

set of functions. The differential wave equation turns into an algebraic eigenvalue problem. The eigenvalues and corresponding eigenvectors are used to acquire the propagation constants by linear combinations of the expansion functions.

In this paper, our work are focused on the derivation and analysis of guided waves equation for unidirectional laminate using the Legendre polynomial. Using the proposed meth, the dispersion curves of different propagation angles with respect to the principal fibre orientation are calculated, and the result solution are compared with the result obtained from Transfer matrix method.

## 2 Formulation of the problem

Consider the general case of a layer structure of infinite extent in the the $x_1$ and $x_3$ directions and thickness H in the $x_2$ direction, as given in Fig. 1.Fig.1 demonstrate the coordinate system for orthotropic layer, where $o-x_1x_2x_3$ and $o-x_1'x_2'x_3'$ represent global and local coordinate system respectively. For unidirectional fiber reinforced composite laminates, the stiffness matrix can be transformed from the local reference coordinates to the global coordinates by[17] :

$$c_{mnop} = \beta_{mi}\beta_{nj}\beta_{ok}\beta_{pj}c'_{ijkl} \tag{1}$$

Where $\beta_{ij}$ indicates the cosine of the intersection angle $\phi$ between the $x_i'$ axis and $x_j'$ axis (i,j=1,2,3).

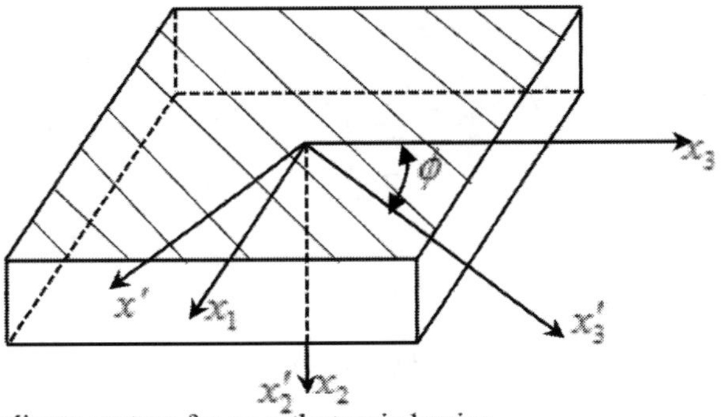


**Figure 1**. The spatial coordinate system for an orthotropic lamina

The wave motion without considering body force, can be written in the following form:

$$\begin{cases} \nabla\cdot(c:\nabla_s u)=\rho\dfrac{\partial^2 u}{\partial t^2} \\ \nabla_s u=\sigma \end{cases} \tag{2}$$

Where $\rho$ stand for the density, $\sigma$ stand for the strain tensor, $u$ stand for the displacement vector. Eq.(2) contain 9 equation and 9 unknowns. The 9 unknowns are three displacement and six stress components. However, for simplicity, this unknown variables are expressed by the vector form

$$u=\begin{bmatrix}u_1 & u_2 & u_3\end{bmatrix}^T e^{-j[kx_1-\omega t]} \tag{3}$$

$$\tau_i=\begin{bmatrix}\sigma_{i1} & \sigma_{i2} & \sigma_{i3}\end{bmatrix}^T e^{-j[kx_1-\omega t]} \tag{4}$$

Substituting Eq. (3-4) into Eq. (2), the harmonic wave term $e^{-j[kx_1-\omega t]}$ is omitted, and the governing equations can be rewritten in term of

$$\frac{\partial\tau_2}{\partial x_2}=-\rho\omega^2[I]u-\frac{\partial\tau_1}{\partial x_1}-\frac{\partial\tau_3}{\partial x_3} \tag{5}$$

Eq.(2) can be rewritten in the form

$$\tau_1=\frac{\partial}{\partial x_1}[D_{11}]u+\frac{\partial}{\partial x_2}[D_{12}]u+\frac{\partial}{\partial x_3}[D_{13}]u \tag{6}$$

$$\tau_2 = \frac{\partial}{\partial x_1}[D_{21}]u + \frac{\partial}{\partial x_2}[D_{22}]u + \frac{\partial}{\partial x_3}[D_{23}]u \tag{7}$$

$$\tau_3 = \frac{\partial}{\partial x_1}[D_{31}]u + \frac{\partial}{\partial x_2}[D_{32}]u + \frac{\partial}{\partial x_3}[D_{33}]u \tag{8}$$

Using the Eq.(7), The first order derivatives of the displacement versus $x_2$ can be expressed as

$$\frac{\partial}{\partial x_2}u = [D_{22}]^{-1}\left(\tau_2 - \frac{\partial}{\partial x_1}[D_{21}]u - \frac{\partial}{\partial x_3}[D_{23}]u\right) \tag{9}$$

At this stage,it is possible to in introduce the state vector $\eta = [u_1 \quad u_2 \quad u_3 \quad \sigma_{21} \quad \sigma_{22} \quad \sigma_{23}]$,the general solution for the state vector can be expressed in the form $\eta = \eta e^{j(\omega t - kx_1)}$ ,governing differential equations for the state vector in term of the displacement and stress quantities can be given

$$\frac{\partial \eta}{\partial x_2} = \bar{A}\eta \tag{10}$$

Where

$$\bar{A} = \begin{bmatrix} [D_{22}]^{-1} & jk[D_{22}]^{-1}[D_{21}] \\ jk[D_{12}][D_{22}]^{-1} & -\rho\omega^2[I] - k^2\left[[D_{12}][D_{22}]^{-1}[D_{21}] - [D_{11}]\right] \end{bmatrix}$$

The relation between $\tau_2$ and $u$ can be obtain from Eq.(7)

$$\tau_2 = [D_{22}]\frac{\partial u}{\partial x_2} - jk[D_{21}]u \tag{11}$$

In order to remove the stress state vector from Eq.(10), Substituting Eq.(11) into the Eq.(10) one obtains

$$k^2[D_{11}]u + jk[D_{12}]\frac{\partial u}{\partial x_2} - \rho\omega^2[I]u + [D_{22}]\frac{\partial^2 u}{\partial x_2^{\,2}} = 0 \tag{12}$$

Considering the flat plate structure as a case, was shown in fig.1. A set of Legendre polynomials are used to represent the unknown displacement in layer. The displacement vector $u$ can be expressed as

$$u = \sum_{n=0}^{N-1}\psi_n P_n(\chi) \tag{13}$$

Where $P_n(\chi)$ is Legendre polynomial, $\psi_n$ is unknown expansion coefficients, and $\chi \in [-1,1]$ being a normalized variable:

$$\chi = \ell(x_2 - x_2^{\,0}), \quad \ell = 2/h \tag{14}$$

Where,h and $x_2^{\,0}$ are the thickness of plate and the coordinate of the plate middle point. Using the Eq.(14),The first and second derivatives of $u$ can be written as

$$\frac{\partial u}{\partial x_2} = \ell\frac{\partial u}{\partial \chi} = \ell\sum_{n=0}^{N-2}p_n P_n(\chi) \tag{15}$$

$$\frac{\partial^2 u}{\partial x_2^{\,2}} = \ell^2\frac{\partial^2 u}{\partial \chi^2} = \ell^2\sum_{n=0}^{N-3}q_n P_n(\chi) \tag{16}$$

with $p_n$ and $q_n$ are expressed by

$$p_m = (2m+1) \sum_{n=m+1,m+3,\cdots}^{N-1} \psi_n \tag{17}$$

$$q_m = \left(\frac{2m+1}{2}\right) \sum_{n=m+2,\ m+4,\cdots}^{N-1} \left(\mathrm{n}(n+1) - m(m+1)\right)\psi_n \tag{18}$$

The introduction of representation Eq.(13) in Eq.(12).Multiplying by Legendre polynomial of order m both sides of this equation, and integrating over $\chi$ from -1 and 1,then using the orthogonal property of Legendre polynomial,gives

$$k^2\left[D_{11}\frac{2}{2m+1}\right]\psi_m + jk\ell\left[D_{12}\frac{2}{2m+1}\right]p_m + \left[-\rho\omega^2\frac{2}{2m+1}I\right]\psi_m + \ell^2\left[D_{22}\frac{2}{2m+1}\right]q_m = 0 \tag{19}$$

Eqs.(19) are quadratic in $k$ ,therefor the total number of unknown is $6N$ ,however we can obtained $3N+3(N-2)$ equation from Eqs.(19),this indicated that the remaining 6 equations are provide by boundary condition.The boundary conditions for the orthotropic plate generally require that: the stress should be zero at the upper and bottom surfaces. The following boundary conditions can be yield

$$\text{at } x_2 = 0: \sum_{n=0}^{N}\left(D_{22}(-1)^{n+1}\ell\frac{n(n+1)}{2} - j(-1)^n k[D_{21}]\right)\psi_n = 0 \tag{20}$$

$$\text{at } x_2 = h: \sum_{n=0}^{N}\left(D_{22}\ell\frac{n(n+1)}{2} - jk[D_{21}]\right)\psi_n = 0 \tag{21}$$

The value of Legendre polynomial and their first derivatives at the end of the interval $[-1,1]$ has to be taken into account for deriving the Eqs.(20-21) :

$$P_n(\pm1) = (\pm1)^n \tag{22}$$

$$\frac{dP_n(\pm1)}{dr} = (\pm1)^{n+1}\frac{n(n+1)}{2} \tag{23}$$

We collected all the relevant equation from Eqs.(19-21),and arranged the unkowns in the column vector $\Omega = [\psi_0,\cdots,\psi_{N-1}]^T$ ,Where $\psi_k = [u_{1k}, u_{2k}, u_{3k}]^T$ ,a generalized eigenvalue problem is obtained:

$$k^2\bar{R}\Omega + jk\bar{S}\Omega + \bar{T}\Omega = 0 \tag{24}$$

Here, $\Lambda = jk\bar{R}\Omega$ are introduced to obtained a linear eigenvalue problem from Eq.(24) . The corresponding equations are as follow:

$$\left(\begin{bmatrix}\bar{S} & -I\\ \bar{R} & 0\end{bmatrix} - \frac{j}{k}\begin{bmatrix}\bar{T} & 0\\ 0 & I\end{bmatrix}\right)\begin{bmatrix}\Omega\\ \Lambda\end{bmatrix} = 0 \tag{25}$$

Where I is the identity matrix. The eigensolutions of $k$ and $\Omega$ represent the wavenumber and the displacement vector associated with the corresponding guided wave mode.

## 3 Numerical examples

In this section, results are shown for hase velocity dispersion curves, the material used as case study is T300/914 .These material properties are shown in table 1.The lamina thickness is 1 mm.the material density is $\rho = 1560 kg/m^3$ . Fig.2 presents the phase velocity results for waves propagating along three different direction of $0^0$ ,45 and 90 with respect to the fiber principal direction. The symmetric and anti-symmetric lamb wave are number as 'Sn' and 'An'(n=0,1,2...),respectively. The SH waves are number as 'SHn'.Figure 2 compares the results obtained from LPM with those obtained by TMM for the case of lamb and SH modes in a single plate. As can be observe in Fig.2 that all the modes were found and the LPM's solution are coincident with those obtained by the TMM's solution.

Table1 Elastic constants of T300/914

Unit: $C_{ij}$ ($10^9$ N/m$^2$)

| $C_{11}$ | $C_{12}$ | $C_{13}$ | $C_{22}$ | $C_{23}$ | $C_{33}$ | $C_{44}$ | $C_{55}$ | $C_{66}$ |
|---|---|---|---|---|---|---|---|---|
| 143.8 | 6.2 | 6.2 | 13.3 | 6.5 | 13.3 | 3.6 | 5.7 | 5.7 |

From fig 2,it can be observe that lamb and SH wave are decoupled for the principal direction ($\phi = 0^0$ and $\phi = 90^0$) and SH0 is almost non-dispersive in the entire frequency range . For the propagation direction at 45 with respect to the direction $x_1$ axis, the lamb and SH wave are coupled and the SH0 mode has some degree of dispersion.it can also be found that the velocity of the lamb waves tend to abate with increasing the propagation angel.

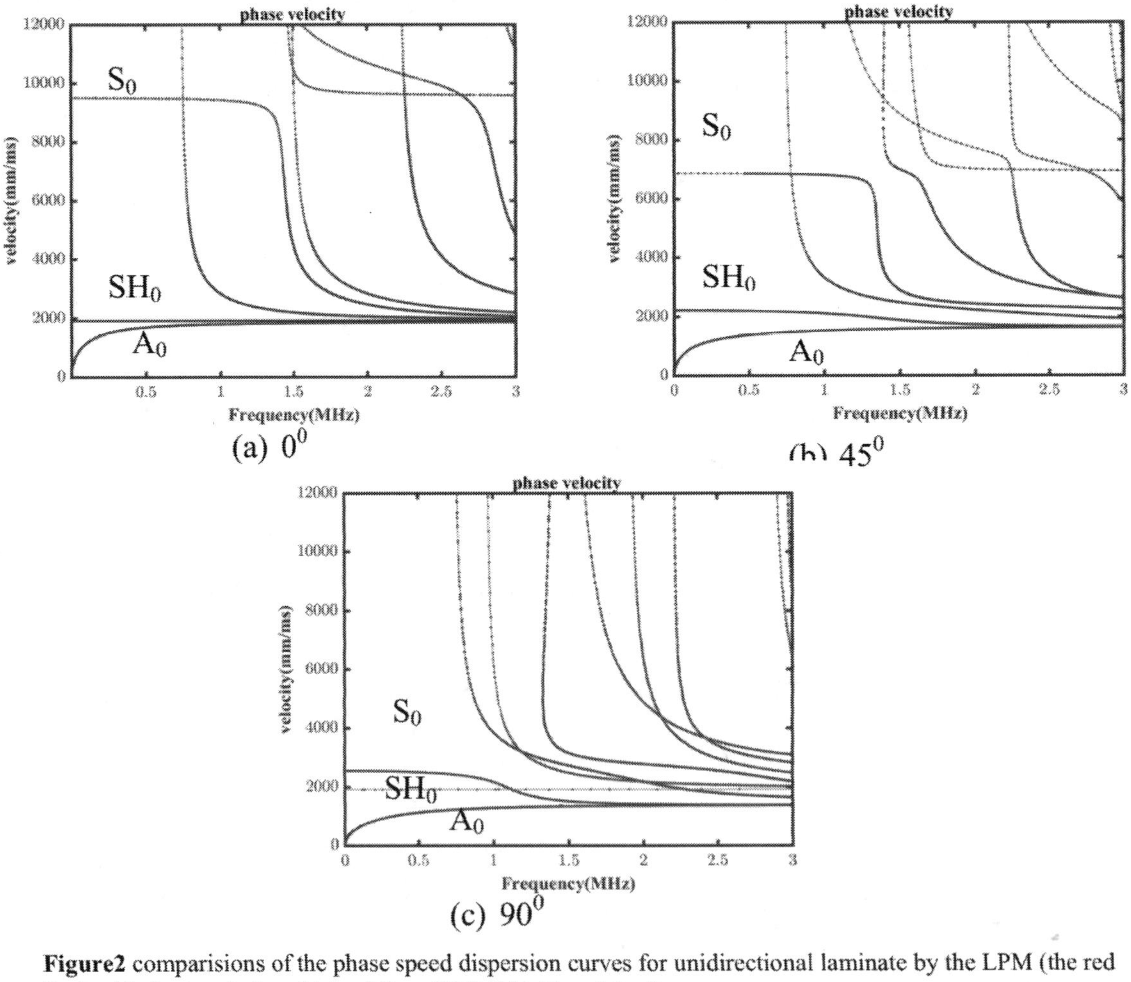


**Figure2** comparisions of the phase speed dispersion curves for unidirectional laminate by the LPM (the red line) with those solution obtained from RMM(the blue dotted)

## 4 CONCLUSIONS

Many different algorithms for solving guided waves propagation in composites layer have been investigated all these years, the article briefly analyzes them. Our work centre on the Legendre polynomial method (LPM) and the efforts for developing an algorithm of the advantages of robust. The Legendre method provide alternative to transfer matrix method. This paper discussed the mathematical formulation about Legendre polynomial method in detail. The cases of composite laminates have also been validated with the results obtained from TMM.

## ACKNOWLEDGMENTS

The reseach was financially supported by the National Natural Science Foundation of China (Grant Nos. 51235001, 51505013, 11372016).

## References

[1] A. Kamal, V. Giurgiutiu, Stiffness Transfer Matrix Method (STMM) for stable dispersion curves solution in anisotropic composites, in, 2014, pp. 906410-906410-906414.
[2] J.L. Rose, C.J. Lissenden, Guided wave mode and frequency selection tips, (2014) 358-364.
[3] W.T. Thomson, Transmission of Elastic Waves through a Stratified Solid Medium, Journal of Applied Physics, 21 (1950) 89-93.
[4] N.A. Haskell, THE DISPERSION OF SURFACE WAVES ON MULTI-LAYERED MEDIA, B Seismol Soc Am, 43 (1951) 86-103.
[5] L. Knopoff, A matrix method for elastic wave problems, B Seismol Soc Am, 54 (1964).
[6] T. Pastureaud, V. Laude, S. Ballandras, Stable scattering-matrix method for surface acoustic waves in piezoelectric multilayers, Appl Phys Lett, 80 (2002) 2544-2546.
[7] Y. Guo, W. Chen, Y. Zhang, Guided wave propagation in multilayered piezoelectric structures, Sci China Ser G, 52 (2009) 1094-1104.
[8] Z. Ma, J. Chen, B. Li, Z. Li, X. Su, Dispersion analysis of Lamb waves in composite laminates based on reverberation-ray matrix method, Composite Structures, 136 (2016) 419-429.
[9] L. Gavric, Computation of Propagative Waves in Free Rail Using a Finite-Element Technique, J Sound Vib, 185 (1995) 531-543.
[10] T. Mazuch, Wave dispersion modelling in anisotropic shells and rods by the finite element method, J Sound Vib, 198 (1996) 429-438.
[11] S. Sorohan, N. Constantin, M. Gavan, V. Anghel, Extraction of dispersion curves for waves propagating in free complex waveguides by standard finite element codes, Ultrasonics, 51 (2011) 503-515.
[12] I. Bartoli, A. Marzani, F.L. di Scalea, E. Viola, Modeling wave propagation in damped waveguides of arbitrary cross-section, J Sound Vib, 295 (2006) 685-707.
[13] J.E. Lefebvre, V. Zhang, J. Gazalet, T. Gryba, Legendre polynomial approach for modeling free-ultrasonic waves in multilayered plates, Journal of Applied Physics, 85 (1999) 3419-3427.
[14] V. Lancellotti, R. Orta, Guided waves in layered cubic media: Convergence study of a polynomial expansion approach, J Acoust Soc Am, 104 (1998) 2638-2644.
[15] J.G. Yu, F.E. Ratolojanahary, J.E. Lefebvre, Guided waves in functionally graded viscoelastic plates, Composite Structures, 93 (2011) 2671-2677.
[16] O. Bou Matar, N. Gasmi, H. Zhou, M. Goueygou, A. Talbi, Legendre and Laguerre polynomial approach for modeling of wave propagation in layered magneto-electro-elastic media, J Acoust Soc Am, 133 (2013) 1415-1424.
[17] C.F. He, H.Y. Liu, Z.H. Liu, B. Wu, The propagation of coupled Lamb waves in multilayered arbitrary anisotropic composite laminates, J Sound Vib, 332 (2013) 7243-7256.
[18] C. Othmani, S. Dahmen, A. Njeh, M.H. Ben Ghozlen, Investigation of guided waves propagation in orthotropic viscoelastic carbon-epoxy plate by Legendre polynomial method, Mech Res Commun, 74 (2016) 27-33.

# Sling tray mechanical anti-swing system simulation and modeling of ship-mounted crane

Han Guang-dong[1, a], Zhang Tong[1], Chen Haiquan[1], Zhang Jinnan[1], Sheng-hai[1] and DUAN Wen-qi[1]

[1]*College of Marine Engineering, Dalian Maritime University, Dalian 116026, Liaoning, China*

**Abstract:** To suppress the swing of the payload, a kind of sling tray anti-swing device was designed, which reduced the length of the lifting rope to reduce the swing of payload. The kinematic equations of the payload were established under the condition of whether anti-swing device was provided, and the dynamic was carried out in Mat lab/Simulink environment. The swing of the payload under different rolling angles was analysed in a comparative manner. The results showed that the in-plane angle reduced 60% and out -plane angle reduced 55% of the payload in the case of the anti-swing device. The simulation results verified the effectiveness of the sling tray anti-swing device.

## 1 INTRODUCTION

Ship-mounted cranes are commonly used in cargo handling, offshore installation and replenishment at sea. However, the ship and payload was swing due to the motion of wave. The swing of payload is unavoidable and may put relevant operations in danger.

There are some methods to restrain payload swing of ship-mounted crane. The kinematics model of payload system is established to study the swing of the payload. It is very important for safe and efficient operation of ship-mounted crane [1]. There are two commonly used methods to study the dynamics of the ship-mounted crane: one is to simplify the effect of the hull on payload to the hanging point of lifting arm [2-4]. The results show that the planar motion of hanging point also make the payload to swing in the space [5-6]. The other one is the ship motion and the swing of payload coupling analysis and develop the multi-body dynamics model [7-9]. JANG built dynamic model of ship-mounted container crane [10], and designed a T-S fuzzy controller to reduce the payload swing. ISMAIL built the three-dimensional model of offshore boom crane [11], designed a second order sliding mode controller which could track to expect trajectory while reducing the payload swing. In addition, the mechanical shock was reduced as well. FANG designed a nonlinear controller based on the dynamic model of offshore boom crane [12], and applied Lyapunov method to prove its stability, and finally carried out experiment to testify its performance. WANG Yang conduct the analysis of the motion characteristics of the anti-swing system under rolling condition of crane sling four rope system [13], And put forward the anti-swing device for a rigid rod.

A sling tray mechanical anti-swing system (STMAS) is proposed in paper aiming to reduce the swing of the payload. An anti-swing test bed is designed, analyzed and built to testify the proposed anti-swing system.

[a] Corresponding author : 18842602951@163.com

# 2 KINEMATICS MODEL OF PAYLOAD SYSTEM

The ship-mounted crane is a mutual coupling rigid body. In addition, the motion of ship-mounted crane is also subject to the interaction of the hull and the wind, wave, flow, with nonlinear characteristics. In paper the ship is looked as an incentive platform, to carry on the roll. The following assumptions are made: the payload could be looked as a point; the quality of rope is negligible and the rope without elastic deformation; the lifting arm of ship-mounted crane as a rigid body without elastic deformation; ship motion in regular wave.

According to these assumptions the kinematics model of the payload system was developed, as shown in Fig. 1.

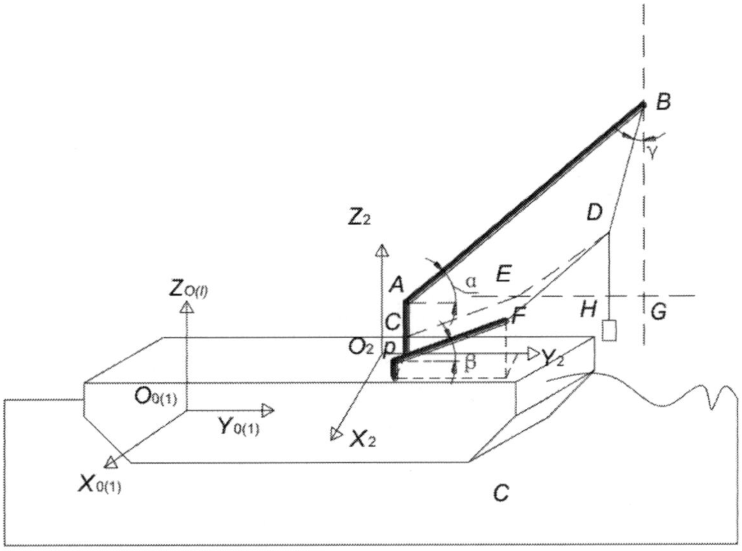


**Figure 1.** The kinematics model of the payload system

The reference coordinate is defined as follows:

($O_0$-$X_0Y_0Z_0$) as the inertial coordinate system; ($O_1$-$X_1Y_1Z_1$) for positioning in the hull on the hull system; ($O_2$-$X_2Y_2Z_2$) is positioning on the base coordinate system origin $O_2$ is the intersection of hull and crane base; the point $p$ is located in the base coordinate system ($O_2$-$X_2Y_2Z_2$) with $O_2$; $\gamma$ is the initial Angle of the lifting rope with the Z axis after put the sling tray mechanical anti-swing system in system.

Through coordinate transform, the coordinate ($O_2$-$X_2Y_2Z_2$) could be represented in the hull coordinate system ($O_1$-$X_1Y_1Z_1$), transformation relation as follow.

$$u_p = \mathrm{T}_{O_2}^{O_1} p^{o_2} \tag{1}$$

In Eq.1 coordinates transform corresponds to the reference coordinate system by superscript and subscript, $u_p$= $[x_p\ y_p\ z_p\ 1]^{\mathrm{T}}$ is the position of $O_2$ in the hull coordinate ($O_1$-$X_1Y_1Z_1$), and $p^{O2}$= $[0\ 0\ 0\ 1]^{\mathrm{T}}$ is the position of crane base original point in base coordinate system, T is coordinate transform matrix between the base coordinate system ($O_2$-$X_2Y_2Z_2$) and the hull coordinate system($O_1$-$X_1Y_1Z_1$).

$$T_{O_2}^{O_1} = \begin{bmatrix} \cos\alpha & -\sin\alpha & 0 & x_{o_2} \\ \sin\alpha & \cos\alpha & 0 & y_{o_2} \\ 0 & 0 & 1 & z_{o_2} \\ 0 & 0 & 0 & 1 \end{bmatrix} \tag{2}$$

Because of the motion of the ship, caused the motion of lifting arm but the lifting arm and cargo are connected by wire rope, only three displacements can be transmitted to the payload to cause the swing of the payload, but the attitude movement is not affected. So only need to derive the translational displacements of ship-mounted crane base, the expression as follow.

$$u_{o_2} = u_p + Hu_{o_1} \tag{3}$$

Assuming that the ship moving under a regular wave, only considering the ship rolling and pitching, it can be assumed that the motion of the ship as follow.

$$u_{o_1} = \begin{bmatrix} 0 & 0 & 0 & \eta_x & \eta_y & 0 \end{bmatrix}^{\mathrm{T}} \tag{4}$$

Without considering the rotation of the crane, the crane as a rigid body, $\alpha = 0$, the position of base of crane can be got, H is the transform matrix.

$$H = \begin{bmatrix} 1 & 0 & 0 & 0 & z_p & -y_p \\ 0 & 1 & 0 & -z_p & 0 & x_p \\ 0 & 0 & 1 & y_p & -x_p & 0 \end{bmatrix} \tag{5}$$

According Eq.1-Eq.5:

$$u_b = \left( 1.5*\eta_y, -1.5*\eta_y + 3.5,\ 3.5*\eta_y + 1.5 \right) \tag{6}$$

The position of point B, E, F can be derived by the coordinates of ub as follows.

$$\begin{cases} B(x_b, y_b, z_b) = \begin{pmatrix} 1.5*\eta_y \\ -1.5*\eta_x + 4.56 \\ 3.5*\eta_y + 2.76 \end{pmatrix} \\ E(x_e, y_e, z_e) = \begin{pmatrix} -1.5*\eta_y - 0.2 \\ -1.5*\eta_x + 3.93 \\ 3.5*\eta_x + 1.85 \end{pmatrix} \\ F(x_f, y_f, z_f) = \begin{pmatrix} 1.5*\eta_y + 0.2, \\ -1.5*\eta_x + 3.93 \\ 3.5*\eta_y + 1.85 \end{pmatrix} \end{cases} \tag{7}$$

The force analysis of the sling tray and the payload was carried out by using the method of Descartes.

(1) Statics analysis of the sling tray

The statics analysis in X, Y, Z direction is

$$\begin{aligned} &F_1 * \left( \left( x_f - x_d \right) / l_{df} + \left( x_e - x_d \right) / l_{de} \right) + \\ &F_2 * \left( \left( x_h - x_d \right) / l_{dh} + \left( x_b - x_d \right) / l_{bd} \right) = 0 \end{aligned} \tag{8}$$

$$\begin{aligned} &F_1 * \left( \left( y_f - y_d \right) / l_{df} + \left( y_e - y_d \right) / l_{de} \right) + \\ &F_2 * \left( \left( y_h - y_d \right) / l_{dh} + \left( y_b - y_d \right) / l_{bd} \right) = 0 \end{aligned} \tag{9}$$

$$\begin{aligned} &F_1 * \left( \left( z_f - z_d \right) / l_{df} + \left( z_e - z_d \right) / l_{de} \right) + \\ &F_2 * \left( \left( z_h - z_d \right) / l_{dh} + \left( z_b - z_d \right) / l_{bd} \right) = 0 \end{aligned} \tag{10}$$

(2) Kinematics analysis of the payload (With the sling tray)

The statics analysis in X, Y, Z direction is

 DOI: 10.1051/matecconf/201710402016

$$m\ddot{x}_h = F_2 * (x_b - x_h)/l_{bh} \tag{11}$$

$$m\ddot{y}_h = F_2 * (y_d - y_h)/l_{dh} \tag{12}$$

$$m\ddot{z}_h = F_2 * (z_d - z_h)/l_{dh} - mg \tag{13}$$

(3) Kinematics analysis of the payload (Without the sling tray)
The statics analysis in X, Y, Z direction is

$$m\ddot{x}_h = F_2 * (x_b - x_h)/l_{bh} \tag{14}$$

$$m\ddot{y}_h = F_2 * (y_b - y_h)/l_{bh} \tag{15}$$

$$m\ddot{z}_h = F_2 * (z_b - z_h)/l_{bh} - mg \tag{16}$$

## 3 THE KINEMATICS SIMULATION OF PAYLOAD SYSTEM

According to the kinematics equations of the payload system, the simulation analysis of the motion of the payload was carried out under the condition of with and without sling tray mechanical anti-swing system in the pitching angles are 5°, 9°, 13°.

### 3.1 The swing of the payload with pitching angle is 5°

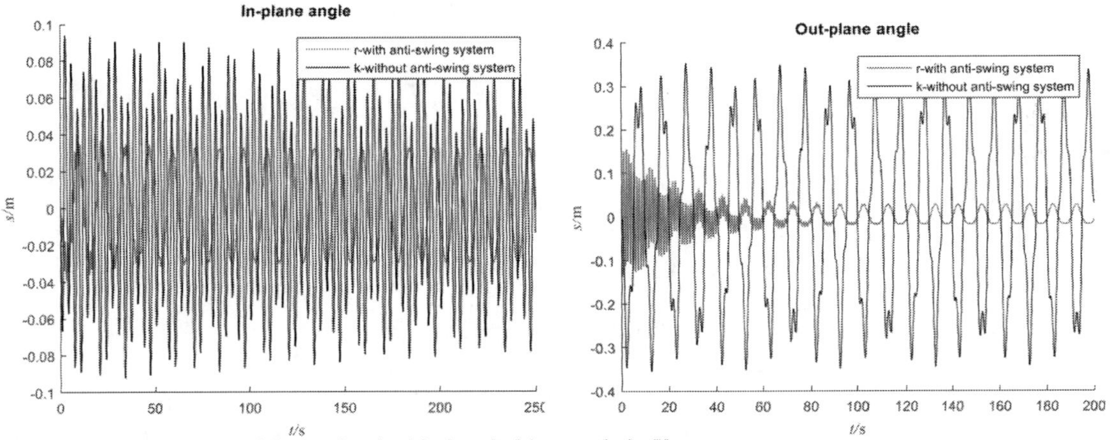


**Figure 2** The swing curve of the payload with the pitching angle is 5°

From Fig. 2 can conclude that the ship pitching angle is 5° without anti-swing system, the maximum in-plane swing of the payload is 0.096m, the maximum out-plane swing of the payload is 0.36m. With the anti-swing device, the maximum in-plane swing of the payload is 0.039m, the range of anti-swing is reduced 59.4%, the maximum out-plane swing of the payload is 0.16m, the range of anti-swing is 55.6%.

### 3.2 The swing of the payload with pitching angle is 9°

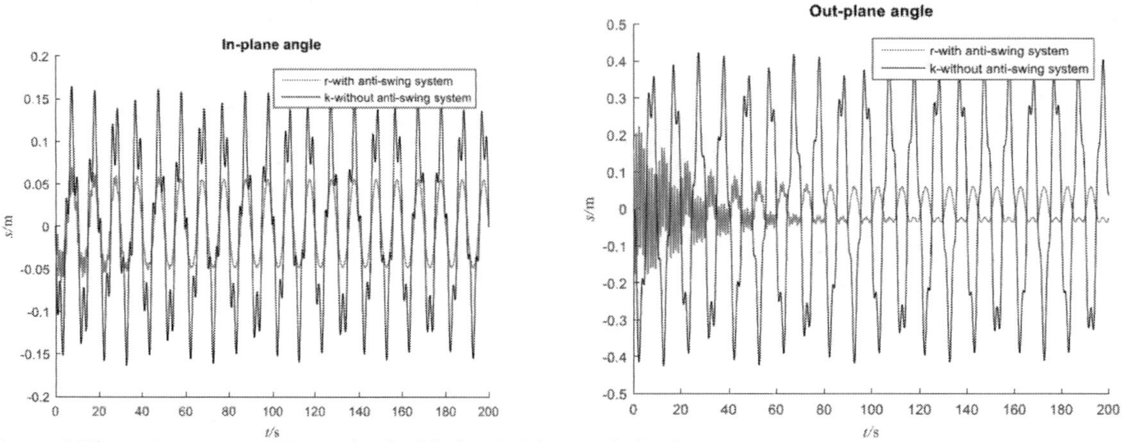


**Figure 3** The swing curve of the payload with the pitching angle is 9°

From Fig. 3 can conclude that the ship pitching angle is 9° without anti-swing system, the maximum in-plane swing of the payload is 0.162m, the maximum out-plane swing of the payload is 0.44m. With the anti-swing device, the maximum in-plane swing of the payload is 0.071m, the range of anti-swing is reduced 56.2%, the maximum out-plane swing of the payload is 0.21m, the range of anti-swing is 52.3%.

### 3.3 The swing of the payload with pitching angle is 13°

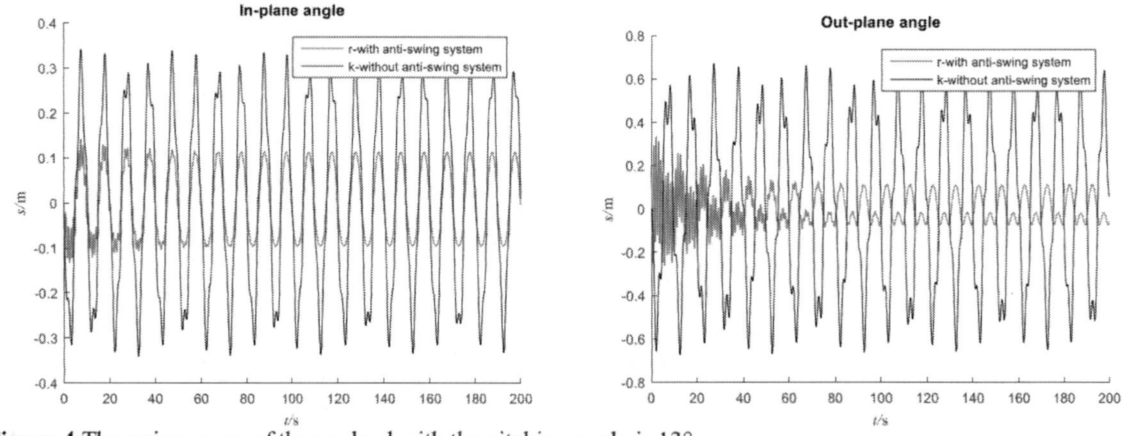


**Figure 4** The swing curve of the payload with the pitching angle is 13°

From Fig. 4 can conclude that the ship pitching angle is 13° without anti-swing system, the maximum in-plane swing of the payload is 0.31m, the maximum out-plane swing of the payload is 0.68m. With the anti-swing device, the maximum in-plane swing of the payload is 0.128m, the range of anti-swing is reduced 58.7%, the maximum out-plane swing of the payload is 0.29m, the range of anti-swing is 57.4%.

## 4 CONCLUSIONS

The kinematics model of the crane payload system is established and the simulation analysis is obtained using Simulink software with the pitching angles are 6°, 11°, 16°. According to the results, the sling tray mechanical anti-swing system can reduce the swing amplitude of the payload about 55%. The effectiveness of the sling tray mechanical anti-swing system is verified, it has a certain significance to improve the safety and efficiency of ship crane.

## ACKNOWLEDGMENTS

The project supported by the Special funds for basic scientific research business of the Central University (3132016345,3132016356); Liaoning Natural Science Foundation of China (2015020132).

## References

1. Wang Xuelin, You Xinyi, Hu Yujin. J. China. M. E. 9, 1077-1082(2010).
2. Henry, R.J, Masoud, Z.N, Nayfeh, A.H, et al. J. V.C, 7, 53-1264(2001).
3. Chin, C, Nayfeh, A.H, Abedl-rahman, E. J. V.C, **7**, 199-220(2001).
4. Chin, C.M, Nayfeh, A.H. J. V.C, **7**, 891-904 (2001).
5. Ellermann, K, Kreuzer, E, Markiewicz, M. J. Mul. Sys. Dyna. **9**, 377-387(2003).
6. Park, KP, Cha, JR, Lee, KY, et al. J. Trans. Korean. S. M.E. **35**,15-120(2011)
7. Miles, J.W, J A. S. A. **20**,21-32(1962).
8. Miles, J.W, J. Phys. D. Nonl. Phe. **11**, 309-323(1984).
9. Patel M, Brown D, Witz J.A. J. Trans. R. I. Nav. Archi. **129**, 103-113(1987).
10. J.J Hoon , K. Sung-Ha. & J.E Tae. J. Cent. South Univ, **19**, 163-167(2012).
11. R. M. T. Raja Ismail, Q. P. *IEEE International Conference on Automation Science and Engineering* (CASE). **17**, 996-1001(2013).
12. Fang Yongchun, Wang Pengcheng, Sun Ning, et al. *IEEE Transactions on Industrial Electronics*, **61**,414-427(2014).
13. Yang Wang, J. China. S. S.T, **35**, 105-108 (2013)

# Modeling and optimization for rotary ultrasonic face milling of carbon fiber reinforced polymers

Muhammad Amin[1, 2,a], Songmei Yuan[1,a], M. Zubair Khan[2], Li Zhen[1], M. Kashif Nawaz[1] and Qi Wu[1]

[1]*School of Mechanical Engineering & Automation, Beihang University, Beijing, Beijing Engineering Technological Research Center of High-efficient & Green CNC Machining Process and Equipment, Beijing 100191, China*
[2]*Department of Mechanical Engineering, Institute of Space Technology, Pakistan*

**Abstract.** Carbon fiber reinforced polymers (CFRP) have got paramount importance in aerospace, and other industries due to their attractive properties of high specific strength, high specific stiffness, high corrosion resistance, and low thermal expansion. However, due to their properties like heterogeneity, anisotropy, and low heat dissipation, the issues in machining like excessive cutting forces and high surface roughness have found. In this research, a cutting force model has developed for rotary ultrasonic face milling of CFRP composites. The experimental machining was carried out on CFRP-T700. From the analysis, it has found that experimental and simulation values of cutting forces have variation/ error below than 10% in the most of the groups of parameters. However, the error found higher in few cases, due to heterogeneity, anisotropy and some other properties of these materials. The formula for contact area of the abrasive core tool improved and an overlapping cutting allowance has applied the first time. The optimal combination of parameters has investigated for cutting force and surface roughness. The developed cutting force model then further validated with pilot experiments and found the same results. So, the model developed in this paper is robust and can be applied to predict cutting force and optimization.

## 1 Introduction and literature review

Carbon fiber reinforced polymer composites have got a wide range of applications in aerospace, defense, and high-performance supporting equipment due to their attractive properties like high specific strength, high specific stiffness, low weight, high corrosion resistance, and low thermal expansion [1-3]. However, their properties like inhomogeneity, anisotropy, and heterogeneity, are the main hindrance for their accurate and defect free machining. Although CFRP materials are designed/ manufactured near-to-net shapes, but some machining processes including face milling are unavoidable. The cutting phenomenon is also complex for such materials and required to be investigated.

Rotary ultrasonic machining (RUM) is a nontraditional machining process which combines the material removal mechanism of diamond grinding and ultrasonic machining. A review of existing literature shows that RUM has many advantages over traditional machining, such as lower cutting forces, fewer surface defects, smaller chipping size, less subsurface damage, and less tool wear [4,5].

[a] Corresponding author: yuansm@buaa.edu.cn, aminbuaa@yahoo.com

Few reports have found on modeling of RUM, namely material removal rate (MRR), tool wear and rotary ultrasonic drilling. Also, some studies have found for rotary ultrasonic face milling (RUFM) of brittle materials like optical glass and ceramics. The RUFM was carried out the first time and found that the cutting forces reduce near to zero after a certain period of time [6]. The theoretical model for RUFM was developed assuming the diamond grit was spherical [7, 8]. The rotary ultrasonic face milling was then carried out with cylindrical diamond core tool for K9 optical glass and a mathematical cutting force model was developed [9]. The rotary ultrasonic face milling was carried out with conical core tool and a cutting force model for ceramic materials was developed [10].

From the literature review, it has found that a few research reports have been found for RUFM of ceramics and other composite materials. The research for rotary ultrasonic face milling of CFRP materials has not reported yet. Also, the optimized cutting force and surface roughness has not investigated till yet. So, there is an essential need to model cutting forces and to optimize cutting force and surface roughness (important for face milling). The excessive cutting forces have adverse effects on properties of composites. In this paper, the mechanistic-based cutting force model is developed to predict the cutting forces in RUFM of CFRP composites. The developed model is improved and the proportionality parameters, $K$ is obtained through designed experiments, calculations, and experimental RUFM. The analysis of variance is carried out. The optimized cutting force and surface roughness are then investigated for the first time in this paper. Conclusions are drawn in the final section.

## 2 Development of cutting force model

The cutting force model has developed by considering single abrasive grit and then applied summation for all active abrasive grits. When a diamond abrasive grit penetrates into the surface of the workpiece material, there is a plastic deformation. With the increase of penetration depth, the median cracks, and the lateral cracks grow/ generate as shown in Figure 1. The extended lateral cracks then induce and peeling off the workpiece material. The assumptions like the diamond abrasive grits/ particles are rigid regular octahedron of the same size and the material removal mode is a rigid brittle fracture, have applied.

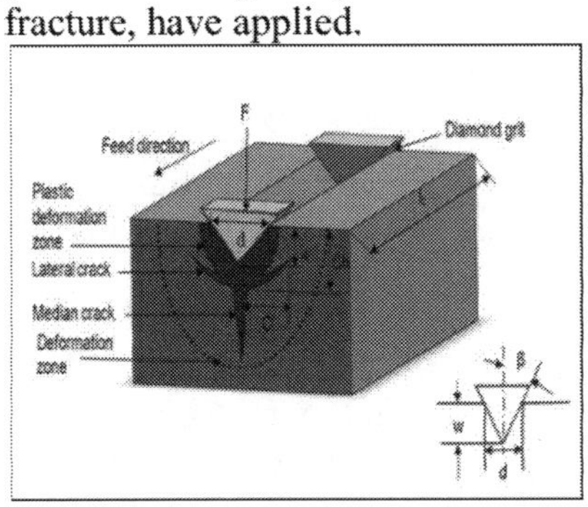


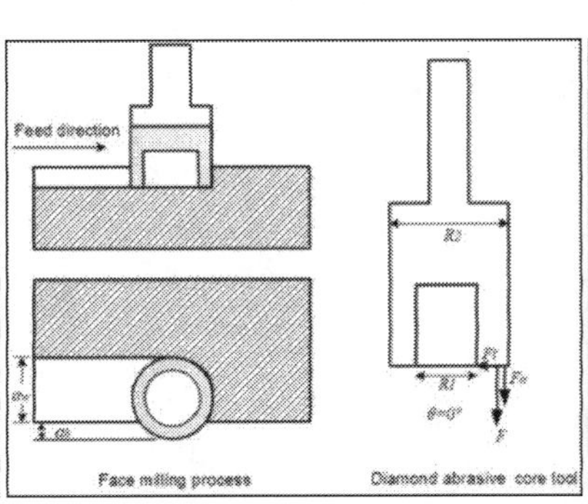

**Figure 1.** Crack generation in material **Figure 2.** Face milling process **Figure 3**. Experimental set up

After considering brittle fracture mechanism/ theories [11-13] and related factors, the developed cutting force model of RUFM for CFRP composites is as follows:

$$F = \frac{K}{\cos\theta}\cdot\left[\frac{C_3.\tan^{26}\beta.K_{IC}{}^{12}.H_v{}^{44}.(1-\nu^2)^6.f.A^{16}.S_g^{32}.f_r{}^{24}.A_0^8}{\cos^8\beta.(R_2+R_1)^{24}.E^{21}.S^{24}.C_\alpha^{32/3}}\right]^{1/35} \quad (1)$$

Where $C_3$ is the dimensionless constant number and has value as follows:

$$C_3 = \left[\frac{(180)^{24}}{(0.051)^8.C_1{}^{16}.C_2{}^{48}.\pi^8}\right] \quad \text{and } C_1=3\times10^{-2} \quad C_2=0.226$$

Equation (1) represents the desired cutting force prediction model for the axial cutting force. The contact area, $A_0$ can be found from Equation (2).

$$A_0 = \pi\left[\left(R_2^2 - R_1^2\right) + R_2(a_p - 2a_b) - a_p.a_b\right] \quad (2)$$

Where, $R_1$ and $R_2$ are the inner and outer radii of the diamond core tool, respectively; $a_p$ is the cutting depth; $a_w$ is the cutting width ($a_w = R_2$ for cylindrical core tool); and $a_b$ (in mm) is the overlapping cutting allowance which is the distance of overlap with the machined surface by previous cutting pass of the tool (the distance which is required to leave uncut for smooth cutting surface and to reduce/ finish scallop height). Practically, it is required to be considered for accurate face milling. Also, for cylindrical core tool (Figure 2), the angle between $F$ and $F_n$ is zero (i.e. $\theta = 0^o$ ). The other parameters related to the model have mentioned in Table 1.

**Table 1.** Parameters/ values related to core tool, material and cutting force model

| Symbol | Nomenclature | Symbol | Nomenclature |
|---|---|---|---|
| $S_g$ | Side length of diamond abrasive grit, *mm* | $\theta$ | Angle between $F$ and $F_n$ , (°) |
| $\beta$ | Half angle of abrasive grit, *45°* | $H_v$ | Vickers-hardness of material ( 0.6 *GPa* ) |
| $C_a$ | Abrasive concentration, *mm* | $E$ | Elastic modulus (53*GPa*) |
| $S$ | Spindle speed, *rpm* | $K_{IC}$ | Fracture toughness, 11.5MPa·m$^{1/2}$ |
| $f_r$ | Feed rate, *mm/min* | $v$ | Poisson's ratio (0.3) |
| $A$ | Ultrasonic vibration amplitude, $1\times10^{-5}$ *m* | $\rho$ | Density (1.8g/cm$^3$) |
| $f$ | Ultrasonic vibration frequency, *1600 Hz* | Core tool | Mesh size 40/45, metal bond |
| $F_n$ | Axial cutting force of an abrasive on the surface of the workpiece, *N(newton)* | $C_a$ | Concentration ( 100) |
| $F$ | Axial cutting force ( found by cutting force Dynamometer), *N* | $R_2$ | Outer radius (6.25mm) |
| $F_{(s)}$ | Axial cutting force simulated from model, *N* | $R_1$ | Inner radius (4.75mm) |

## 3 Experimental setup and conditions

The actual experimental setup has shown in Figure 3. The setup has composed of three parts: Ultrasonic vibration system, CNC vertical machining center, and diamond core tool. The ultrasonic vibration system has an ultrasonic spindle and an ultrasonic generator. The CNC vertical machining center (Model: VMC 0850B, Shenyang, China) has fitted with ultrasonic vibration device/ attachment (developed by Tianjin University, China) having the ultrasonic spindle. The cutting force has measured with the dynamometer (9257B, Kistler, Switzerland). The workpiece material used in experiments was CFRP-T700 with dimensions 96 x 40 x 5mm.The properties of workpiece material and the core tool have reported in Table1. The average grit size 385μm has applied [14]. After machining, the specimens have then inspected with the non-contact 3D optical profiler (model: CCI MP by Taylor Hobson, Japan) to find roughness average (i.e., a surface roughness ($S_a$) over the area and also known as the arithmetic mean height of surface) and then recorded in Table 2. Firstly, the experiments have designed by single factor experiment array with 3 factors. The level of each factor is selected by the theoretical calculations, previous experiments and keeping in view the higher values of MRR for practical/ industrial applications. For the second group, the experiments have designed on the basis of full factorial design with 2-levelof parameters, ($S$, $f_r$, and $a_p$) by applying Minitab16.

## 4 Results and discussion

The cutting force value is the mean value of maximum values in a stable stage that has obtained through measurement in graphical form through Dynoware software. The graphical cutting force data then transformed to numerical data through MATLAB software and recorded in column 5 of Table 2. The simulation values and measured values of axial cutting force have found a close match with the minimum value of the factor $\sum(F_{(m)} - K' * F_{(s)})^2$. This factor has differentiated with respect to *K*, putting the values for each experiment, summation of the values for all experiments, and then the value of *K* is obtained. It is the relationship of the workpiece material and properties (geometry, material, etc.) of cutting tool. The value of *K* has found 0.036. The cutting force data obtained by applying the cutting force model has recorded in column 6 of Table 2.

The developed cutting force model is more close to practical machining as compared to the existing models proposed by Zhang [9], and Zhang [10]. The contact area calculations have improved in this paper and the overlapping cutting allowance, $a_b$ have incorporated for the results of developed model match to the practical machining situations. The maximum values of feed rate and cutting depth ($f_r$ = 180 mm/min and $a_p$ = 0.8mm) have applied to develop the cutting force model whereas the values applied by Zhang [9] are lower at considerable level ($f$ = 12mm/min and $a_p$ = 0.08mm). Also, the higher values of machining parameters are required to increase the *MRR* as, $MRR = f_r.a_p.a_w$. The cutting force has found decreased with the increase of spindle speed while found increased with the increase of feed rate and cutting depth. The same has reported by Zhang [10].

The research for rotary ultrasonic face milling of CFRP, particularly CFRP-T700 composites has been carried out first time in this paper. However, the cutting force models have also reported for RUFM of K9 optical glass and C/Sic materials [9-10].

The cutting force has found decreased with the increase of *S* while it has found increased with the increase of $f_r$ and $a_p$ (Figure 4). Also, the measured and simulated (from the model) values of cutting force values have a close match (nearly equal). However, a variation/ error of 2.043 N (19.81%) for Exp. 12 (Table 2) has observed. From graphs of Figure 5, the surface roughness, $S_a$ has found increased with the increase of *S* up to 4000 rpm and then found decreased rapidly up to 5000 rpm. Also, $S_a$ has found increased with the increase of $f_r$ and $a_p$. Although, some variations from uniform curved path has been found which are due to the properties like anisotropy, and heterogeneity of CFRP material.

**Table 2.** Cutting force and roughness average data related to parameters

| **Exp. No** | ***S* *(rev/min)*** | **$f_r$ *(mm/min)*** | **$a_p$ *(mm)*** | **Measured axial force ($F_{(m)}$) *(N)*** | **Simulated axial force ($F'_{(s)}$ *without K*) *(N)*** | **Simulated axial force ($F_{(s)}$ *with K*) *(N)*** | **$S_a$ *(μm)*** |
|---|---|---|---|---|---|---|---|
| 1 | 2000 | 60 | 0.5 | 16.921 | 453.232 | 16.316 | 2.683 |
| 2 | 2500 | 60 | 0.5 | 14.980 | 388.927 | 14.001 | 2.746 |
| 3 | 3000 | 60 | 0.5 | 13.104 | 343.220 | 12.355 | 2.954 |
| 4 | 3500 | 60 | 0.5 | 10.677 | 308.792 | 11.116 | 3.393 |
| 5 | 4000 | 60 | 0.5 | 9.892 | 281.773 | 10.143 | 3.610 |
| 6 | 4500 | 60 | 0.5 | 9.215 | 259.911 | 9.356 | 2.794 |
| 7 | 3000 | 60 | 0.5 | 12.404 | 343.220 | 12.355 | 2.525 |
| 8 | 3000 | 90 | 0.5 | 15.845 | 453.232 | 16.316 | 3.245 |
| `9 | 3000 | 120 | 0.5 | 20.842 | 552.069 | 19.874 | 2.914 |
| 10 | 3000 | 150 | 0.5 | 22.080 | 643.348 | 23.160 | 3.114 |
| 11 | 3000 | 180 | 0.5 | 25.918 | 729.024 | 26.244 | 3.307 |
| 12 | 3000 | 60 | 0.5 | 10.312 | 343.220 | 12.355 | 2.200 |
| 13 | 3000 | 60 | 0.6 | 12.312 | 345.688 | 12.444 | 3.307 |
| 14 | 3000 | 60 | 0.7 | 13.309 | 348.098 | 12.531 | 3.172 |
| 15 | 3000 | 60 | 0.8 | 13.875 | 350.454 | 12.616 | 3.425 |

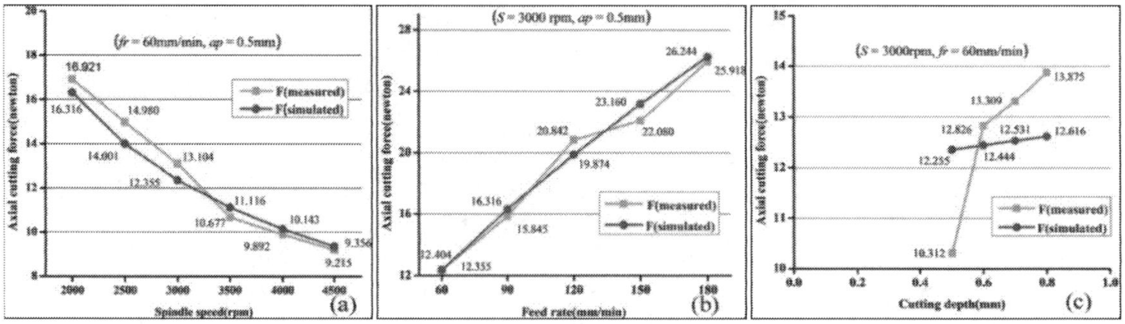


**Figure 4.** Relationship of cutting force and machining parameters

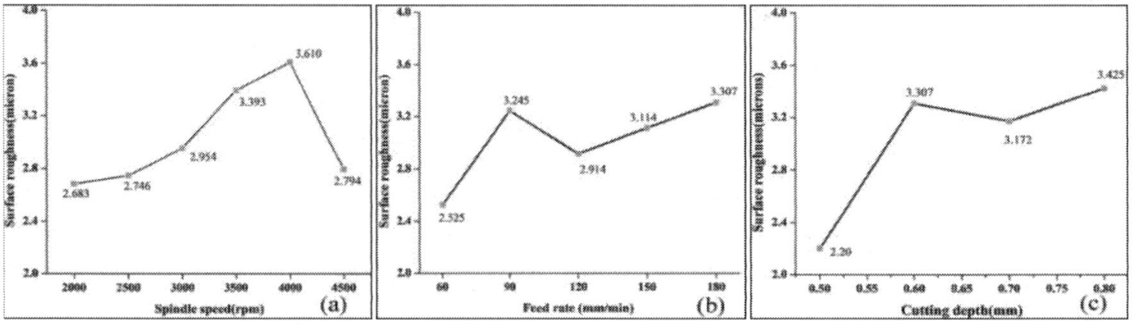


**Figure 5.** Relationship of surface roughness and machining parameters

## 4.1 Further Analysis and Optimization

The second group of experiments was carried out to know whether the error is random or due to some issues in cutting force model. The variation more than 10 % has found in only two experiments (Exp.1 as 23.20% and for Exp.5 as 16.90 %) shown in Table 4. Such variations are mainly due to heterogeneity and anisotropy of CFRP composites. Also, these variations can rise due to uneven material properties and dislocations of fibers during RUFM. Other factors contributing to this behavior are their inhomogeneity and varying thermal behavior. So, the error up to limited level is expected and required to be accepted due to nature of the material.

The analysis of variance was carried out (Table 3) and found that $S$ and $f_r$ have a significant effect on $F$ while $a_p$ has a non-significant effect for RUFM of CFRP composites. From ANOVA for $S_a$, a significant and marginally significant effect has found by $S$ and $a_p$ respectively, while $f_r$ has shown a non-significant effect. From the analysis of contour plots (Figure 6), $F$ found decreased with the increase of $S$, while it increased with the increase of $f_r$. The optimal value of $F$ (less than 11N) can be found with $S$= 5000 rpm, $f_r$=120 mm/min and $a_p$=1.2 mm. The contour plots (Figure 7) showed that $S_a$ increased with the increase of cutting depth and decreased with the increase of spindle speed. The optimal value of $S_a$ ( less than 2.5 µm) can be found with 5000 rpm of $S$, 110 mm/min of $f_r$ and 2.2 mm of $a_p$. The curvilinear profile for contour plots for $F$ and $S_a$ shows the quadratic model fitting.

**Table 3.** ANOVA for axial cutting force and surface roughness

| Group | Parameter | DoF | Sum of squares | Adj-mean | F-value | P-value | Finding |
|---|---|---|---|---|---|---|---|
| Cutting force | $S$ | 1 | 32.176 | 32.176 | 11.90 | 0.026 | Significant |
| | $f_r$ | 2 | 212.242 | 212.242 | 78.49 | 0.001 | Significant |
| | $a_p$ | 3 | 0.112 | 0.112 | 0.04 | 0.848 | Non-significant |
| Surface roughness | $S$ | 1 | 3.759 | 3.759 | 23.94 | 0.008 | Significant |
| | $f_r$ | 1 | 0.078 | 0.078 | 0.50 | 0.518 | Non-significant |
| | $a_p$ | 1 | 0.436 | 0.436 | 2.78 | 0.171 | Marginally significant |

**Table 4.** Cutting force and surface roughness data

| Exp. No | $S$ *(rev/min)* | $f_r$ *(mm/min)* | $a_p$ *(mm)* | Axial force $(F_{(m)})(N)$ | Axial force $(F_{(s)}$ *without* $K)(N)$ | Axial force $(F'_{(s)}$ *with* $K)(N)$ | $S_a$ *(μm)* | Error $\frac{(F'_{(s)}-F_{(m)})}{F_{(m)}}\times100\%$ |
|---|---|---|---|---|---|---|---|---|
| 1 | 5000 | 100 | 1.2 | 10.500 | 359.377 | 12.937 | 2.826 | +23.20 |
| 2 | 3500 | 100 | 0.6 | 15.188 | 441.472 | 15.892 | 3.500 | +4.63 |
| 3 | 5000 | 200 | 1.2 | 22.015 | 578.057 | 20.810 | 2.493 | -5.47 |
| 4 | 5000 | 200 | 0.6 | 19.887 | 556.039 | 20.017 | 2.563 | +0.65 |
| 5 | 3500 | 100 | 1.2 | 14.133 | 458.953 | 16.522 | 3.901 | +16.90 |
| 6 | 3500 | 200 | 0.6 | 24.875 | 710.107 | 25.560 | 3.627 | +2.75 |
| 7 | 3500 | 200 | 1.2 | 27.213 | 738.226 | 26.576 | 4.820 | -2.34 |
| 8 | 5000 | 100 | 0.6 | 12.960 | 345.688 | 12.444 | 2.482 | -4.00 |

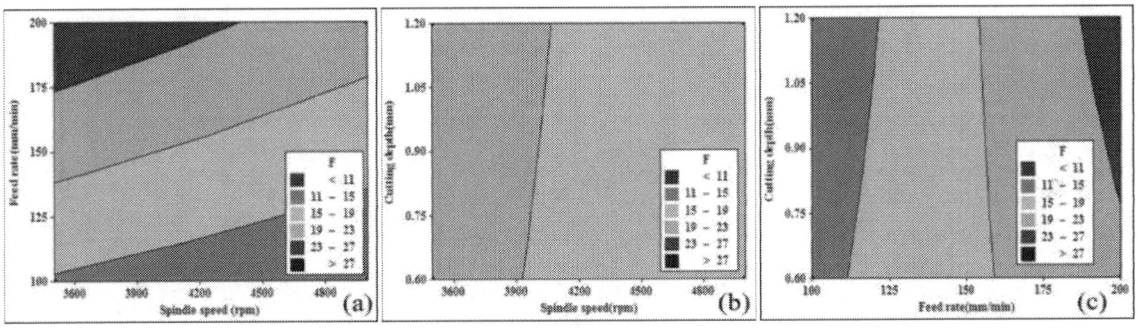


**Figure 6.**Surface plot for cutting force

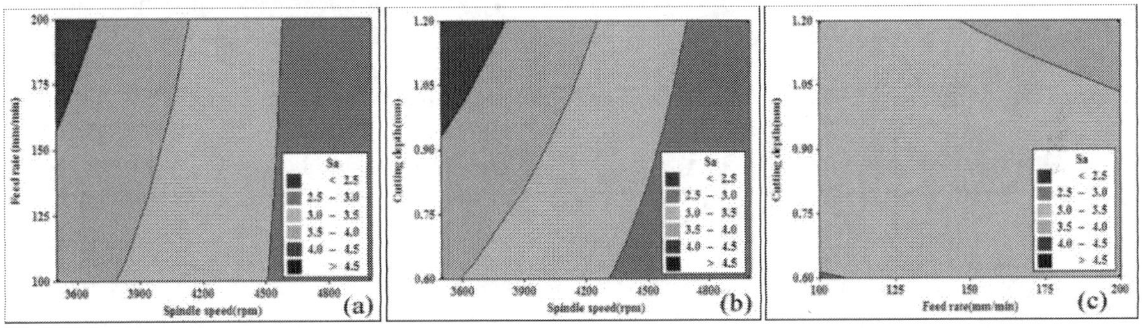


**Figure 7.**Surface plots for surface roughness

# 5 Conclusions

The salient features and outcomes of this research work have been recorded beneath:

**1.** The mechanics based cutting force prediction model has been developed for RUFM of CFRP-T700 composites. The improved formula for contact area, $A_0$ have been proposed in the paper and an overlapping cutting allowance, $a_b$ has incorporated for accurate machining. From the graphical/ numerical analysis, the measured and the simulated values of cutting forces have found a close match (nearly equal) and the percentage variation found less than 10%. However, in the few cases, the variation was recorded more than this value. The cause of this variation is due to the heterogeneity, anisotropy and some other properties of CFRP composites. So, the developed cutting force model is accurate/ robust and can be applied for finding cutting forces.

**2.** The significantly higher values of machining parameters have applied ($f_r$=200mm/min $a_p$=1.2mm, $S$=5000 rpm) for the first time in order to enhance *MRR* and to close to practical machining. From

ANOVA, $S$ and $f_r$ found significant for $F$ while $a_p$ has found non-significant. From ANOVA for $S_a$, $S$ has found significant while $a_p$ has found marginally significant. The $F$ and $Sa$ have found decreased with the increase of spindle speed however both have found increased with the increase of feed rate and cutting depth.

**3.** The optimal value of $F$ (less than 11N) have found with $S$= 5000 rpm, $f_r$=120 mm/min and $a_p$=1.2 mm. Also, the optimal value of $S_a$ ( less than 2.5 μm) can be found with 5000 rpm of $S$, 110 mm/min of $f_r$ and 2.2 mm of $a_p$. The curvilinear profile for contour plots for $F$ and $S_a$ shows the quadratic model can be adequately fitted.

The developed cutting force model in this paper can be used for prediction/ minimizing of cutting forces for rotary ultrasonic face milling of CFRP composites and input resources can be saved by the increase process efficiency and quality of the product.

## Acknowledgment

This research is financially supported by National High Technology Research and Development Program of China under program No. 863 with Grant No.2013AA040105. The authors are indebted to this financial support to accomplish this research.

## References

1. R. M. Jones, *Mechanics of composite materials. 2nd edn.Taylor & Francis, Inc.,Philadelphia* (1999).
2. A. K. Kaw, *Mechanics of composite materials. 2nd edn. Taylor & Francis Group* (2006).
3. J.Y. Sheikh-Ahmad, *Machining of polymer composites. Springer science.* doi:10.1007/978-0-387-68619-6 (2009).
4. J. Liu, D. Zhang, L. Qin, and L. Yan, Feasibility study of the rotary ultrasonic elliptical machining of carbon fiber reinforced plastics (CFRP). Int J Mach Tool Manuf., **53**, 141-150 (2012).
5. D. Geng, D. Zhang, Y. Xu, F. He, D. Liu, and Z. Duan, Rotary ultrasonic elliptical machining for side milling of CFRP: Tool performance and surface integrity. Ultrasonics, **59**, 128-137(2015).
6. Z. J. Pei, P. M. Ferreira, and M. Haselkorn, Rotary ultrasonic drilling and milling of ceramics. The design for manufacturability and manufacture of ceramic components symposium, Am cera Soc., 96 th Annual meeting (1994).
7. Z. J. Pei, and P. M, Ferreira, An experimental investigation of rotary ultrasonic face milling. Int J Mach Tool Manuf., **39**, 1327–1344(1999).
8. D. J. Feng, F. L. Zhao, Z. G. Xu, and D. M. Guo, Mathematic model of material removal rate for ultrasonic milling. China Mech Eng., **17(13),** 1399-1403(2006).
9. C. Zhang, J. Zhang, and P. Feng, Mathematical model for cutting force in rotary ultrasonic face milling of brittle materials. Int J Adv Manuf Technol., **69**,161–170 (2013).
10. C. Zhang, S. Yuan, M. Amin, and F. Fan, Development of cutting force model based on brittle fracture for C/SiC using rotary ultrasonic facing milling. Int J Adv Manuf Technol., doi: 10.1007/s00170-015-7894-4(2015).
11. D. B. Marshall, B. R. Lawn, and A. G. Evans, Elastic/plastic indentation damage in ceramics: the lateral crack system. J Am Cera Soc., **65 (11),** 561–566 (1982).
12. B. R. Lawn, A. G. Evans, and D. B. Marshall, Elastic/plastic indentation damage in ceramics: the median/radial crack system. J Am Cera Soc., **63 (9–10),** 574–581(1980).
13. J. A. Collins, Failure of materials in mechanical design. Wiley, New York (1981).
14. www.inlandcraft.com/uguides/lapidary_grit_chart.pdf. Accessed 15 September 2016.

# Simulation of flow around a slender body at high angles of attack

Osama Obeid[1,a] and Ibraheem Alqadi[1]

[1]*Aeronautical Engineering Department, King Abdulaziz University, Jeddah, Saudi Arabia*

**Abstract.** LES of the flow around an ogive-cylinder body at high angles of attack were carried out to investigate the possibility of the development of asymmetric wake-vortex without the introduction of artificial perturbations. The study investigated the effect of grid resolution and scheme bias on the solution. The numerical solution was found to be sensitive to the bias in the numerical scheme. The simulation was carried for angles of attack $\alpha = 30°, 40°, 50°, 55°$, and $60°$. The simulation at $\alpha = 30° - 40°$ produced symmetric wake-vortex. At $\alpha = 50°$, the wake-vortex is also symmetric but with vortex separation. At $\alpha = 55°$, the wake-vortex becomes asymmetric. At $\alpha = 60°$, the wake-vortex is highly asymmetric with vortex separation and breakdown. It was concluded that asymmetric flow around slender bodies at high angles of attack can be simulated in the absence geometrical or flow perturbations.

## 1 Introduction

When slender bodies fly at high angles of attack, a wake-vortex system is generated on the leeward side. At small angles, the wake-vortex is symmetric and attached to the body. However, as the angle of attack is increased, the wake-vortex becomes asymmetric. As a result, side force is generated. This presents control difficulties. The lack of understanding of this phenomenon is due to the difficulties encountered when performing either experimental or numerical studies. The obstacles in experimental studies include model imperfections, incoming flow conditions, and limitations of available measurements techniques. On the other hand, high fidelity numerical studies are demanding. DNS for such flows is not feasible given the flow resolution requirements. Large Eddy Simulation (LES) provides an alternative to DNS. In LES, the Navier-Stokes equations are solved to resolve large-scale turbulent structures while modelling the effects of small-scale structures. Most of previous numerical studies were performed using RaNS and/or URaNS in which the introduction of artificial perturbations [1,2], or geometrical imperfections [3], is necessary to generate flow asymmetry. Likewise, some of the available LES studies [4] had to add artificial perturbations as well whereas others argue that asymmetry can be captured without them. On the other hand, experimental studies are limited by the resolution and accuracy of available measurement techniques. Experimental measurement, conducted by Zilliac, Degani, and Tobak [5], showed four different flow pattern; symmetric flow on the leeward side for ($\alpha < 30°$), steady asymmetry flow within the range of ($30° > \alpha > 50°$), nearly bi-stable range ($50° > \alpha > 65°$), and bluff body vortex shedding for

[a] Corresponding author: oobeid@stu.kau.edu.sa

($\alpha > 65^\circ$). In addition, a peculiar behaviour is observed where the side force direction changes with model revolution [5,6]. Further experimental
studies revealed a number of factors that influence this asymmetry such as Mach number [7,8], Reynolds number [7], nose bluntness, and fineness ratio [8].

From a numerical point of view, the separated, unsteady, asymmetric flow around a slender body at high angle of attack poses several challenges. The main complexity arises from the unsteady viscous flow behind the wake-vortex where nonlinearity effects are dominant. Also, the behaviour of the wake-vortex is complex where the flow exhibits vortex interaction, separation, and breakdown. With such complexity, a simulation would require resolving flow behaviour. Degani and Schiff [1], Murman [2], and Degani et al [9] performed computational studies using Thin-Layer RaNS. However, the simulations showed that flow asymmetry did not develop in the absence of geometrical or flow perturbations. The objective of this study is to investigate the development of asymmetric wake-vortex around slender bodies at high angle of attack using Large Eddy Simulation without the introduction of neither geometrical nor flow perturbations. A grid-refinement study was performed to investigate the sensitivity of the solution to grid resolution and numerical scheme. The study was carried out for at angle of attack $\alpha = 30^\circ$ and $60^\circ$. Statistical analysis was performed to investigate time averaged solution. In addition, mean normal and side force coefficients ($C_y, C_z$) were computed and analysed.

## 2 Computational Setup

Calculations were carried out using Large-Eddy-Simulation approach in which three-dimensional time-dependent Navier-Stokes and continuity equations are solved numerically after being spatially filtered for all motions with a scale larger than the mesh size of the chosen numerical grid. The small-scale, mainly dissipative motion, is simulated by a sub-grid scale (SGS) model. The filtered equations read in tensor notation:

$$\frac{\partial \overline{u}_i}{\partial x_i} = 0 \tag{1}$$

$$\frac{\partial \overline{u}_i}{\partial t} + \frac{\partial \overline{u}_i \overline{u}_j}{\partial x_j} = -\frac{1}{\rho}\frac{\partial \overline{p}}{\partial x_i} + \frac{\partial\left(2\nu \overline{S}_{ij}\right)}{\partial x_j} - \frac{\partial \tau_{ij}}{\partial x_j} \tag{2}$$

The equations are written in compact index-notation. i and j varies from 1 to 3. $\overline{u}_i$ represents velocity vector components and $\overline{p}$ is pressure. The fluid density is $\rho$ and the kinematic molecular viscosity is $\nu$. The spatially filtered quantities are denoted with overbar. The last term in Navier-Stokes equation is the gradients of the sub-grid-scale stresses $\tau_{ij} = \overline{u_i u_j} - \overline{u}_i \overline{u}_j$ that is modelled using the eddy-viscosity concept:

$$\tau_{ij} = 2\nu_t \overline{S}_{ij} + \frac{1}{3}\delta_{ij}\tau_{kk} \ , \quad \overline{S}_{ij} = \frac{1}{2}\left(\frac{\partial \overline{u}_i}{\partial x_j} + \frac{\partial \overline{u}_j}{\partial x_i}\right) \tag{3}$$

For larger scale motions, spatial filtering is used on the three-dimensional time-dependent Navier-Stokes equations before being solved numerically. Dynamic Smagorinsky model, developed by Germano et al [10] and enhanced by Lilly [11], will be used to compute the eddy viscosity of the sub-grid scale.

$$\nu_t = C_S \Delta^2 \left|\overline{S}\right| \text{ Where } \overline{S} = (2\overline{S}_{ij}\overline{S}_{ij})^{\frac{1}{2}} \text{ and } \Delta = (\Delta x \Delta y \Delta z)^{\frac{1}{3}} \tag{4}$$

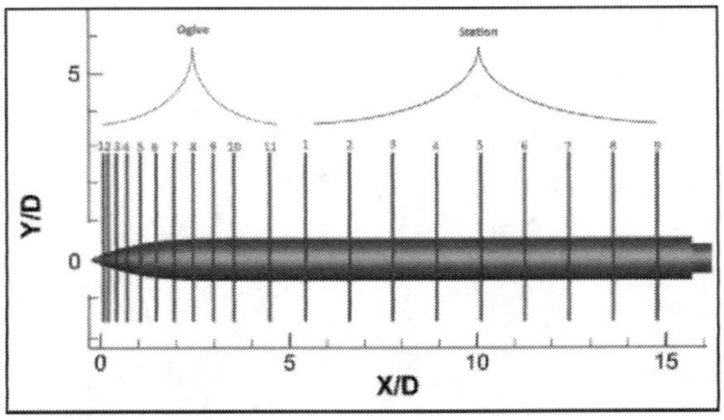


**Figure 1**: Stations.

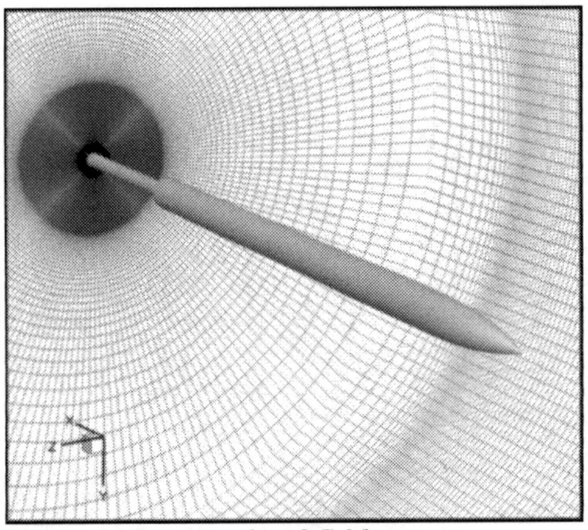

**Figure 2**: Computational Grid

The coefficient $C_S$ is computed with the Dynamic Smagorinsky-Lilly procedure where a second filter is implemented with twice the width Δ based on the grid spacing. The difference between the two filtered solutions is used to compute $C_S$ dynamically. For a physical solution, the value of $C_S$ is limited to ($0.0 < C_S < 0.23$). The geometry used in the study is identical to the model used by [12] which consisted of fineness ratio of 3.5 and 12.5 for tangent ogive and cylindrical afterbody, respectively. A model holder was added at the end of the cylinder with a length of 11.5D and a diameter of 0.5D. The overall length of the model is about 34.5cm and the diameter is $D$ = 2.16cm for the cylindrical afterbody. The time history of the solution is collected at 11 stations along the Ogive and 9 stations along the cylinder (figure 1). The Ogive profile matching the one used in the experiment model is computed as $y = D/ \ (1-(1-x/L))^{2.75}$ . $L$ is set equal to 3.544$D$. The length of the ogive ($L$) is taken slightly more than 3.5$D$ in order to compensate the loss in length after adding a small hemisphere to simulate the nose. Since it is not possible from a practical point of view to manufacture a perfectly sharp tip. Hence, a hemisphere will be added at the nose with a radius of 0.025$D$, where $D$ = 2.16cm. The computational domain is setup such that the domain boundaries are 6.5$D$ upstream of the nose, 11.5$D$ downstream of the cylinder base, and 15$D$ in the radial direction away from the model. The base computational grid consisted of 8.7 million hexahedral cells distributed as follows: 220 in the longitudinal direction, 194 in the radial direction, and 197 in the azimuthal direction (figure 2). The initial base grid spacing in the radial direction (r), the steam-wise direction (x), and the azimuthal direction ($\theta$) are $\Delta r^+ = 10$ , $\Delta x^+ = 9$ , and $\Delta\theta^+ = 80$ wall units. It should be noted that the above estimated wall spacing is for attached turbulent boundary layer. However, the separation in this case is laminar and transition to turbulence takes place in the separated shear layer. The computations were carried out using Ansys-Fluent V17.0. The pressure-based solver uses SIMPLE scheme for pressure-velocity coupling, the Least Square approach for gradients evaluation, and 2nd order pressure interpolation scheme. The solution is advanced in time using implicit 2nd order time marching scheme. For spatial discretization, two spatial discretization schemes were used to assess the effect of scheme bias on the solution; 2nd order Central scheme and 2nd order Bounded Central scheme.

**Table 1.** Grid size (in millions) for each grid refinement level.

| | **1**: $\Delta x$ | **2**: $\Delta x/2$ | **3**: $\Delta x/4$ |
|---|---|---|---|
| **A**: $\Delta\theta$ | 8.70 | 13.87 | 24.21 |
| **B**: $\Delta\theta/2$ | 11.26 | 17.88 | 31.16 |
| **C**: $\Delta\theta/4$ | 19.25 | 34.43 | 52.88 |

## 3 Results and Discussion

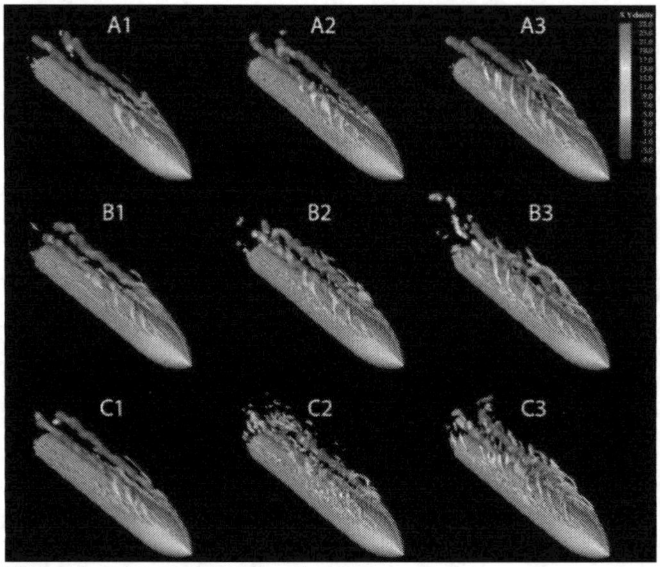


**Figure 3**. $\lambda$-2 criterion (Bounded)

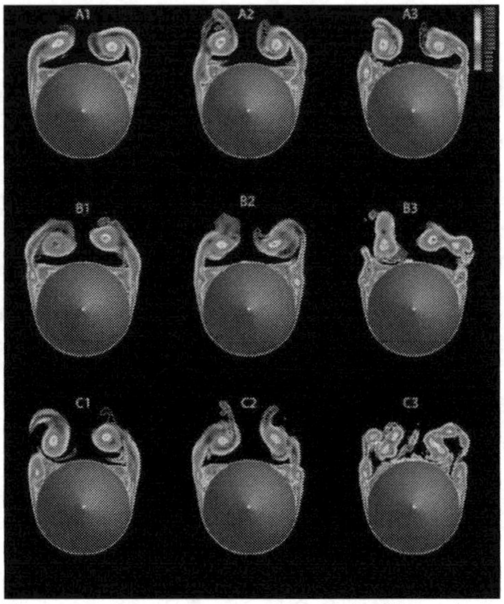


**Figure 4**. Instantaneous *x*-vorticity (Bounded)

The Time step was set to $10^{-4}$ seconds with a maximum of 20 error-reduction sub-iterations per time-step and a maximum residual of $10^{-5}$. Extrapolation was used at the out-flow boundary while velocity components were specified at the inflow boundary. The density and viscosity are held constant at $1.225\ \mathrm{kg.m^{-3}}$ and $1.784\times 10^{-5}\ \mathrm{kg.(m.s)^{-1}}$, respectively. The freestream velocity is set such that the Reynolds number, based on model diameter $D$ = 2.16 cm, is $\mathrm{Re}_D = 26{,}000$ . Two levels of grid refinement were performed in both the azimuthal and the axial directions as shown in Table 1. For the grid refinement study, calculations were performed at $\alpha = 30°$. The solutions were advanced 18 flow-through times (7000 time steps) to allow for flow field development. Flow statistics is then collected over 14000 time steps. Additionally, instantaneous data of the solution at 20 stations along the model, shown in figure 1, were also saved.

Figure 3 shows instantaneous iso-surfaces of $\lambda$-2 criterion for the grid configurations listed in Table 2. The figure illustrates the increase in the number of captured flow scales with grid refinement. Contour plots of the instantaneous x-vorticity on the *y-z* plane at Ogive-11 are depicted in figure 4 for Bounded Central scheme. In case A1, the wake-vortex system is captured as two large vortices. As the grid is refined, the captured wake-vortex system consists of several smaller vortical structures. It is noted that the refinement in the axial direction (cases A1, A2, and A3) has a stronger effect on the solution compared with the effect of refinement in the azimuthal direction (cases A1, B1, and C1). However, the combined effect of refinement in both directions (case C3) produces higher resolution of the wake-vortex system. Table 2 shows a comparison of the solution using Bounded Central scheme with the standard central scheme. There is a noticeable difference between the two spatial schemes as the grid is refined, especially in case C3. The Bounded Central scheme demonstrate multiple small vortical structures for case C3 whereas the Central scheme shows the primary wake-vortex accompanied with unsteady small structures.

Figure 5 depicts the normal and side force coefficients of all investigated grids along the model stations. For the first 6 station, up to Ogive-6, the side and normal forces are virtually identical for all cases. Starting form Ogive-1, the normal force coefficient ($C_y$) grows reaching a maximum at Ogive-6. Downstream of Ogive-6, $C_y$ starts to decrease but with disparity between the different cases. Oscillations in the force coefficient appear distinctively for the Bounded Central scheme. However, the oscillation diminishes with grid refinement. This applies to refinements from "1" to "2" in axial direction. Further refinement,"3", increases the fluctuation. Similarly, the side force coefficient, $C_z$, decreases as the grid is refined in the azimuthal direction, "A" to "B", for Bounded Central scheme. Additional refinement in the azimuthal direction, "B" to "C", or in the longitudinal direction, "1" to

"2" and "2" to "3", increases the side force coefficient downstream Ogive 9. In general, the side force increases for all grids downstream Ogive 9. Also, the direction of side force varies with respect to grid size. Cases A1, A2, B2, and B3 exhibit negative $C_z$ whereas A3, B1, C1, C2, C3 have a positive $C_z$.

**Table 2.** Comparison of instantaneous $x$-vorticity.

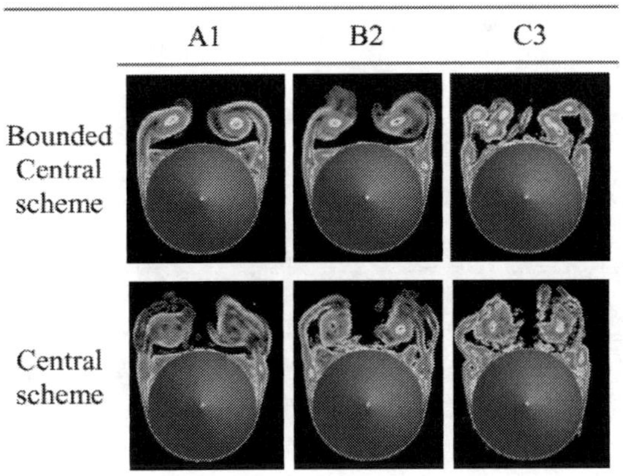


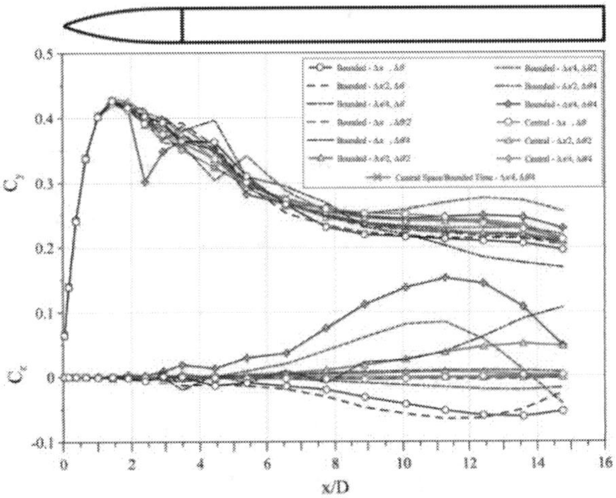


**Figure 5**. Variation of normal and side forces with grid resolution and spatial scheme

The maximum side force coefficient was obtained close to Station 6 for case C3 with a magnitude of $C_z = 0.152$ (62% of $C_y$). Moreover, using the Central scheme shows marginal variation in normal and side force coefficients with respect to grid refinement, cases A1, B2, and C3. The direction of the side force coefficient changes slightly as the grid is refined. For case A1 using Central scheme, the side force coefficient was negative downstream Ogive 9. B2 refinement decreased the magnitude of side force coefficient. As the grid is refined further, C3, the side force coefficient slightly increases to become positive. The maximum side force coefficient for the Central scheme was obtained at Station 6 for C3 grid with a magnitude of $C_z = 0.00715$ which is 2.9% of $C_y$ at that location. While the Central scheme produced almost zero side force, symmetric wake-vortex, the Bounded central scheme produced considerable side force, asymmetric wake-vortex. Additionally, the Central scheme produces consistent solution behaviour with grid refinements, while the Bounded Central scheme exhibit large and inconsistent behaviour with grid refinement. The difference in scheme behaviour can be attributed to the bias built in the Bounded scheme. Figure 6 displays instantaneous iso-surfaces of $\lambda$-2 criterion and stream traces emitted from the nose at $\alpha = 40°$, $50°$, $55°$, and $60°$ using the Central scheme. The solution produced symmetric and attached wake-vortex at $\alpha = 40°$. At $\alpha = 50°$, the wake-vortex is also symmetric but with vortex separation. At $\alpha = 55°$, the asymmetry started to develop along the cylinder with asymmetric vortex separation. However, the wake-vortex at $\alpha = 60°$ is highly asymmetric with vortex separation starting near the nose with subsequent vortex breakdown further downstream. The results illustrate, numerically, the development of asymmetric wake-vortex in the absence of geometrical imperfections or flow perturbations.

## 4 Conclusion

LES of the flow around an Ogive-cylinder body at high angles of attack were carried out. A parametric study was performed to investigate the influence of grid resolution and numerical scheme on the simulation. For Bounded Central scheme, the refinement in the axial direction increased the oscillations in the normal force and the magnitude of side force. On the other hand, the Central scheme produced consistent variation in normal and side force with grid refinement. The study also illustrated the development of asymmetric wake-vortex in the absence of artificial forcing or perturbations.

## Acknowledgments

This work was supported by (KACST) under the NSTIP program - Project. No. (8-SPA135-3). Computations were performed on Aziz Supercomputer at KAU HPC Centre (http://hpc.kau.edu.sa/). The authors would like to acknowledge the technical support provided by the centre.

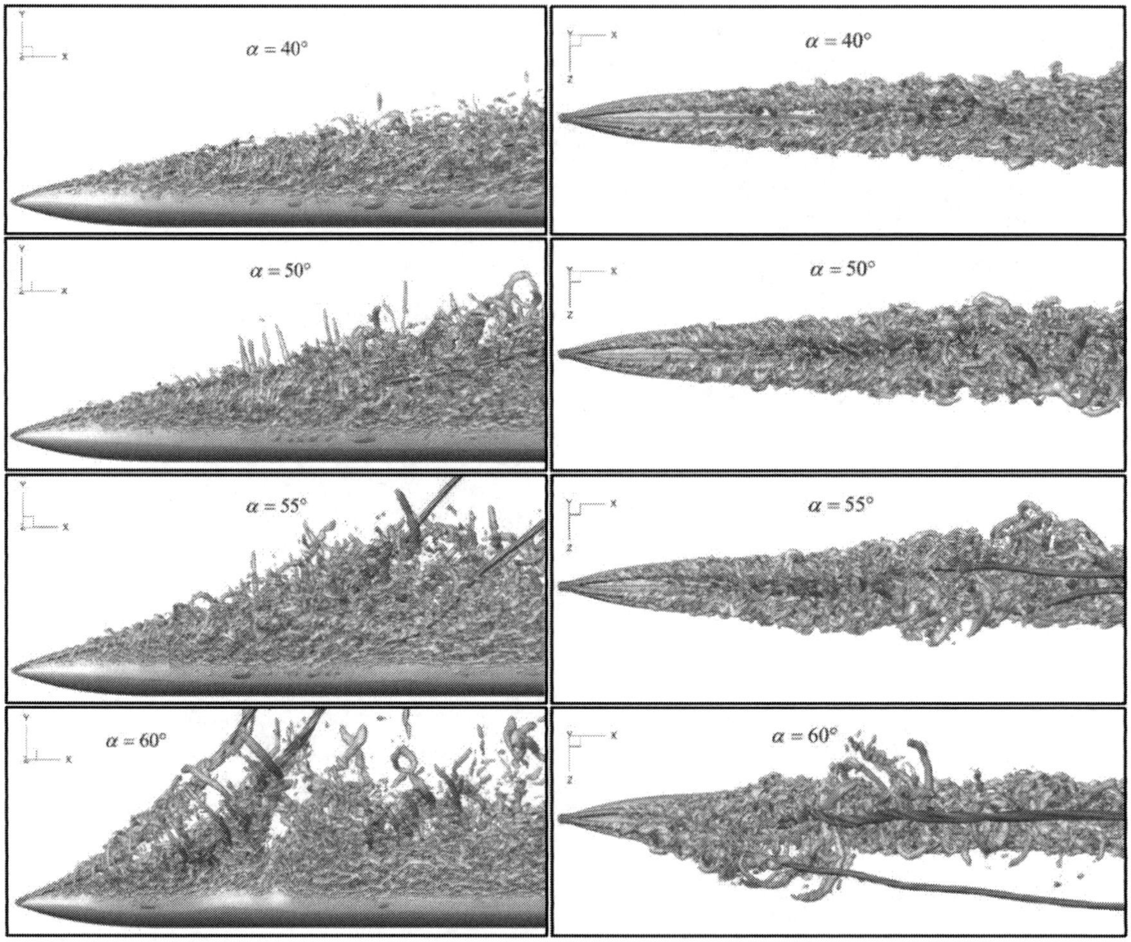


**Figure 6**: Instantaneous -2 isosurface and streamtubes for various angles of attack.

## References

[1] D. Degani and L. Schiff, *AIAA journal,* **29**, 344 (1991)
[2] S. M. Murman, *AIAA Journal,* 2000-4103 (2000)
[3] P. Kumar and J. K. Prasad, *Journal of Spacecraft and Rockets*, **53**, 195 (2016)
[4] B. Ma, Y. Huang, and T. Liu, *Chinese Journal of Aeronautics,* **27**, 772 (2014)
[5] G. G. Zilliac, D. Degani, and M. Tobak, *AIAA journal*, **29**, 667 (1991)
[6] P. C. Dexter and B. L. Hunt, *AIAA, 19th Aerospace Sci. Mtg* (1981)
[7] E. R. Keener, G. T. Chapman, L. Cohen, and J. Taleghani, *NASA-TM-X-3437* (1977)
[8] A. B. Wardlaw and A. M. Morrison, *Journal of Spacecraft and Rockets*, **13**, 589 (1976)
[9] D. Degani and Y. Levy, *AIAA journal*, **30**, 2267 (1992)
[10] M. Germano, U. Piomelli, P. Moin, and W. H. Cabot, *Physics of Fluids A: Fluid Dynamics (1989-1993)*, **3**, 1760 (1991)
[11] D. K. Lilly, *Physics of Fluids A: Fluid Dynamics (1989-1993)*, **4**, 633 (1992)
[12] D. Degani and G. Zilliac, *AIAA journal*, **28**, 642 (1990)

# Research on Learning from Demonstration of Mobile Robot with Autonomous Navigation

LI Yan-fei[1,a] and ZHANG Wen-zhi[2]

[1,2]*School of Mechanical Engineering· Inner Mongolia University of Technology, Hohhot 010051, China*

**Abstract:** A new method of batch learning from Demonstration is presented, in order to solve the problem of mobile robot independent navigation. According to the actual situation, the model of the Learning form Demonstration is given, and the neural network is used to realize the robot's learning. Considering that the single artificial neural network cause dimension disaster, we designed the Learning from Demonstration model which is the coexistence of multi-neural network and is dynamic switching. The simulation results demonstrate that mobile robot independent navigation is realized.

## 1 Introduction

Mobile robot should be able to perceive changes in the surrounding environment and in accordance with changes in the environment appropriate to adjust their action path and behavioral strategies [1]. In the field of military, mobile robot technology has been applied to a variety of advanced unmanned early warning aircraft, demining robots; In the civil field, domestic mobile, entertainment, medical and other types of mobile robots more and more people in the field of vision. In short, the mobile robot has a very broad space for development and application prospects. However, navigation is a necessary problem to be solved by the mobile robot, which determines the action set of the mobile robot from the initial point to the target point, and avoids the collision with the obstacle [2,3]. The existing algorithms include grid method, potential force method and fuzzy control method. These algorithms must be designed by the professionals according to the surrounding environment of the robot, and the environment changes will affect the navigation and obstacle avoidance of the mobile robot. And even the need to rewrite the control procedures by experts, bringing expensive human and material resources [4,5]. Aiming at the existing navigation algorithms of mobile robots, A navigation controller for mobile robot based on batch demonstration learning is proposed. According to the frame of demonstration and the actual situation of the mobile robot, a mobile robot model based on demonstration learning is designed. And the neural network learning algorithm is used to compensate the non-linear term between the environment state and the action in the model. Using the control method proposed in this paper, a two-wheeled mobile robot is used to simulate an arbitrary path in an obstacle-free environment in order to realize autonomous navigation.

## 2 Demonstrate learning model of Mobile robot

### 2.1 Frame of Batch of learning from demonstration

In batch learning, all presenter sample data are collected prior to learning, and the learning update itself often uses the mathematical properties of the strategy evaluation value M. The batch learning process is shown in Fig1. From human brain

[a] Corresponding author: 839765543@qq.com

 DOI: 10.1051/matecconf/201710402019

strategies to collect large amounts of human state space data [6]. The human state space is mapped into a common task space by the human brain's task space operator. Through the theoretical analysis of the strategy evaluation value M, the effective data in the general state space is selected and imported into the updating operator U, and finally the robot control strategy is derived through the robot task space operator.

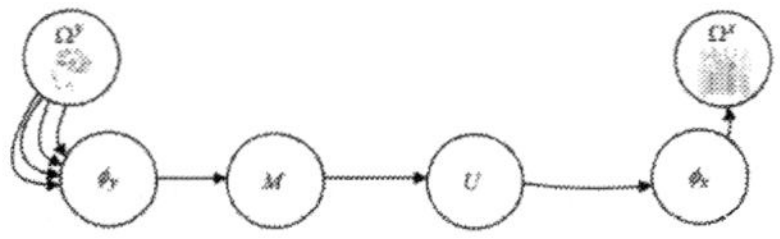

**Figure 1.** Batch learning

### 2.2 Overall Model of Mobile Robot Based on Batch Demonstration Learning

The essential problem of mobile robot based on batch demo learning is to solve the nonlinear mapping between environment states and actions. Combining the frame structure of batch demonstration learning and the actual situation of mobile robot, the whole model of mobile robot based on batch demonstration learning is designed. As shown in Figure 2.

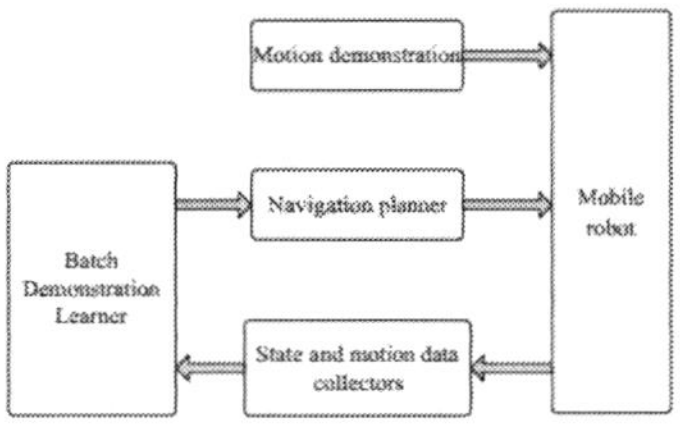


**Figure 2.** Demonstrates learning the mobile robot model

### 2.3 Design of Navigation Planner

In the overall model of the mobile robot, the navigation planner is the most critical module for the robot to realize self-navigation. Its role is to achieve the robot state of the environment and the implementation of the nonlinear mapping between the actions. There is no fixed pattern for this many-to-one mapping, so it is almost impossible to find the formulas between them. At present, artificial neural network (ANN) has been widely used in the development of nonlinear models. It is especially suitable for applications where input and output are not well defined. It is feasible to apply it to the navigation planner in Fig2. Considering the complexity of the navigation planning function, if a single neural network is used to complete the function of the navigation planner, the artificial neural network will be too large, the large ne In view of this feature, the navigation planner is divided into several smaller planners, and a classifier is added to select the corresponding subordinate planner to control the robot sailing movement according to the state of the robot. According to this idea, the navigation planner can be designed as shown in Figure 3 the overall structure.

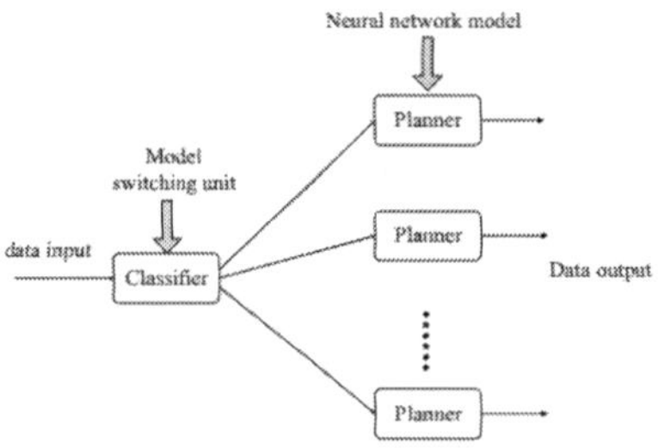


**Figure 3.** Learning model overall framework

 DOI: 10.1051/matecconf/201710402019

In the above diagram, each planner is implemented with a small-scale neural network to form the structure of the multi-neural network. Each neural network can use the same structure, but the neural network training using different data sets, of course, the network structure can also be used in different forms of structure. Although each neural network is not perfect, it can only generalize some types of robot environment states and motion maps, but through the model switching unit constructed before the mulch-neural network, the dynamics of each neural network model in the robot running process conversion. Dynamic conversion to make the performance of each neural network perfect play to achieve the proper function of the navigation planner. In order to determine the number of planners and planner functions in Figure 3, it is necessary to analyze the state information obtained by sensors on the actual robot, the characteristics of navigation target points and so on. The robot used to test the learning effect of the mobile robot is a two-wheeled robot equipped with three distance detection sensors with an angular spacing of 20° between the three sensors, which are located in front of the robot, front left and front right respectively. The state of the machine can be divided into eight states according to whether the three sensors detect an obstacle: NNN, NNE, NEN, ENE, NEE, ENE, EEN, and EEE, where E is the detected state Existing, N indicates that no obstacle has been detected (Nothing). In accordance with the overall structure of the design idea, each small planner uses an artificial neural network to replace. The entire navigation planner consists of eight neural networks and a classification switching unit. And represent the obstacle distance values detected by the three sensors on the mobile robot, respectively, and is the steering angle of the mobile robot, which can be used as the input signal of the eight neural networks. Thus, the navigation planner shown in Figure 3 can be further refined to form a detailed plan of the navigation planner as shown in Figure 4. The module switching unit dynamically triggers one of the eight neural networks according to the output value of the three sensors, and outputs the control parameter for controlling the steering of the robot.

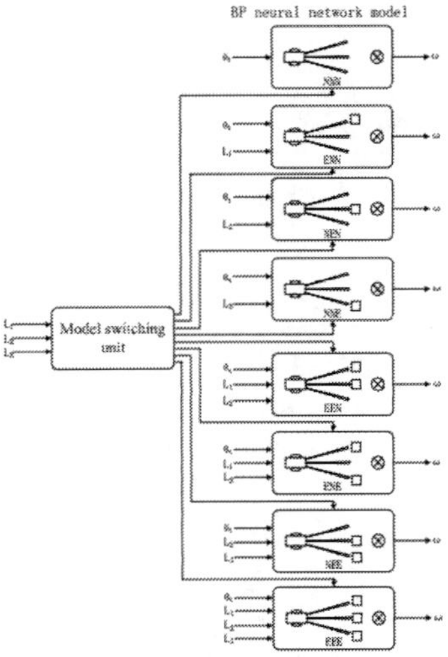


**Figure 4.** Detailed block diagram of the navigation planner

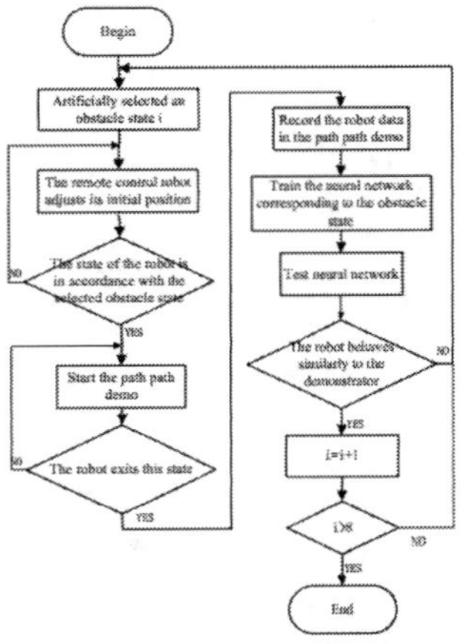


**Figure 5.** Demonstrate learning flow chart

In order to make the mobile robot perform the task of demonstration learning, the weights of the internal nodes of each neural network in Figure 4 must be updated by using the data extracted by the state and action data collectors in Figure 3 after the demonstrator is finished. Can be used BP algorithm, the field of intelligent control is widely used in a kind of neural algorithm [7]. It can store and generalize this complex input-output mapping relationship and control the output precision of the network by training the steepest descent learning rule under the condition that the complex input-output mapping relationship is difficult to be expressed by mathematical function [8, 9].

## 3 Simulation experiment

### 3.1 Demonstrate learning process

In the MATLAB simulation process, the virtual mobile robot sensor distribution as shown in Figure 5, mobile robot, obstacle and target sphere position of any of the three placed. The learning flow is shown below.

### 3.2 Analysis of learning result

With the human hands demonstrate the behavior of the ongoing remote control, eight kinds of obstacle status will produce a large number of presentation data. Using this data, the corresponding neural network is trained repeatedly, and finally eight neural network models are obtained which accord with the demonstrator behavior. Figure 6 shows 2D neural network control graphs without obstacle state, 3D neural network control diagram with single obstacle and double obstacle.

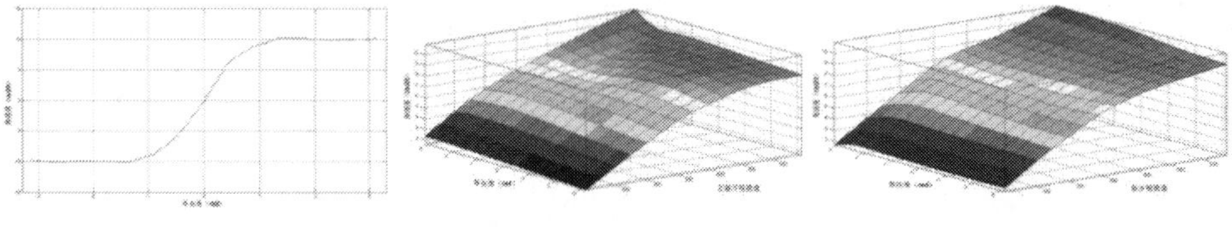

a) NNN state b) ENN state c) NEN state

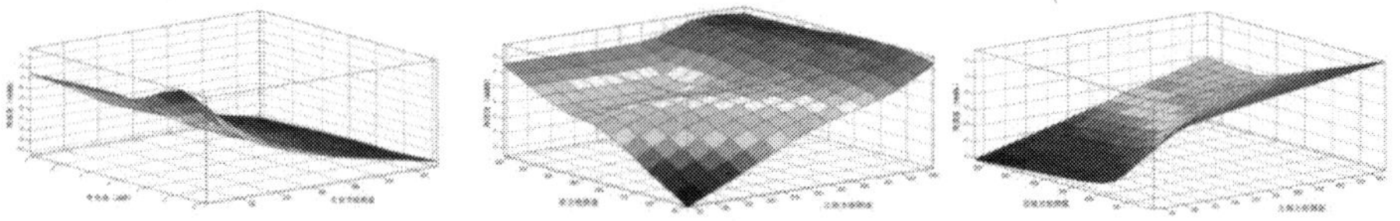

d) NNE state e) EEN state f) ENE state

**Figure 6.** Curve and surface diagram of neural network control

The data sources in each image in Figure 6 are extracted from the trained neural network controller, which can reflect the performance of the demonstration learning controller. Some typical data points are selected from the figure to illustrate:

As can be seen from the figure a, the guide angle is -3.14 and the angular velocity output is -10; the guide angle is 3.14 and the angular velocity output is 10; It is shown that the NNN neural network can produce a large angular velocity when the moving direction of the mobile robot and the deviation angle of the target sphere are large.

From Figure b, c can be seen in the case of a single obstacle, the sensor detects the value of 700, the guide angle of any given angular velocity output in the vicinity of 10; d is similar, it is clear that the closer the mobile robot is to the obstacle, the larger the angular velocity of the single obstacle neural network.

Can be seen from the e graph, the left front detection value and the front detection value as long as there is a 400, the angular velocity output in the vicinity of 10; It is shown that the angular velocity of the EEN neural network increases as the mobile robot approaches the obstacle.

As can be seen from the f graph, the left front detection value and the right front detection value are 500, the angular velocity output in the vicinity of 0. It is shown that the ENE neural network can control the mobile robot to move along the obstacle.

### 3.3 Simulation and experimental platform testing

In order to test the performance of the self-navigation controller, the obstacle in 3D virtual scene of the simulation platform is rearranged and two kinds of complicated test environments are designed. The simulation results are shown in Figure 7.

**Figure 7.** Virtual Robots in a Static Environment and in a G - Type Environment

DOI: 10.1051/matecconf/201710402019

After testing the performance of the demonstration learning control model, the subsystem is built on the platform of the two-wheeled mobile robot, and the obstacle environment is arranged. The demonstration experiment is completed on the experimental platform. As shown in Figure 8, the robot successfully avoids a plurality of obstacles and reaches a predetermined target point.

Experimental results on the experimental platform show that the self-navigation control strategy of mobile robot based on batch demonstration learning has better navigation performance.

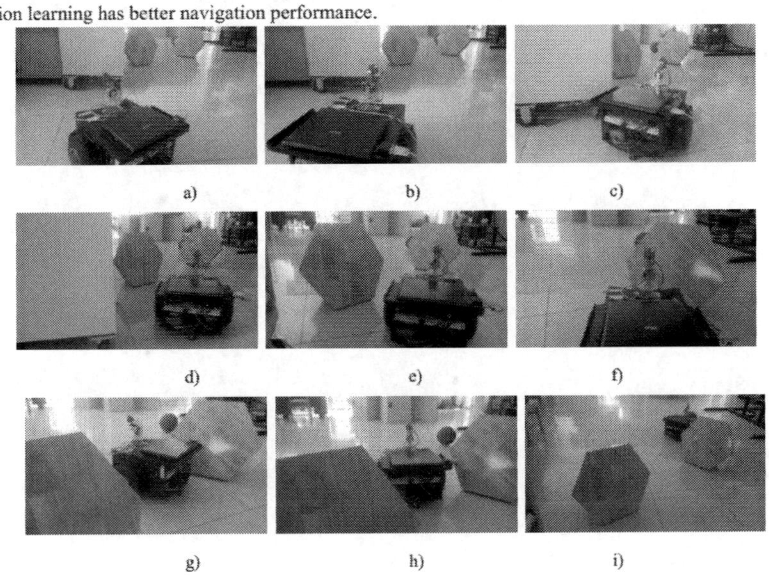

**Figure 8.** Mobile robot test platform path navigation test

## 4 Conclusions

Aiming at the characteristics of mobile robot demonstration learning, artificial neural network is proposed to realize the learning action of mobile robot to demonstrator. Considering single artificial neural network to realize complex navigation demonstration learning will make neural network too complex or cause neural network convergence Neural network, and each neural network is relatively simple, only to achieve a certain state of the mobile robot learning action, the robot in its mobile In the process, according to their work status changes at any time to switch to the appropriate neural network, at any time there is only one neural network in working condition. Experiments show that the neural network with this structure has a faster convergence rate. The learning model is simulated by the simulation platform, and the test result is good. Finally, the test results on the real robot show that the self - navigation control method based on batch demonstration learning is feasible.

## References

1. C.Mericli, Manuela Veloso, H. Levent Akin. A. IEEE-RAS,(2010).
2. H.Bener Suay, Sonia Chernova. A comparison of two algorithms for robot learning from demonstration[C]. Conference Proceedings - IEEE International Conference on Systems, Man and Cybernetics. IEEE Press,(2011).
3. C.Sun, Wei He, Weiliang Ge, Cheng Chang. Adaptive Neural Network Control of Biped Robots[C]. Man and Cybernetics: Systems. IEEE Press,(2016).
4. H.Niu, Niu Wang, Nan Li. The adaptive control based on BP neural network identification for two-wheeled robot [C] .World congress on intelligent control and automation. IEEE Press,(2016).

5. Santiago Morante, Juan G. Victores, Carlos Balaguer. Automatic demonstration and feature selection for robotlearning [C] .Humanoid Robot. IEEE-RAS,(2015).
6. Sonia Chernova, Manuela Veloso. Multi-thresholded approach to demonstration selection for interactive robot learning [C] . Proceedings of the 3rd ACM/IEEE International Conference on Human-Robot Interaction. IEEE Press,(2008) .
7. S.Jia, Quan Qiu, Junmin Li; You Li, Yue Cong. BP neural network based localization for a front-wheel drive and differential steering mobile robot [C] . Information and Automation. IEEE,(2015).
8. Maria Koskinopoulou, Stylianos Piperakis, Panos Trahanias. Learning from Demonstration facilitates Human-Robot Collaborative task execution [C] . Human-Robot Interaction.ACM/IEEE,(2016).
9. Hiroaki Arie, Takafumi Arakaki, Shigeki Sugano,et al. Imitating others by composition of primitive actions: A neuro-dynamic model [J] . Robotics and Autonomous Systems,(2012).

MATEC Web of Conferences **104**, 02020 (2017) DOI: 10.1051/matecconf/201710402020
*IC4M & ICDES 2017*

# Non-intrusive uncertainty quantification in structural-acoustic systems using polynomial chaos expansion method

Mingjie Wang[1], Qibai Huang[1,a], Shande Li[1], Lin Li[1] and Zhifu Zhang[1]

[1]*State Key Laboratory of Digital Manufacturing Equipment and Technology, Huazhong University of Science and Technology, Wuhan 430074, P R China*

**Abstract.** A framework of non-intrusive polynomial chaos expansion method (PC) was proposed to investigate the statistic characteristics of the response of structural-acoustic system containing random uncertainty. The PC method does not need to reformulate model equations, and the statistics of the response can be evaluated directly. The results show that compared to the direct Monte Carlo method (MCM) based on the original numerical model, the PC method is effective and more efficient.

## 1 Introduction

It is well known that traditional methods are conducted with to analysis the structural-acoustic system determined parameters. However, a structural-acoustic model inevitably contains some uncertain factors due to uncertain load conditions, uncertain materials and so on. Thus, it is necessary to investigate the structural-acoustic system considering uncertainties.

The MCM [1] is widely used due to its robust characteristic and the ease of implementation, while its main shortcoming is time consuming to obtain accurate results. Therefore, the MCM is usually employed to validate other approaches. Perturbation stochastic method is an effective method to deal with random uncertainty [2]. However, this method may be only limited for the problems with small uncertainty level of parameters. The polynomial chaos expansion method has been previously applied in uncertain analysis among different fields, such as fluid dynamics [3], structural dynamics [4] and acoustic fields [5,6] and so on. Sarkar et al. [4] employed the PC method to investigate the mid-frequency vibration of linear random systems with random uncertainty. LePage et al. [6] studied the sonar system performance in uncertain environments based on the PC method. The simulation results agree very well with the results estimated by MCM with less computational costs. The PC method is suitable for quantifying the uncertainty in the systems with uncertainties in initial conditions, inputs, or model parameters [7].

However, up to now, to the best of the authors' knowledge, the relevant research on the non-intrusive methods for quantifying the uncertainty in the structural-acoustic problems is still not reported. In this study, the PC method is adopted to study uncertain propagation in the structural-acoustic system containing random uncertainty. A 2D structural-acoustic model containing random uncertainty is investigated.

[a] Corresponding author : D201377175@hust.edu.cn

## 2 Theory and formulation

### 2.1 Structural-acoustic dynamic model containing random uncertainty

The dynamic equation of a structural-acoustic system can be written as

$$\mathbf{ZU} = \mathbf{f} \tag{1}$$

where $\mathbf{f}$ denotes the force vector, $\mathbf{U}$ represents the response vector, $\mathbf{Z}$ denotes the structural-acoustic dynamic matrix, which can be given by

$$\mathbf{Z} = -w^2\mathbf{M} + jw\mathbf{C} + \mathbf{K} \tag{2}$$

where $\mathbf{K}$, $\mathbf{M}$ represent the structural-acoustic stiffness matrix, mass matrix, respectively. Rayleigh damping $\mathbf{C} = \alpha\mathbf{M} + \beta\mathbf{K}$ is considered. And $\alpha$, $\beta$ represent the Rayleigh damping coefficients. $w$ is the angle frequency.

Assume that the statistics information about uncertain parameters is known, and the uncertain parameters are treated as random variables. The vector of random variables $\boldsymbol{\xi} = (\xi_1, \xi_2, \ldots, \xi_n)$ represents all the random parameters. The dynamic equation (Equation (1)) of the structural-acoustic system with random variables can be given by

$$\mathbf{Z}(\boldsymbol{\xi})\mathbf{U}(\boldsymbol{\xi}) = \mathbf{f}(\boldsymbol{\xi}) \tag{3}$$

Then the response of the structural-acoustic system considering random uncertainty can be calculated by the following equation

$$\mathbf{U}(\boldsymbol{\xi}) = \mathbf{Z}(\boldsymbol{\xi})^{-1}\mathbf{f}(\boldsymbol{\xi}) \tag{4}$$

### 2.2 The polynomial chaos expansion method for random variables

In this section, we assume that the structural-acoustic model contains normally distributed random variables. The response $U$ can be expressed by an expansion of multidimensional Hermite polynomials function $\boldsymbol{\xi} = (\xi_1, \xi_2, \ldots, \xi_n)$ as follows

$$U(\boldsymbol{\xi}) = \sum_{k=0}^{\infty} u_k \psi_k(\boldsymbol{\xi}) \tag{5}$$

where $u_k$ represent the coefficients to be calculated, $\psi_k(\boldsymbol{\xi})$ are the Hermite polynomial chaos of order $k$. In the process of numerical computation, the terms given by Equation (5) must be truncated. Assume the number of the truncated terms is $s$, $U(\boldsymbol{\xi})$ can be approximated by

$$U(\boldsymbol{\xi}) \approx \sum_{k=0}^{s-1} u_k \psi_k(\boldsymbol{\xi}) \tag{6}$$

where the number of the truncated terms $s = (n+r)!/n!r!$, $n$ denotes the number of random dimensions, and $r$ represents the highest order of the polynomials $\psi$. The coefficients $u_k$ in Equation (6) can be obtained as follows

$$u_k = \frac{\langle U(\boldsymbol{\xi}), \psi_k \rangle}{\langle \psi_k^2 \rangle} = \frac{1}{\langle \psi_k^2 \rangle} \int U(\boldsymbol{\xi}) \psi_k(\boldsymbol{\xi}) w(\boldsymbol{\xi}) d\boldsymbol{\xi} \tag{7}$$

The numerical integration in Equation (7) can be obtained by the tensor product quadrature method, i.e.

$$\int U(\boldsymbol{\xi}) \psi_k(\boldsymbol{\xi}) w(\boldsymbol{\xi}) d\boldsymbol{\xi} \approx \sum_{q=1}^{N} U(\boldsymbol{\xi}_q) \psi_k(\boldsymbol{\xi}_q) w(\boldsymbol{\xi}_q) \tag{8}$$

where $N$ represents the total number of collocation points. The collocation points are produced by the roots of the Hermite polynomial of order $r+1$. Then the coefficient $u_k$ can be calculated via the following equation

$$u_k \approx \frac{1}{\langle \psi_k^2 \rangle} \sum_{q=1}^{N} U(\xi_q) \psi_k(\xi_q) w(\xi_q) \tag{9}$$

By varying $k$ from 0 to $s-1$, the coefficients vector can be obtained

$$[u_0,\ldots,u_{s-1}] = [U(\xi_1),\ldots,U(\xi_N)]\,\mathbf{A}(\xi) \tag{10}$$

$$\mathbf{A}(\xi) = \begin{bmatrix} a_0(\xi_1) & \cdots & a_{s-1}(\xi_1) \\ \vdots & \ddots & \vdots \\ a_0(\xi_N) & \cdots & a_{s-1}(\xi_N) \end{bmatrix}, \quad a_k(\xi_q) = \frac{1}{\langle \psi_k^2 \rangle} \psi_k(\xi_q) w(\xi_q) \tag{11}$$

After obtaining the coefficients vector, the mean $\mu$ and the standard variance $\sigma$ of $U(\xi)$ can then be calculated via the following equation

$$\mu(U(\xi)) = u_0, \quad \sigma(U(\xi)) = \sqrt{\sum_{k=1}^{s-1} u_k^2 \langle \psi_k^2 \rangle} \tag{12}$$

## 3 Numerical study

A 2D structural-acoustic problem taken from Wang and Huang [8] is adopted to investigate the performance of the PC method. The uncertain parameters, listed in Table 1, are assumed to be independent of each other.

**Table 1**. Values of the uncertain parameters

| parameters | $E_a$ (m/s) | $\rho_a$ (kg/m$^3$) | $E_s$ (kg/m$^3$) | $\rho_s$ (Pa) |
|---|---|---|---|---|
| values | $\mu_{E_a} = 7.0\times10^{10}$<br>$\sigma_{E_a} = 7.0\times10^{8}$ | $\mu_{\rho_a} = 2700$<br>$\sigma_{\rho_a} = 54$ | $\mu_{E_s} = 2.11\times10^{11}$<br>$\sigma_{E_s} = 2.11\times10^{9}$ | $\mu_{\rho_s} = 7850$<br>$\sigma_{\rho_s} = 157$ |

First, we choose the order of PC as 2. Second, 70 samples are employed to evaluate the response of original system. Third, obtain the coefficient vector using Equation (10). Four, after building the PC model, Equation (12) is used to calculate expectation and standard variance of the response. Based on the original numerical model, the solutions yielded by the direct MCM with $10^4$ samples are regarded as the reference solutions.

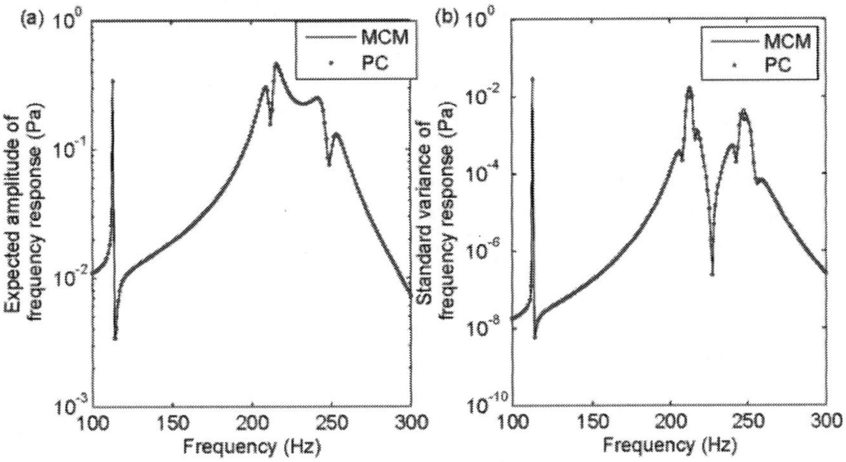


**Figure 1**. Statistic characteristics of response at node S: expectation (a) and standard variance (b)

**Table 2**. The relative errors of statistic characteristics of the response at node S

| Frequency (Hz) | Expectation | | | Standard variance | | |
|---|---|---|---|---|---|---|
| | MCM (Pa) | PC (Pa) | Relative error (%) | MCM (Pa) | PC (Pa) | Relative error (%) |
| 100 | 1.09E-02 | 1.09E-02 | 0.00 | 1.78E-08 | 1.75E-08 | 2.15 |
| 120 | 1.00E-02 | 1.00E-02 | 0.00 | 2.62E-08 | 2.55E-08 | 2.37 |
| 140 | 1.57E-02 | 1.57E-02 | 0.01 | 9.06E-08 | 8.83E-08 | 2.48 |
| 160 | 2.41E-02 | 2.41E-02 | 0.01 | 3.92E-07 | 3.82E-07 | 2.64 |
| 180 | 4.49E-02 | 4.49E-02 | 0.01 | 3.18E-06 | 3.09E-06 | 2.86 |
| 200 | 1.37E-01 | 1.37E-01 | 0.03 | 1.13E-04 | 1.09E-04 | 3.21 |
| 220 | 3.48E-01 | 3.48E-01 | -0.01 | 4.80E-04 | 4.72E-04 | 1.71 |
| 240 | 2.47E-01 | 2.47E-01 | 0.03 | 5.19E-04 | 5.09E-04 | 1.97 |
| 260 | 8.60E-02 | 8.60E-02 | 0.03 | 6.42E-05 | 6.17E-05 | 3.96 |
| 280 | 2.16E-02 | 2.16E-02 | -0.01 | 2.69E-06 | 2.65E-06 | 1.57 |
| 300 | 7.22E-03 | 7.22E-03 | -0.01 | 2.52E-07 | 2.48E-07 | 1.75 |

As shown in **Figure 1** (a) and **Figure 1** (b), the results obtained by the PC are similar to those obtained by MCM. To study the performance of the PC method in detail, the relative errors of statistic characteristics of response are listed in Table 2.From.

Table 2, we can see that the maximum relative error between PC and MCM is 3.96%. Furthermore, the relative errors of standard variance are larger than the counterpart of expectation, which means that expectation are less sensitive to uncertain parameters compared to standard variance.

**Table 3.** Execution time of the PC and MCM

| MCM | PC | | |
|---|---|---|---|
| Total execution time (s) | Execution time of determining **U**(s) | Execution time of determining statistic characteristics (s) | Total execution time (s) |
| 431,142.85 | 3018.14 | 3.89 | 3022.03 |

The execution time of the MCM and PC to calculate statistic characteristics of the response are shown in Table 3. We can see that the total execution time of the MCM method was considerably longer than that of the PC method. In fact, it only takes about 3022.03 s in the PC method, much less than the 431,142.85 s required by MCM. Furthermore, in the PC method, it takes the largest fraction of the execution time to obtain the response matrix **U**, and the statistic characteristics of the response can be calculated by only extra 3.89 s.

Based on the Equation (6), $10^4$ MCM simulations are conducted to evaluate the PDF of the response. Based on the original numerical model, the direct MCM with total $10^4$ samples is regarded as the referenced solution. The response PDF for frequency 100Hz, 200 Hz at node S using the direct MCM and the PC are plotted in Figure 2 (a) and Figure 2 (b), respectively. We can see that the overall shapes of the probability distribution mapped by the PC method practically coincided with the direct MCM results.

DOI: 10.1051/matecconf/201710402020

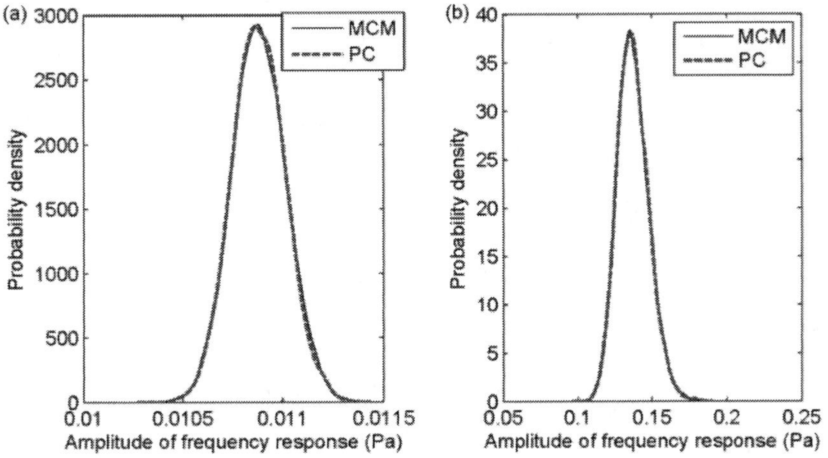


**Figure 2**. The response probability density of node S calculated by MCM and PC: (a) 100Hz, (b) 200Hz

Statistic characteristics of the response evaluated at nodes along the line $y = 0.15m$ for frequency 100Hz, 200 Hz for are shown in Figure 3 and Figure 4, respectively. The results show that the results yielded by PC method match with solutions obtained by MCM perfectly.

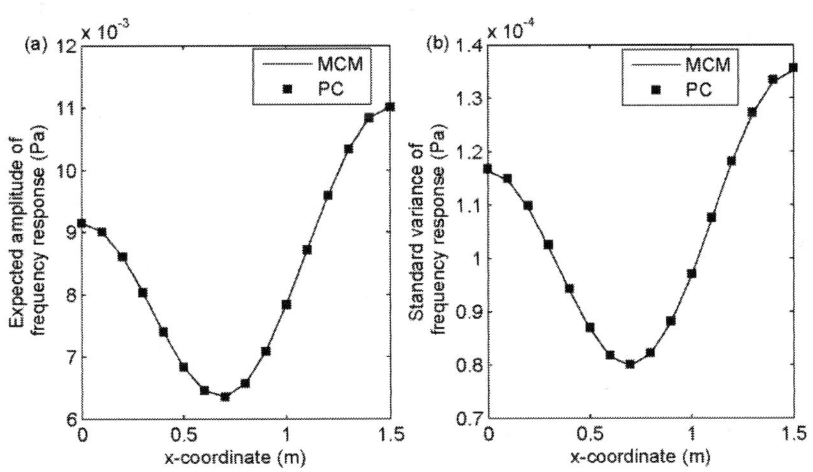


**Figure 3**. Statistic characteristics of the response along the line $y = 0.15m$ (100Hz): expectation (a) and standard variance (b)

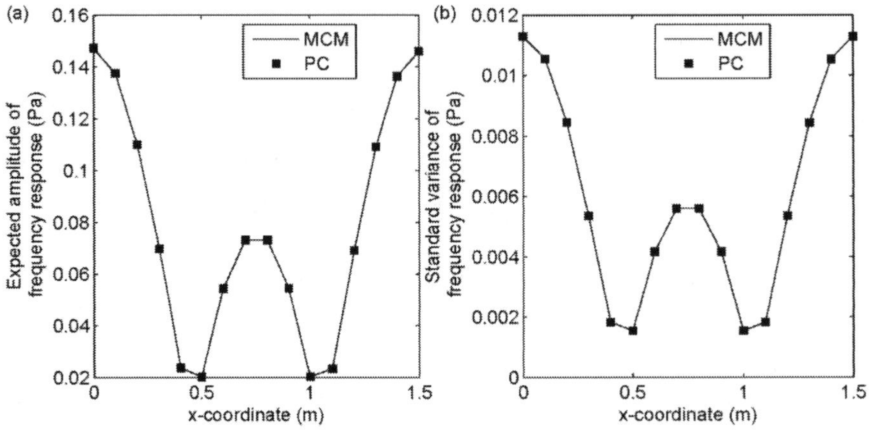


**Figure 4**. Statistic characteristics of the response along the line $y = 0.15m$ (200Hz): expectation (a) and standard variance (b)

## 4 Conclusion

The PC method is implemented to study uncertain propagation in the structural-acoustic problem containing random uncertainty. Based on the PC model, statistic characteristics of response of the structural-acoustic problem containing random uncertainty can be evaluated. The numerical results show that the PC method solutions match well with those yielded by the MCM. Furthermore, the computational effort of the PC method is significantly lower than that of the MCM.

## Acknowledgments

The research was sponsored by the National Nature Science Foundation of China (No.51575201).

## References

[1] M. Shinozuka. Monte Carlo solution of structural dynamics. Comput. Struct. **2** : 855-874(1972).
[2] G. I. Schuëller. Developments in stochastic structural mechanics. Arch. Appl. Mech. **75** : 755-773(2006).
[3] O. M. Knio, O. P. L. Maître. Uncertainty propagation in CFD using polynomial chaos decomposition. Fluid Dyn. Res. **38** : 616-640(2006).
[4] A. Sarkar, R. Ghanem. Mid-frequency structural dynamics with parameter uncertainty. Computer Methods in Applied Mechanics & Engineering. **191** : 5499-5513(2002).
[5] M. Wang, Z. Wan, Q. Huang. A New Uncertain Analysis Method for the Prediction of Acoustic Field with Random and Interval Parameters. Shock & Vibration. **2016** : 1-16(2016).
[6] K. D. Lepage. Estimation of Acoustic Propagation Uncertainty Through Polynomial Chaos Expansions. International Conference on Information Fusion. **2006**:1-5(2006).
[7] Z. K. Nagy, R. D. Braatz. Distributional uncertainty analysis using power series and polynomial chaos expansions. J. Process Contr. **17** : 229-240(2007).
[8] M. Wang, Q. Huang. A new hybrid uncertain analysis method for structural-acoustic systems with random and interval parameters. Comput. Struct. **175** : 15-28(2016).

# Prediction of Residual Stress Field in Milling Planes by Establishing Bivariate Mathematical Models

Fengyun Wang[1, a], Kuanmin Mao[1,a], Shanguo Wu[2], Yikang Du[1] and Xiaobo Mao[1]

[1] *School of Mechanical Science and Engineering, Huazhong University of Science and Technology, Wuhan 430074, China*
[2] *Yantai Chengxin Steel Grating Co. Ltd, Yantai 264006, China*

**Abstract.** The knowledge of the distribution of residual stresses in milling planes is of paramount importance to understand the surface qualities of engineering components, research on manufacturing processes and improve the assembly qualities of machines. For this purpose, a prediction method for the residual stress fields was proposed. Residual stresses of some surface discrete points of a rectangular specimen were measured. Based on the measurements, the residual stress field was predicted by establishing bivariate mathematical models. The obtained results are close to the experimental technique and superior to the previous average method. The predicted stress fields are almost symmetrical along the length or width. The proposed method avoids simulating the process of residual stresses generation and has a unique potential application for engineering components of which the generation process of residual stresses can hardly be simulated in detail.

## 1 Introduction

Residual stresses are special self-equilibrium stresses remained within engineering with no thermal gradients or external forces. They can influence the mechanical properties of engineering components such as the distortion, structural stability and fatigue life [1]. The surface residual stress, as a critical parameter of evaluating the surface quality and assembly quality, can directly affect the working performance and working life [2-3]. So for design, manufacturing and structural integrity assessment, it is necessary to know the residual stress distribution of surface residual stress as far as possible.

Residual stresses are inevitably existed in processes such as manufacturing, remanufacturing and using of mechanical components. The process of residual stresses generation is hard to be simulated. To date, surface residual stress could be measured by experimental methods[4-8] such as X-ray diffraction (XRD). However, due to the time consuming, expensive and limits of testing on some discrete points, the measured stress could only reflect the local stress characteristics. It is hard to find out the distribution of residual stress when the machine part is bigger just using experimental methods. Previous studies have ignored the initial residual stresses in the finite element analysis, considered initial residual stresses using average stress or just researched on the distribution of residual stresses in a straight line recently [9-11]. However, little information has been focused on establishing bivariate

[a] Corresponding author : kmmao4546@sina.com, fywang0213@sina.com

regression mathematical functions to predict plane residual stress fields.

The purpose of this paper is to obtain the distribution of residual stresses in a milling surface of 316 stainless steel shown in Figure.1 by establishing bivariate mathematical models. Based on the theoretical analysis of surface residual stress, the prediction method is first described. Furthermore, the computational procedure is shown. In addition, the predicted residual stresses of the obtained mathematical models are compared not only with the experimental data but also with the mean. Finally, the obtained residual stress field is analyzed. With the help, it would be easy to know the distribution of surface residual stress field and has a better understanding of the surface qualities of engineering components.

# 2 Methodology

## 2.1 Theoretical background

According to the elastic-plasticity theory, the stress components satisfy the equilibrium equations and boundary condition as follows:

$$\partial\sigma_x/\partial x+\partial\tau_{xy}/\partial y+\partial\tau_{xz}/\partial z+f_x=0\ ,\ \partial\sigma_y/\partial y+\partial\tau_{xy}/\partial x+\partial\tau_{yz}/\partial z+f_y=0 \text{ and } \partial\sigma_z/\partial z+\partial\tau_{yz}/\partial y+\partial\tau_{xz}/\partial x+f_z=0 \quad (1)$$

$$l\sigma_x+m\tau_{xy}+n\tau_{xz}=\overline{f_x}\ ,l\tau_{xy}+m\sigma_y+n\tau_{yz}=\overline{f_y} \text{ and } l\tau_{zx}+m\tau_{yz}+n\sigma_z=\overline{f_z} \quad (2)$$

where $f_x$, $f_y$ and $f_z$ represent the body force , $\bar{fx}$ , $\bar{fy}$ *and* $\bar{fz}$ denote the surface force, $\sigma_x$, $\sigma_y$ and $\sigma_z$ are the residual stress component in the *x*, *y* and *z* directions, respectivelly; *l*, *m* and *n* denote the three directions cosine of the normal to the boundary plane; $\tau_{xy}$, $\tau_{yz}$ and $\tau_{xz}$ are the shear stress on the *xy*, *yz* and *xz* planes, respectively.

Residual stresses are special stresses of a mechanical component when the surface forces are zero. In the plane parallel to the *x-y* coordinate plane, $l=m=0$ and $n=1$. Thus, it could be concluded that $\sigma_z=\tau_{xz}=\tau_{yz}=0$ from (2). Furthermore, (1) could be deduced to the follows:

$$\partial\sigma_x/\partial x+\partial\tau_{xy}/\partial y+f_x=0 \text{ and } \partial\sigma_y/\partial y+\partial\tau_{xy}/\partial x+f_y=0 \quad (3)$$

Ignoring the body forces, the solution of the differential equations (3) is $\sigma x=\partial^2\varphi/\partial^2 y$, $\sigma y=\partial^2\varphi/\partial^2 x$ and $\tau_{xy}=-\partial^2\varphi/\partial x\partial y$. No matter what the function $\varphi$ is, the three stress components could satisfy (3). The three stress components could be described as functions of *x* and *y* as follows:

$$\sigma_x=f_x(x,y)\ ,\ \sigma_y=f_y(x,y) \text{ and } \tau_{xy}=f_{xy}(x,y) \quad (4)$$

## 2.2 Description of the proposed method

According to (4), the three residual stress components could be described as the function of two planar natural coordinates. Thus, bivariate mathematical functions for predicting residual stress components could be established using measurements of surface points. To obtain the suitable mathematical functions, residual stress components of some points on the surface could be first measured and then be used to establish or check the bivariate mathematical functions.

# 3 Experimental procedures and results

## 3.1 Sample and experimental setup

The experimentation was performed on a rectangular specimen. The geometry of the specimen and the test instrument are shown in Figure 1. The dimensions of the specimen are: length=30mm, width=10mm, and height=5mm. The material of the sample is 316 stainless steel. Residual stresses were tested by using *XRD* technique. The test instrument used was the Proto XRD residual stress test instrument.

To obtain the stress components $\sigma_x$, $\sigma_y$ and $\tau_{xy}$, measurements were made with Mn_K-Alpha radiation in the length, width and rotation 45° direction in the gage region limited by a collimator diameter of 1mm. The $\sin^2 \psi$ method was used with seven beta angles at maximum beta angle interval from -25° to 25°. The hkl-311 diffracting planes were selected. The bragg angle was 152.8°.

Because it is rather difficult to test the residual stresses of the points on the boundary line accurately, eleven points on the upper surface were measured as shown in Figure 1. The origin of the coordinate system was located at the center of the surface. The x-axis was along the length direction while the y-axis was along the width direction. The 1st to 9th points were the quadrant points that divided the length and width directions into four equal parts. Meanwhile, the coordinates of the 10th and 11th points were (-11,25, 3.75) and (-3.75,1.25) , respectively.

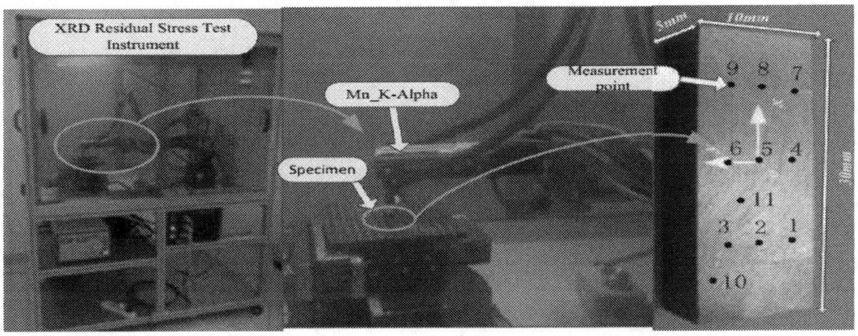


**Figure 1.** Experimental setup and test points.

### 3.2 Experimental results

The obtained residual stresses of the 11 test points are shown in Figure 2. The maximum stress and minimum stress of the longitudinal residual stress $\sigma_x$, transversal residual stress $\sigma_y$ and shear stress $\tau_{xy}$ are 252.50MPa and -77.59MPa, 796.27MPa and 513.04MPa, -49.50MPa and -268.59MPa, respectively. It can be seen that the fluctuation ranges are large. Meanwhile, the means of the three stress components are 61.77MPa, 623.42MPa, -169.11MPa, respectively.

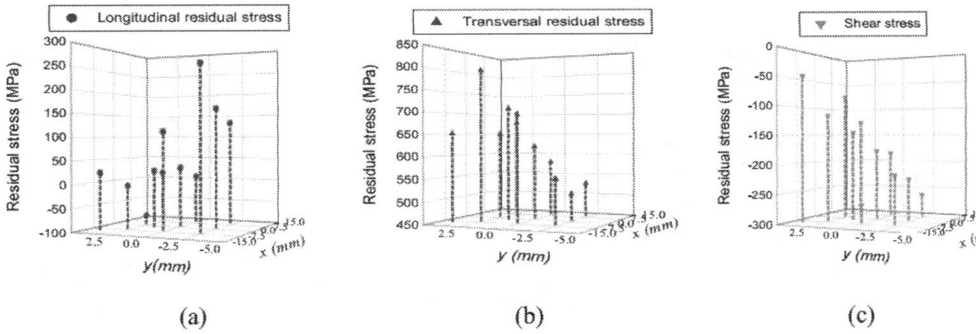


**Figure 2.** Experiments of (a) $\sigma x$, (b) $\sigma_y$, and (c) $\tau_{xy}$ of 11 test points.

## 4 Establishment of bivariate mathematical models of the sample

## 4.1 Choice of modelling points and checking points

To find suitable mathematical models for predicting the residual stress field, some points would be used to establish mathematical models, while another would be used to check the obtained models. These points are shown in Figure 3. The 1st to 9th points were the modelling points to establish the mathematical models. Meanwhile, the 10th and 11th points were the checking points to validate the models.

Similarly to residual stresses in the planes parallel to the *x-y* coordinate plane, some of residual stress components of points in planes parallel to the *x-y* or *y-z* coordinate plane could also be deduced. The stress components $\sigma_y = \tau_{xy} = \tau_{yz} = 0$ in the planes parallel to the *x-z* plane, while $\sigma_x = \tau_{xy} = \tau_{xz} = 0$ in the planes parallel to the *y-z* plane. Thus, in the *x-y* plane, the stress components $\sigma_y = \tau_{xy} = 0$ on the boundary lines parallel to the *x*-axis, while both of the $\sigma_x = \tau_{xy} = 0$ on the boundary lines parallel to the *y*-axis.

To avoid checking residual stresses of points on boundary lines and improve the prediction accuracy, some points on the boundary lines could be selected to establish the mathematical models. The selected boundary points of the three residual stress components are shown in Figure 3. Ten points on boundary lines parallel to the *y*-axis, ten points on boundary lines parallel to the *x*-axis and sixteen points on boundary lines parallel to the *x*-axis or *y*-axis were used to establish the model of $\sigma_x$, $\sigma_y$ and $\tau_{xy}$ respectively.

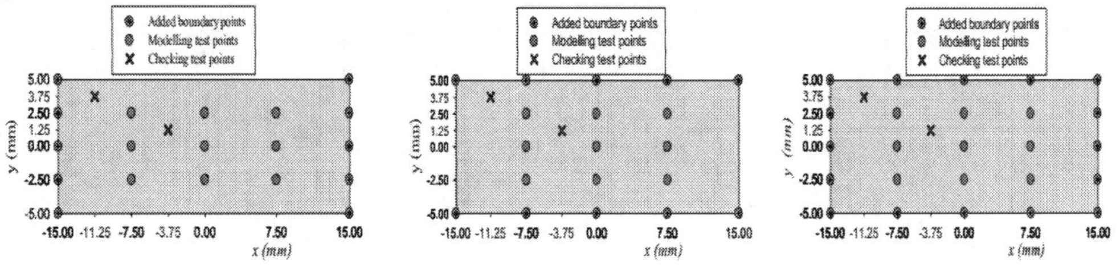


(a) (b) (c)

**Figure 3.** Choice of points to establish or validate models of (a) $\sigma_x$, (b) $\sigma_y$ and (c) $\tau_{xy}$.

## 4.2 Establishment of bivariate mathematical models

Based on the test points and boundary points, An automatic curve fitting and equation estimating software Table Curve 3D was used to find suitable bivariate mathematical models. By importing the data (x, y, $\sigma_x$), (x, y, $\sigma_y$) and (x, y, $\tau_{xy}$) of the mentioned modeling points into Tablecurve 3D v4.0.01 software, respectively, the obtained data graphs were shown in Figure 4. Each graph contains a 3D scatter plot or interpolated surface of the currently selected XYZ data.

Usually functions with bigger coefficients of determination $r^2$ have better agreement between predicted and measured capillary pressure curves. According to the sequence of coefficients of determination $r^2$ from large to small, it was found that surface-fit graphs of Chebyshev X, Y Bivariate Polynomials Order 4 shown in Figure 5 were similar with graphs in Figure 4. The coefficients of determination $r^2$ of the three stress components σx, σy and τxy were 0.98507968, 0.99426621 and 0.91892214, respectively. Thus, the following Chebyshev *X, Y* Bivariate Polynomials Order 4 was chosen as the mathematical model of the surface residual stress components:

$$f(x,y) = a + bT_1(x') + cT_1(y') + dT_2(x') + eT_1(x')T_1(y') + fT_2(y') + gT_3(x') + hT_2(x')T_1(y') + iT_1(x')T_2(y') + jT_3(y') + kT_4(x') + lT_3(x')T_1(y') + mT_2(x')T_2(y') + nT_1(x')T_3(y') + oT_4(y') \qquad (5)$$

where $x'=x/15$, $y'=y/5$, $T_n(x')=\cos(n\cdot\mathrm{acos}(x'))$, $T_n(y')=\cos(n\cdot\mathrm{acos}(y'))$, $f(x, y)$ denotes the residual stress component $\sigma_x$, $\sigma_y$ or $\tau_{xy}$, and *a-o* are the fitting coefficients. The values of the coefficients *a-o* in (5) are listed in Table 1.

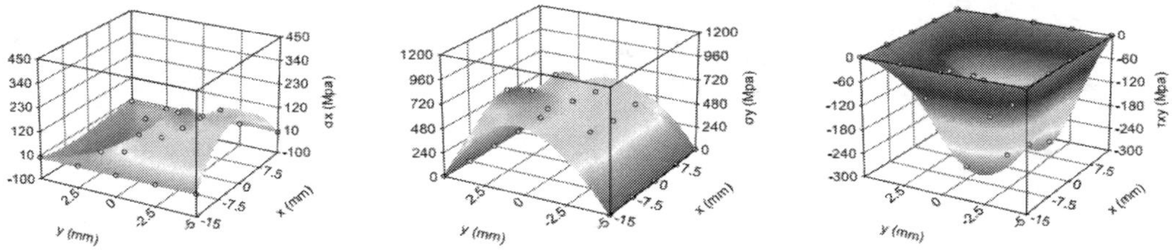


(a) (b) (c)

**Figure 4.** Data graphs of (a) $\sigma_x$, (b) $\sigma_y$ and (c) $\tau_{xy}$ obtained by importing the coordinate and stress components of mentioned modeling points into Tablecurve 3D v4.0.01 software.

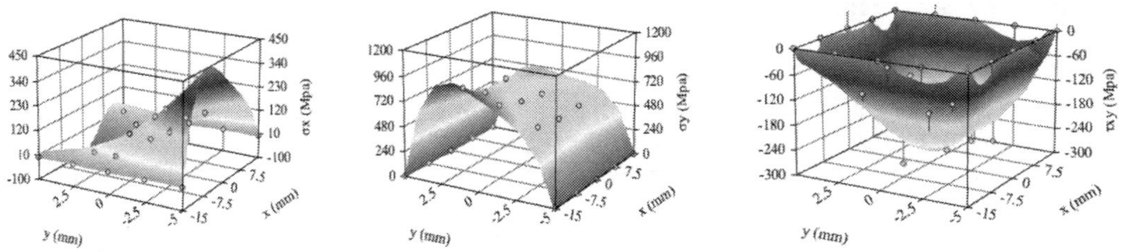


(a) (b) (c)

**Figure 5.** Surface-fit graphs of (a) $\sigma_x$, (b) $\sigma_y$ and (c) $\tau_{xy}$ displaying the importing data along with the surface for the fitted Chebyshev X, Y Bivariate Polynomials Order 4.

**Table 1.** Coefficients of residual stress model in (5).

| Coefficient | Value in $\sigma_x$ | Value in $\sigma_y$ | Value in $\tau_{xy}$ |
|---|---|---|---|
| a | 37.385844 | 178.49217 | 10.088706 |
| b | 10.719248 | 16.697397 | 0.2395292 |
| c | 43.101709 | 23.707313 | 1.3812165 |
| d | 18.446365 | 17.984915 | 6.1213734 |
| e | 16.6463 | 10.533959 | 0.1313418 |
| f | 14.241578 | 138.78497 | 7.9157676 |
| g | 10.719248 | 0.1822267 | 0.2395292 |
| h | 42.995814 | 0.4381588 | 0.8206999 |
| i | 0.6049351 | 16.879624 | 0.3774382 |
| j | 0.1058953 | 23.707313 | 1.3812165 |
| k | 18.548416 | 1.153549 | 3.6882823 |
| l | 16.6463 | 0 | 0.0938155 |
| m | 13.654984 | 17.292786 | 7.5745897 |
| n | 3.25E-19 | 10.533959 | 0.0938155 |
| o | 0.9776567 | 39.24578 | 0.6976252 |

# 5 Results and discussion

## 5.1 Comparison of experiments, mean and predicted stresses

Figure 6 displays the comparison of experiments of the mentioned 11 test points, mean of experiments and predicted stresses of the obtained mathematical models. The solid lines are composed of experiments, while the dashed lines are of the means and the predicted stresses. It can be noted that the means have a significant difference with experiments. The differences between them are bigger than 40MPa and the maximum difference is up to nearly 200MPa. However, the predicted stresses are close to experiments, and the differences between them are basically less than 20Mpa. Thus, the proposed method is superior to the average method.

DOI: 10.1051/matecconf/201710402021

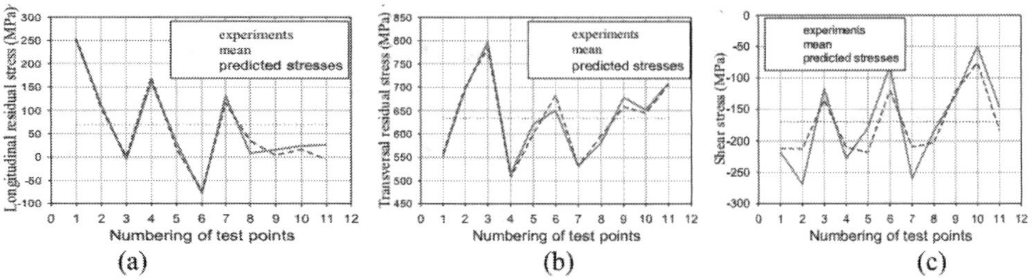


**Figure 6.** Comparison of experiments, mean of eperiments and predicted stresses of (a) $\sigma_x$, (b) $\sigma_y$ and (c) $\tau_{xy}$.

### 5.2 Analysis of predicted residual stress field

The predicted residual stress fields of the obtained mathematical models are shown in Figure 5. It can be seen that the stress fields are almost symmetrical along the width or length. The longitudinal residual stress $\sigma_x$ is symmetrical along the length, the transversal residual stress $\sigma_y$ is symmetrical along the width, and the shear stress $\tau_{xy}$ is symmetrical along the length and width. Meanwhile, the maximum stress or minimum stress exists in the symmetry axis or the boundary lines. The maximum stresses of $\sigma_x$, $\sigma_y$ and $\tau_{xy}$ are near 450MPa, 1200MPa and 0MPa, while the minimum stresses are near -100MPa, 0MPa and -300MPa. Fluctuation ranges are about 550MPa, 1200MPa and 300MPa. It indicates that it is of supreme importance to explore such a method to establish a mathematical model to predict the surface residual stress field.

## 6 Conclusions

To reconstruct the milling surface residual stress field, a prediction method by establishing bivariate mathematical functions based on measurements and boundary points is proposed. The results show that the prediction errors are basically less than 20Mpa and the distributions of the three stress components are almost symmetrical along the length or width. The proposed method is favorable to a significant underestimation of the residual stress field of milling surface. The Chebyshev X, Y Bivariate Polynomials Order 4 was the suitable mathematical model in the specimen. Whether there is a unified function that could display residual stress fields of different materials and structures is worthy to be studied.

## Acknowledgements

This work was supported by the High-End CNC Machine Tools and Basic Manufacturing Equipment Technology Major Projects (Grant no. 2013ZX04005-011).

## References

1. I. Samardzic, T. Vuherer, D. Maric, P. Konjatic, Metalurgija. 54, 527 (2015).
2. F. Gunnberg, M. Escursell, M. Jacobson, J. Mater. Process. Tech. 174, 82 (2006).
3. M. Liu, J.-i. Takagi, A. Tsukuda, J. Mater. Process. Tech. **150**, 234 (2004).
4. M. Birkholz, C. Genzel, T. Jung, J. Appl. Phys. 96, 7202 (2004)
5. G. S. Schajer, Exp. Mech. 50, 245 (2010).
6. M. Ya, F. L. Dai, H. M. Xie, J. Lu, Acta. Mech. Sinica. 19, 567 (2003).
7. Y. Y. Santana, J. G. La Barbera-Sosa, M. H. Staia, Surf. Coat. Technol. 201, 2092 (2006).
8. M. Ya, P. Marquette, F. Belahcene, J. Lu, Mater. Sci. Eng. A. 382, 257 (2004).
9. M. Achintha, D. Nowell, K. Shapiro, P. J. Withers, Surf. Coat. Technol. 216, 68 (2013).
10. A. M. Korsunsky , G. M. Regino, J . Appl. Mech-T. Asme. 74, 717 (2007).
11. A. T. DeWald , M. R. Hill, Exp. Mech. 46, 473 (2006).

# Biomechanical research of the femur finite element model combined with different material assignment methods

Wei Sheng [1], Aimin Ji [1, a] and Changsheng Chen [1, 2]

[1]*College of Mechanical & Electrical Engineering, Hohai University, 213022 Changzhou, Jiangsu, China*
[2]*Changzhou Orthmed Medical Instrument Co., Ltd., 213022 Changzhou, Jiangsu, China*

**Abstract.** The effects of different material assignment methods on femur biomechanical finite element analysis were explored, which provided guidance for material assignment in the study. According to different material assignment methods, the study was divided into the control group and experimental groups, the femur of the control group was assigned to 180 kinds of materials depend on the greyvalue, and the material assignment method of experimental group including the femur that assigned to 120, 60 and 10 kinds of materials depend on the greyvalue, the simplified femur model that assigned to two materials and the femur model that assigned to one material. Six groups different femoral finite element models was axial loaded and solved, and attain maximum equivalent stress of each group. The results showed that the experimental results of the material assignment schemes in the experimental group that including the femur that assigned to 120, 60 and 10 kinds of materials, and the simplified femur that assigned to two kinds of materials were consistent with the control group, while the experimental results of the femur that assigned to one material has obvious different with the control group. The analysis results show that when carries on femoral biomechanics finite element analysis, the femur evenly assigned to 10 kinds of materials depend on the greyvalue or simplified femoral model that assigned to two kinds of materials can be used, which can ensure the effectiveness and simplicity of the femoral material assignment.

## 1 Introduction

With the development of computer technology, finite element analysis of orthopedic biomechanics is becoming more and more popular. Determining the material properties is the prerequisite of finite element analysis of bones, and the prevailing approach is to assigned one uniform material. Zhang Junwen, etc. used uniform material assignment to study the effects that the size of the axial load on the biomechanical properties of the femur [1]. Wu Zehai, etc. also used uniform material assignment to research finite element analysis of the allogeneic bone plate combined with different screws in the treatment of femoral shaft fracture [2]. Because previous studies showed that there is a linear relationship between the CT image greyvalue and the bone density ρ [3], bone density and femoral unit material properties in empirical formula [4], so it can be regarded as a continuum, and the material assignment can be made according to the mutual relationship. Li Peng, etc. assigned 10 different material properties to the mandible depend on the CT value, efficient and accurate

[a] Corresponding author : wei_321_yue@163.com

established the human mandible three-dimensional finite element model [5]. Fei Wanghua took the fresh pig femoral stem as the research object, established finite element model of fresh femoral stem based on CT images, which assigned to 802 kinds of different materials [6].

This study mainly start form the material assignment method depend on the greyvalue, the femur were assigned to 180 kinds of different material properties in the control group, and the biomechanical properties of the femur could be reduced by the assignment method of the control group. Regarding the control group as a reference, to explore whether it is can well reflect the biomechanical properties of the femur, when the femur were assigned to 120, 60 and 10 kinds of materials. At the same time, the material assignment method depend on the greyvalue must be conducted material assignment for body grid. When researching the femoral fracture internal fixation system, grid is relatively complex, femoral assignment is easy to get wrong. Therefore, this paper proposes a simplified femur model that assigned to two kinds of materials.

## 2 The establishment of the geometric model

Femoral data were obtained from a Philips / Brilliance 64-slice spiral CT scanner and the data were stored in DICOM format. DICOM standard format has been developed into an international standard for medical imaging information technology, which not only retains the real information of human femur, but also can conveniently communicate and exchange with different medical image devices. The DICOM data is directly imported into the MIMICS software, to determine the orientation of each image and obtain four engineering views. The threshold of the femur is set in the range of 226-2433, which distinguishes bones from other tissues. After the image editing, regional growing and 3D calculation, the femur model was obtained.

The femur model was imported into the Geomagic Studio to form the point cloud data. After the point processing, the polygon processing and the shape processing, the ideal curved surface model was formed. The NURBS surface model was transformed into a CAD model to complete the reconstruction of the femur. The femur geometry model can be directly applied to the analysis that the material assignment of the femur based on the greyvalue and femoral uniform assignment directly, femur geometry model is shown in figure 1.

**Figure 1.** The femur geometry model

By researching the characteristics of femur tissue, bone marrow was found in the central part of the femoral shaft. A 10 mm diameter hollow cylinder was set up in the femur model to simulate the bone marrow similar to the elasticity of water. And as the femoral cortical bone part of the very thin, a cancellous bone can be simulated by setting a ring cylinder with a thickness of 10 mm on the outside of the hollow cylinder, the rest are regard as cortical bone, the simplified femur model can be assigned to two materials in the subsequent processing. In summary, six group experimental geometrical models were established, and six groups femur models were introduced as shown in Table 1.

**Table 1.** Introduction of the femur models of different group

| Group | Introduction |
|---|---|

| | |
|---|---|
| Group 1 (control) | The femur was evenly assigned to 180 materials by the greyvalue |
| Group 2 (experimental) | The femur was evenly assigned to 120 materials by the greyvalue |
| Group 3 (experimental) | The femur was evenly assigned to 60 materials by the greyvalue |
| Group 4 (experimental) | The femur was evenly assigned to 10 materials by the greyvalue |
| Group 5 (experimental) | The simplified femoral model was assigned to two materials |
| Group 6 (experimental) | The femur was uniformly assigned to one material |

# 3 The establishment and solving of the finite element model

## 3.1 Meshing and material assignment

When the finite element model of material assignment depend on greyvalue is established, the surface grid meshing of the three-dimensional model of the femur is completed by MIMICS. The parameters are set according to the default value, and the surface grid is saved as Ansys element format. Then, the surface grid file is imported into Ansys, and complete volume grid meshing based on surface grid. The nodes, elements and PREP7 files were written based on the volume grid respectively. In the FEA module of MIMICS, read nodes, elements and PREP7 files, select solid 92 as the element type, and use material function to assign the material value depend on greyvalue of the CT image, the concrete expression as shown in follow equations[7]:

$$Den = -13.4 + 1017 * Gra \tag{1}$$

$$E\text{-}Mod = 388.8 + 5925 * Den \tag{2}$$

$$Poi = 0.3 \tag{3}$$

Where, *Den*, *Gra*, *E-Mod* and *Poi* are the density, grey value, e-modulus and poisson's ratio of the femur.

When the finite element model was set up in group 5 and 6, femur and simplified femur model were respectively imported into the finite element analysis software Ansys to set the element type and material property directly, and to complete the division of the volume grid meshing. the element type and Material properties are set as shown in Table 2.

**Table 2.** The setting of element type and material properties [2,8]

| Group | Elastic Modulus [MPa] | | Poisson's ratio | Element type |
|---|---|---|---|---|
| Group5 (experimental) | Cortical bone | Cancellous bone | 0.3 | Solid 92 |
| | 15000 | 1100 | | |
| Group6 (experimental) | 12000 | | 0.3 | Solid 92 |

### 3.2 Loading with solving

A loading node was created above the femoral head, and a rigid region was established between the node and nodes included in the femoral head surface. Considering the axial force of the femur under normal standing and walking, the axial load of 600N and 1000N was applied to the loading node, and the displacement of the distal face end of the femur was restrained. The default solver is used to calculate the finite element model, and the corresponding equivalent stress contour map is obtained.

## 4 Results

According to the finite element analysis model established, the biomechanical analysis of 12 groups experiments was completed. The results of the experiment are shown in Figure 2. The maximum equivalent stress in the control group was 13.4431Mpa, when standing by one leg, the group 6 has a difference with the control group, but the other groups are close to the control group, the maximum equivalent stress contour map in the control group is shown in Figure 3. In the walking condition, the maximum equivalent stress of the control group was 21.845MPa, and the performance of the experimental group was consistent with standing by one leg, the maximum equivalent stress of the control group was shown in Figure 4.

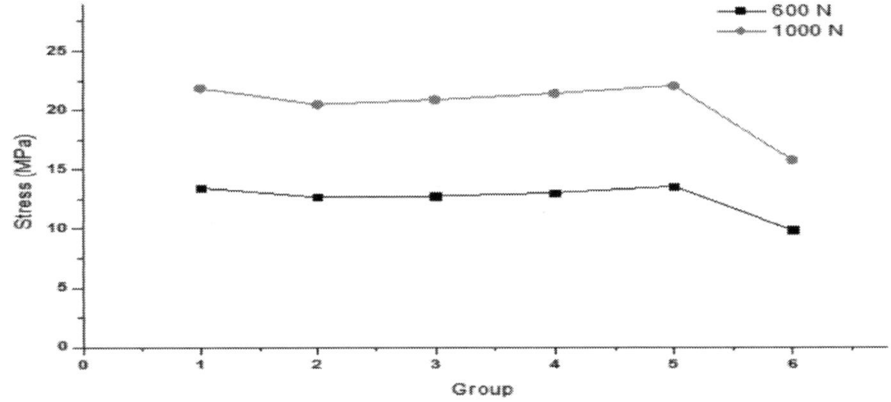


**Figure 2.** The maximum equivalent stress of each group femur under the different axial loading

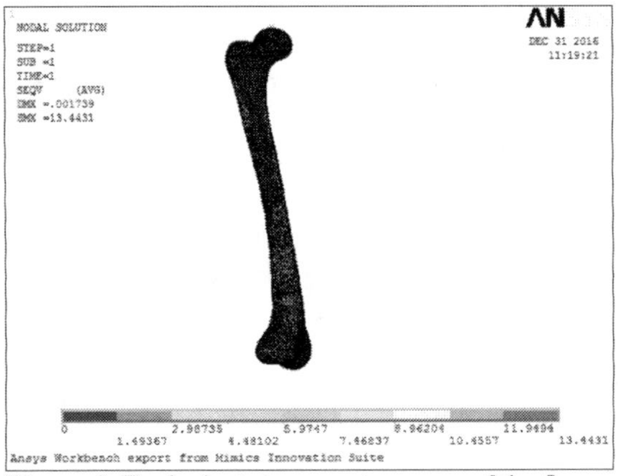


**Figure 3.** Equivalent stress contour map of the femur in the control group under the 600N axial loading

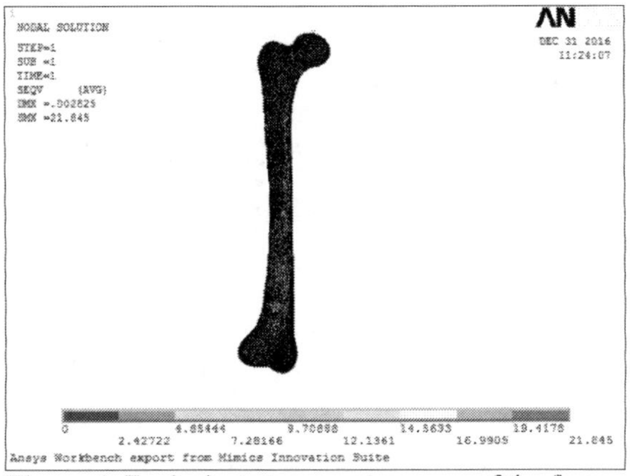


**Figure 4.** Equivalent stress contour map of the femur in the control group under the 1000N axial loading

## 5 Discussion

In this study, finite element method was used to evaluate femur biomechanics of different material assignment. Finite element method was applied to the study of orthopedic biomechanics since 1972, and has made great achievements in the orthopedic and orthopedic medical devices biomechanics research. At the same time, the finite element method has the advantages of cost-saving, time-saving, high-efficiency, high simulation and repeatability, which has an irreplaceable role in other biomechanical experimental methods, and be used in bone biomechanics research increasingly to provide guidance for clinical research[9].

When using finite element method to conduct the orthopedic biomechanical analysis, the bone material assignment is one of the most important step. The goal of femur material assignment is not only to guarantee the accuracy of model material assignment, but also to realize the simplicity and economic of assignment operation. This research is aimed at this goal, through the biomechanical finite element analysis for the femur that using different meterial assignment methods, to choose the most simplicity and economic meterial assignment method based on the validity of the assignment.

The analysis of the research results show that the biomechanical effects of the femur evenly assigned to 120, 60, or 10 kinds of materials depend on the greyvalue in the experimental group were consistent with the femur assigned to 180 kinds of materials in the control group, no matter how much the axial load is. So if you want to assign femur's material depend on the greyvalue, the femur could be assigned to 10 kinds of materials directly. When the simplified femur model was assigned to two kinds of materials in the experimental group, it was consistent with the biomechanical effect of the control group. But when the femur model was assigned to one unified material in the experimental group, it has a quite different biomechanical effect with the control group. When there is no greyvalue data or hoping to find a more convenient material assignment method, the simplified femoral model that assigned to two materials can be used to ensure the biomechanical properties of the femur.

In this study, under the premise of guaranteeing the accuracy of the research results, the finite element analysis model is simplified boldly, which can ensure the simplicity and generality of the research. The only major axial force in the external loading was considered, but the actual situation of fracture stress is often affected by muscle and ligament and other soft tissue effects [10]. Although there is little need to be improved in the whole finite element analysis process, but the analysis results has a quite consistent with the real situation. In the following study of this topic, we will gradually improve the shortcomings of the study, and strive to provide a more convenient and effective research methods.

## 6 Conclusions

Based on the finite element analysis of biomechanics, the following conclusions can be drawn from the research and discussion:

(1) The finite element analysis (FEA) method has been successfully applied in this study, the data obtained were matched to the relevant human femoral experiments, and the validity of the results of finite element analysis can also ensure the validity of the conclusion of the study.

(2) When conducting finite element analysis of the femur and needing to assign material to the femur depend on the greyvalue, to evenly assign 10 kinds of materials to the femur is a good choice, which can ensure the simplicity and the validity of the material assignment at the same time.

(3) When the researchers were unable to obtain the CT data of the femur or were unable to perform the operation of the MIMICS software, the study shows that the use of simplified femoral model and the assignment method can also be effective in the material assignment of femur and research of corresponding finite element analysis.

## Acknowledgements

This work is financially supported by the central university basic scientific research business expenses special funds (No. 2016B49614), the innovation training project of graduate student in jiangsu province (No. KYLX16_0708) and the special fund project of the transformation of scientific and technological achievements in jiangsu province (No. BA2012061).

## References

1. Zhang Junwen, Deng Shuangcheng, Zheng Haiyang, et al. The Establishment and Static Analysis of the Finite Element Model of the Human Femur[J]. Journal of Beijing Institute of Petro-chemical Technology, **20(3)**, 45-47(2012).
2. Wu Zehai, Wang Yongqing, Dong Limin, et al. Finite Element Analysis of Allogeneic Bone Plates Combined with Different Screw in the Treatment of Femoral Shaft Fractures[J]. Journal of Practical Orthopaedics, **20(10)**, 910-913(2014).
3. Rho JY, Hobatho MC, Ashman RB. Relations of Mechanical Properties to Density and CT Numbers in Human Bone[J]. Med Eng Phys, **17(5)**, 347-355(1995).
4. Wirtz DC, Schiffers N, Pandorf T, et al. Critical Evaluation of Known Bone Material Properties to Realize Anisotropic FE-simulation of the Proximal Femur[J]. J Biomech, **33(10)**, 1325-1330 (2000).
5. Li Peng, Liang Rui, Shen Longduo, et al. Establishment of Three-dimension Finite Element Model of Human Mandible with MIMICS and ANSYS Software[J]. J Pract Stomatol, **28(6)**, 709-713(2012).
6. Fei Wanghua. Finite Element Analysis and Experimental Verification of Fresh Bone CT Images[D]. Nanjing University of Science and Technology, (2008).
7. Zhang Guodong, Liao Weijing, Tao Shengxiang, et al. Methods for Material Assignment of Finite Element Analysis with Femurs[J]. Journal of Rehabilitative Tissue Engineering Research, **13(43)**, 8436-8441(2009).
8. Gao Yalei. Finite Element Analysis and Experimental Verification of Titanium Plate Fixation for Femoral Fracture System[D]. Tianjin University of Technology, (2014).
9. Zhang Yan, Yang Tieyi, Liu Yishu, et al. Stress Distribution of Proximal Screw Fixation of Single and Double Cortex Different Combinations in Less Invasive Stabilization System-distal Femur Treatment: A Finite Element Analysis[J]. Chinese Journal of Tissue Engineering Research, **16(4)**, 581-584(2012).
10. Zhou Xingheng, Wang Yongqing, Dong Limin, et al. Finite Element Analysis of Proximal Tibial Fractures Fixation with Interlocking Intramedullary Nail and Locking Compression Plate[J]. Journal of Practical Orthopaedics, **20(11)**, 995-998(2014).

MATEC Web of Conferences **104**, 02023 (2017) DOI: 10.1051/matecconf/201710402023
*IC4M & ICDES 2017*

# Watermarking for Fused Deposition Modeling by Seam Placement

Felix W. Baumann[1,a] and Dieter Roller[1]

[1] *Institut für Rechnergestützte Ingenieursysteme University of Stuttgart Universitätsstr. 38 70569 Stuttgart, Germany*

**Abstract.** With the increased usage and application of 3D-printing or Additive Manufacturing (AM) the question arises of how content providers or creators can ensure their intellectual property on such model data. Similar to other digital media such object information that is represented in a number of file formats is easy to copy and reproduce lossless. This research contributes by a proposition of a watermarking schema for Fused Deposition Modeling (FDM) type 3D-printers. This system embeds information into the 3D-printed object without alterations to the structure or geometry by altering the entry points of each layer in a specific manner. With such a watermarking schema employed objects can be embedded with additional information such as a serial number or other traceable information.

## 1 Introduction

For commercial scenarios in 3D-printing where the creator of a digital model wants to ensure that objects 3D-printed are traceable to the customer it is essential to embed such information in a way that can't be removed easily. In the currently existing 3D-Model marketplaces, such as TurboSquid [1] try to prevent the Intellectual Property (IP) misuse by contractual obligations, others like i.materialise [2] and Shapeways [3] do not allow the user to download the acquired model but only the direct 3Dprinting through the service itself.

In case of limiting the use of a digital model by contract the unlicensed use of the digital model requires some form of identification or information within the model to prosecute the misconduct.

For future use cases where marketplaces for digital models allow the user to download the model it is also required to employ some form of tracing information when the service provider wants to ensure the legitimate usage of the digital models.

With this work a proposal for the addition of such information to digital models that are 3Dprinted is provided by a watermarking schema. This watermarking schema is intended for FDM 3Dprinting and tested in an experiment to possess the required features for identification. The watermarking schema does not require additional materials or components and does not alter the geometrical and structural composition and features of the object.

This work is an addition to Baumann et al. "Streaming Provisioning for Additive Manufacturing – Securing IP", which is submitted to the Journal of Smart and Sustainable Manufacturing Systems and is currently under review. Fragments from this work are used for this article. This work is structured as follows: In the next section (2) the existing and related literature is examined and discussed for the

[a] Corresponding author : felix.baumann@informatik.uni-stuttgart.de

applicability to the problem discussed within this work. In Sect.3 the conceptual background to the watermarking schema is presented. In Sect. 4 the watermarking schema is explained with Sect. 5 presenting information on the detection and decoding process required. The Sect. 6 presents the results of an experimental implementation of the watermarking schema. This work concludes with a brief summary in Sect. 7.

## 2 Related Work

Research on the security of digital models in AM is currently sparse.

This research is required to avoid problems [4] encountered in the video and music industry by the improvements in compression technology and the widespread use of broad-band Internet connections [5–7]. One distinction between the video and music sector and the AM sector is that with the latter, physical objects are the end-product and thus only duplication in previous steps can occur, as digital replication from analogue objects is non-error free [8].

Existing research on watermarking on 3D meshes is performed for more than 15 years. In Benedens [9], the author presents the required and recommended properties of 3D watermarking technologies as well as an overview of contemporary algorithms. The use case for this work is the Virtual Reality Modeling Language (VRML) data of the 3D Web, which is still in development. The author identifies Ohbuchi, Masuda, and Aono [10] as the first publication on 3D watermarking.

Research on the perceptual quality of such mesh-bashed watermarking algorithms is presented in the work of Gelasca et al. [11]. The authors develop their watermark quality comparison on the surface roughness which they identify as indicative of the quality of the object and the watermarking algorithm. Chou and Tseng [12] provide a review on 3D watermarking techniques with their work. In their work they identify and present six robust watermarking algorithms and six fragile watermarking methods.

Identification of objects not relying on surface watermarking or the integration of additional objects or material into an additively manufactured object is presented in the work of Okada et al. [13]. The authors describe a method to extract information on internal structural properties like cavities in objects by the utilisation of far-infrared light. A similar method is introduced by Willis and Wilson [14] which is called InfraStructs and relies on the detection of mixed-material detection using Terahertz radiation.

The book by Yu et al. [15] has a separate chapter dedicated to 3D model watermarking. The authors provide detailed information on the requirements for such systems, attack scenarios, restrictions and applications. In this work eight different classification schemes are presented. Despite providing a thorough work on 3D watermarking, the authors do not explicitly research the application in 3D-printing but only mention 3D-printing as a means to generate objects.

## 3 Theory

In order to watermark a specific object 3D-printed with a FDM type 3D-printer without requiring additional or embedded material like a Radio-frequency identification (RFID) chip and in the moment of 3D-printing is possible to alter the geometrical structure of the object or alter the deposition strategy that can leave traces in the finished object. Alterations to the structure of an object are described in a number of publications on 3D mesh watermarking, see Yu et al. [15]. Structural alterations on the surface are prohibitive in cases that require the verbatim reproduction of the model data. As an alternative to these methods, internal structural changes for watermarking purposes are possible, as described in Willis and Wilson [14].

The approach proposed in this work is based on the alteration of the layer entry points of the model. In all 3D-printing methods that are based on slicing and that rely on the tracing of the structural elements of the objects, i.e., that are not manufacturing one layer in its totality, one point is selected as a starting point for this layer. The starting point can be located inside of the object or on its boundary. In either case, one point is the first point that is 3D-printed on the boundary of the layer of

the slice of the object. In case of an internal starting point, the strategy to 3D-print this layer can be altered without consequences to have a starting point at the boundary.
In case of FDM 3D-printing this starting point of a layer is travelled to by the printhead in a non-extruding move with retracted filament. By a miniscule increase of material extruded at this point, the starting point can be identified in the completed object. See Fig. 1 for a depiction of a visible stacking of starting points on an object 3D-printed with yellow acrylonitrile butadiene styrene (ABS) filament. In this figure, all the starting points for every layer are positioned at the same X- and Y-coordinates by the slicing software.

**Figure 1.** Object with Stacked and Visible Layer Start Points

As these starting points can be identified in the completed object and are free to place on the boundary of the object by the slicing software without alterations to the structure, geometry or topology, it is possible to encode information by controlled placement of these starting points and the consecutive altered layer deposition path.

## 4. Watermarking Schema

In the slicing software MiracleGrue [16] for Makerbot 3D-printers and Slic3r [17] the default setting for the starting or insertion point of a new layer is always at the same position for layers. This leads to a seam like structure that is visible on the 3D-printed object. By employing the following schema for the placement of the starting points it is possible to encode one of eight different symbols per layer.

In the proposed schema, the object's layers are viewed from above and are virtually centred around a centre point. The starting point is then located at any point of the layer's boundary that lies within the segment that reflects the value of the information that is to be encoded for this layer. See Fig. 2 for an illustration of the segmentation. This segmentation can be extended to employ more encoding capabilities by shrinking the areas of the separate segments. The segmentation can also be reduced by expanding the area of the separate segments down to two segments. In Fig. 2 the value 2 is encoded with this layer as the starting point, depicted by the green rectangle, is placed in the second segment. The segments are indicated on a circle with a dotted line and separated by dashed lines. The object's layer boundary or perimeter is indicated by the blue line.

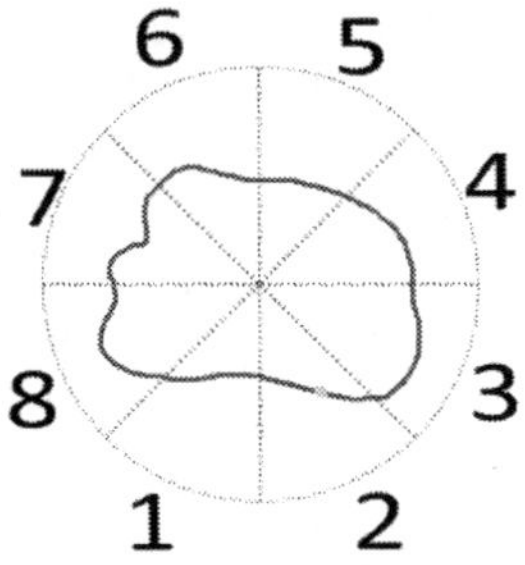


**Figure 2.** Watermarking Coding Schema with Starting Point Encoding Segment 2

The visibility of the entry point can be enhanced by either extruding a small amount of extra material at this location or by a miniscule displacement to generate a hole.

The path for the extrusion is to be adapted for the specific entry point and is under the responsibility of the utilised slicing software. Each slicing software employs a specific strategy for optimal path planning for each layer with respect to the layer below [18].

As a negative consequence of the application of this watermarking schema it is possible for the object's printing duration to be extended, as the selected starting point might be farthest from the last point of the layer below, thus increasing the travel time of the printhead.

To minimize the effects of stacked starting points in consecutive layers for the encoding of the same information the starting point must not be placed at the same X- and Y-coordinates. This is achieved by randomisation of the starting point within the respective segments with thresholding around the previous layer's starting point and margins at the segments borders. In Fig. 3 this pattern is depicted. In this figure, the value for segment 8 is encoded in two consecutive layers, with the red dots on the boundary representing the margin at the segment borders that are unavailable for the selection of the starting point. The yellow circle with the black border around the green starting point represents the thresholding boundary around the starting point which is unavailable for the selection of the starting point in the consecutive layer.

In order to identify the order of the segments on the 3D-printed object the first three layers are to be identified with markings in segment 1. For error identification purposes an adapted ISBN10 checksum [19] is utilized. This checksum (see Eq. 1) is added to the information that is to be embedded in the object at every tenth place. In Fig. 1 the indication pi is for the $i^{th}$ place of the information.

$$p_{10} = ((\sum_{i=1}^{9} i * p_i)\ mod\ 7) + 1 \quad (1)$$

The watermarked information is to be repeatedly adjusted with markings in the segment 2 for two consecutive layers after the checksum is placed every tenth layer.

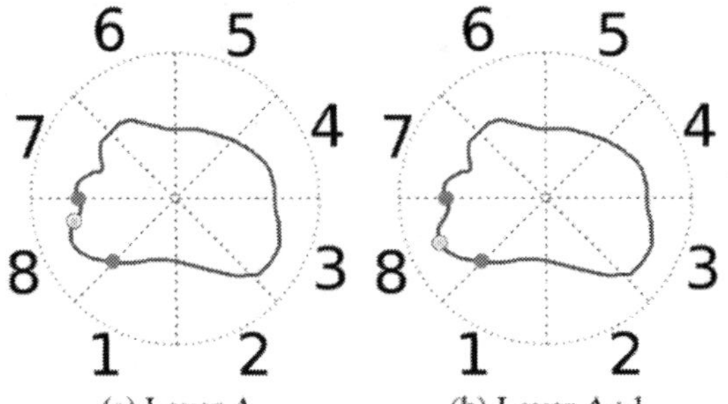


**Figure 3.** Encoding of the Segment 8 in Two Consecutive Layers

The information to be embedded as a watermark is to be repeated for all layers of the object when the first iteration of the watermark is embedded. Consecutive iterations must not be identified with the initial three layer marking in segment 1.

If the information is not a multiple of 9, then the remaining layers to fill 9 layers for the checksum are to be identified by repetition of a marker in the segment of the last layer that contained information with an additional marking in the following segment. The layers indicated by this dual marking are not to be considered for the checksum.

As an example the information 352 662 631 334 476 1 is to be encoded in the following way. The object to be manufactured is 8.7 mm in height and the layer size is set to 0.3 mm. The information is stored in the layers starting with layer 1 at height 0.0 as follows: *1 1 1* 3 5 2 6 6 2 6 3 1 **7** *2 2* 3 3 4 4 7 6 1+2 1+2 **4** 3 5 2 6. In this listing, **bold digits** indicate the checksum digits, underlined digits indicate the missing information places and *italicised digits* indicate the placement markers.

## 5 Detection and Decoding

For the detection of the watermark at least four digital images of the object must be available. These images must cover all sides of the object and must be of sufficient resolution to detect separate layers and the starting points. Common layer heights of FDM 3D-printed objects range between 0.1 mm and 0.4 mm. A camera with 12 Megapixel resolution yields an image of 4200x2800 pixel (px) size. For an object of 10 cm height photographed in its entirety this results in approximately = 42 px per mm. As the common layer height is as described above, this means that one layer is approximately represented by 4 pxs.

The decoding can be performed manually by visual inspection or software supported. From digital images the orientation of the virtual segmentation can be deduced by the initial marking of the first three layers in segment 1. On identification of the first segment, the other segments can be identified from digital images as they are placed in a counter clockwise orientation.

In digital photo manipulation software, e.g., GIMP [20] is then possible to highlight or otherwise mark these entry points which can then be transcribed back to the original values. The repeated reposition markers after the checksum every tenth layer are further indicators that ease the decoding of the contained information.

## 6 Implementation

A prototype is implemented as a proof of concept by script that manually generates a tower with side lengths of 20 mm at variable heights. Information of 8 bits can be stored per layer with a selected layer height of 0.3 mm. The system is tested on a MakerBot Replicator 2Xwith ABS 1.75 mm Yellow, ABS filament from REC, Moscow, Russia.

The software to generate the G-code (G-Code) is a GNU Bash [21] script, available from the authors upon request.

See Fig. 4 for the result of an object created containing information embedded on its surface structure. In this figure the starting points are highlighted by red circles and the layers are indicated by blue lines. The numbers next to the indicative circles for the starting points indicate the respective layer. The figure depicts the front side with segments 1 (left-hand side) and 2 (right-hand side) visible. The information contained in this example is xxx1x2x1x2x1xxxxx (with x indicating values not visible from the figure, i.e., on other sides). The image was acquired using a Canon LiDE 210 [22] scanner at a resolution of 1200 dots per inch (dpi). Due to brevity reasons only one side is depicted. Future implementations will embed the watermarking schema into the Slic3r software.

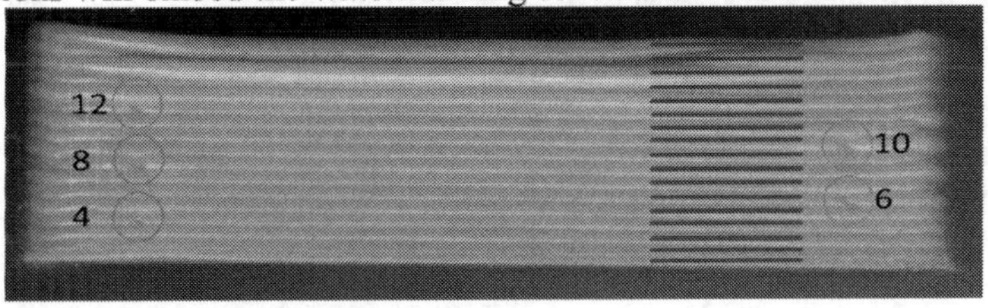


**Figure 4.** Experiment Result with Marked Starting Points

## 7 Summary

In this work, an approach to digitally ensure the traceability of a FDM manufactured object by embedding a watermark without the alteration of both the structure and geometry, is presented. The watermarking schema does not require additional materials or components and is readily available to any object that is built on an FDM type 3D-printer.

## References

1. TurboSquid, 3d models for professionals :: Turbosquid, http://www.turbosquid.com
2. i.materialise, Online 3d printing service | i.materialise, https://i.materialise.com
3. Shapeways, Shapeways - 3d printing service and marketplace, https://www.shapeways.com
4. G. McDonald, C. Roberts, Journal of Product & Brand Management , **3**, 55 (1994)
5. S. Bhattacharjee, R.D. Gopal, G.L. Sanders, Communications of the ACM, **46**, 107 (2003)
6. M. Fetscherin, M. Schmid, Comparing the Usage of Digital Rights Management Systems in the Music, Film, and Print Industry, in Proceedings of the 5th International Conference on Electronic Commerce (ACM, New York, NY, USA, 2003), ICEC '03, pp. 316–325, ISBN 158113-788-5, http://dx.doi.org/10.1145/948005.948047
7. R.D. Gopal, G.L. Sanders, S. Bhattacharjee, M. Agrawal, S.C. Wagner, Journal of Organizational Computing and Electronic Commerce, **14**, 89 (2004)
8. Q. Jiang, X. Feng, Y. Gong, L. Song, S. Ran, J. Cui, Computers and Geotechnics, **73**, 210 (2016)
9. O. Benedens, IEEE Computer Graphics and Applications, **19**, 46 (1999)
10. R. Ohbuchi, H. Masuda, M. Aono, Watermarking three-dimensional polygonal models, in MULTIMEDIA'97 Proceedings of the fifth ACM international conference on Multimedia (1997), pp. 261–272, ISBN 0-89791-991-2, http://dx.doi.org/10.1145/266180.266377
11. E.D. Gelasca, T. Ebrahimi, M. Corsini, M. Barni, Objective evaluation of the perceptual quality of 3D watermarking, in IEEE International Conference on Image Processing 2005 (2005), Vol. 1, pp. 1–4, ISSN 1522-4880, http://dx.doi.org/10.1109/ICIP.2005.1529732
12. C.M. Chou, D.C. Tseng, International Journal of Computer Science and Network Security, **7**, 328 (2007)
13. Okada, P. Silapasuphakornwong, M. Suzuki, H. Torii, Y. Takashima, K. Uehira, Nondestructively reading out information embedded inside real objects by using far-infrared light (2015), Vol. 9599, pp. 1–7, http://dx.doi.org/10.1117/12.2189486
14. K.D.D. Willis, A.D. Wilson, ACM Transactions on Graphics (TOG) - SIGGRAPH 2013 Conference, **32**, 1 (2013)
15. F. Yu, H. Luo, Z. Lu, P. Wang, 3D Model Watermarking (Springer Berlin Heidelberg, Berlin, Heidelberg, 2010), pp. 305–370, ISBN 978-3-642-12651-2, http://dx.doi.org/10.1007/978-3-642-12651-2_5
16. Makerbot Industries, Miraclegrue (beta), http://makerbot.github.io/Miracle-Grue (2012)
17. Ranellucci, H.B. Andersen, N. Dandrimont, M. Hindness, P. Ledvina, Y. Sapir, M. Sheldrake, G. Hodgson, Slic3r – g-code generator for 3d printers, http://slic3r.org (2011)
18. F. Baumann, H. Bugdayci, J. Grunert, F. Keller, D. Roller, Computer-Aided Design and Applications, **13**, 14 (2016)
19. ISBN-Agentur für die Bundesrepublik Deutschland, Tech.rep. (2012), http://www.german-isbn.org/downloads/ISBN-Handbuch.pdf
20. T.G. Team, Gimp – gnu image manipulation program, https://www.gimp.org (1996)
21. Free Software Foundation, Bash – gnu project – free software foundation, https://www.gnu.org/software/bash (2007)
22. Canon Europe, Canon canoscan lide 210 – canoscan flatbed photo & document scanners – canon europe, http://www.canon-europe.com/scanners/flatbed-scanners/canoscan_lide_210 (2012)

# The Design of a Bistable Composite Connection Method Applied on the Low Volume Production of a Breast Board Table Connection

Arnout Debyser[1], Bert Boute[1,2,a] and Jan Detand[1]

[1]*Ghent University, Faculty of Engineering and Achitecture Department of Industrial Systems Engineering and Product Design, Graaf Karel de Goedelaan 5, B-8500 Kortrijk, Belgium*
[2]*Ghent University, Department of Radiotherapy and Experimental Cancer Research, Faculty of Medicine and Health Sciences, St. Pietersnieuwstraat 33, 9000 Gent, Belgium*

**Abstract.** Bistable morphing composite laminates have been studied over the recent years because of their exquisite applications in energy harvesting and aerodynamic performance. A bistable laminate has two passive, stable positions that require an actuating bending moment for snap through. Different techniques have been developed to obtain bi-stability in composite laminates by using thermally or mechanically induced strain. Little study has been done around the more accessible method of laminate bi-stability through elastic buckling. This paper presents the design process of a buckling bistable composite connection method with human actuation.

## 1 Introduction

The use of morphing composite laminates in different applications have been studied over the recent years. Bistable laminates possess two passive stable positions, convertible by a snap-through actuator. Once settled in an equilibrium state, the bistable structure will remain there without demanding continuous power. Different techniques to obtain bi-stability in composite laminates were studied over the recent years. The following techniques are the most common:

- Asymmetrical stacking with thermally induced strain[1]
- Pre-curved asymmetric Laminates [2]
- Pre-stressed symmetric laminates [3]
- Variable Angle Tow -steering [2]
- Viscoelastic pre -stress [4]
- Hybrid composites through thermal mismatch [5]

Nowadays, the applications of bistable composite laminates are concentrated on piezoelectric energy harvesting [6] and aerodynamic optimization in for example wind turbine blades [7]. In these applications, surrounding vibration and wind are the actuators for the snap-through.

The current methods to create bi-stability are often complex and require specified machinery. Little study has been done on the more accessible technique of creating bi-stability in laminates through elastic buckling. Moreover, the applications of bistable laminates are limited to high-tech

[a] Corresponding author : bert.boute@ugent.be

products. In this paper, a concept will be presented where a bistable composite laminate is used as a load-carrying connection method.

The design will be applied on the table connection of a new type of breast board, developed and researched by Ghent university in collaboration with the university hospital[8]. A breast board is a prone patient support device for breast and regional lymph node radiotherapy. The device is mounted on the treatment couch and the upper body part is hanging over the table, resulting wide range of desirable beam paths (Figure 1).

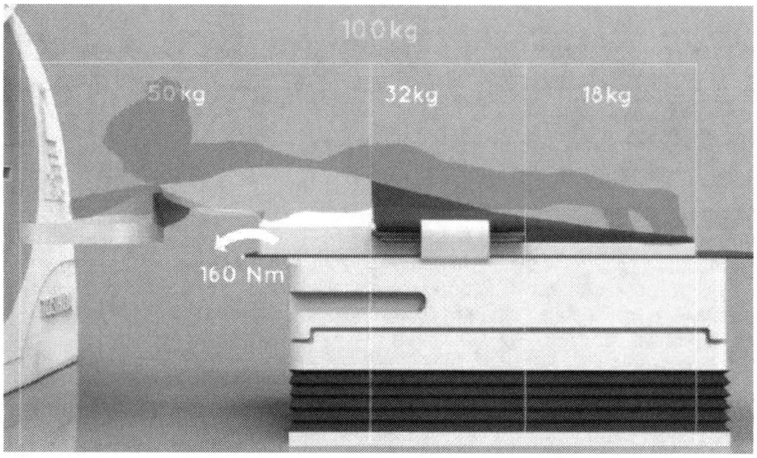


**Figure 1.** Prone crawl breast board with the applied forces. The clamps are subjected to a moment of 160 Nm, or 80Nm each.

The current table connection is a 3D-printed part that has to be applied manually around the table and has the following issues:

- Not safe - when the connection is not applied, the breast board can fall when the patient's weight is applied;
- Not strong enough - the current connection fails too easily because of applied moment;
- Not user friendly - There is no specific feedback when the connection is closed.

This paper presents the design process and person-interaction of a bistable buckling composite connection method. The concept will be applied on the table connection of a breast-board. The usage of fibre reinforced plastic can increase the load capacity of the connection, while the bistable behaviour improves the usability and safety.

## 2 Design process

### 2.1 Design methodology

Four main categories in the design process of a composite part will influence the characteristics of the bistable plate: shape, material, construction and production (Figure 2). Firstly, we will propose a connection design based on the buckling of a laminate. Secondly, we will determine the materialization, construction and production of this laminate. Every decision in each category, influences the other categories. Thirdly, we will perform user tests to study the person interaction. At last, the integration of the bistable connection in the breast board body will be discussed.

 DOI: 10.1051/matecconf/201710403001

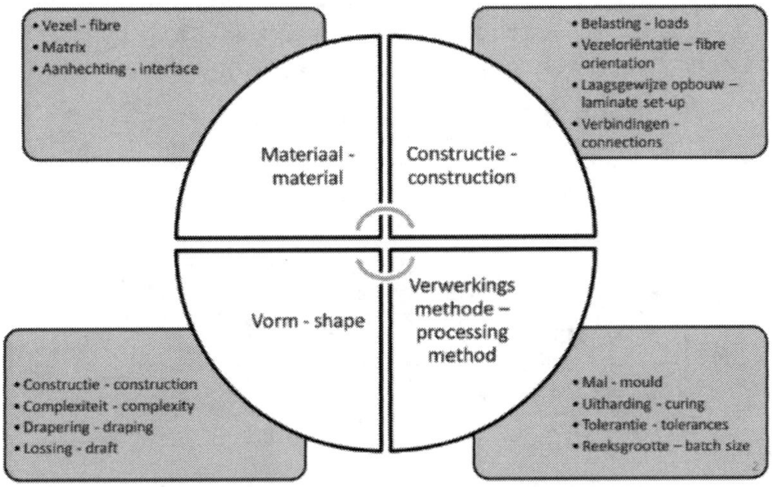


**Figure 2.** A composite design methodology[9].

## 2.2 Design principle

The concept of the proposed design is based on the simple principle of a buckling beam that is clamped between two points, because of pre-stress P (Figure 3). An actuation force F can snap-through the beam towards a second stable state.

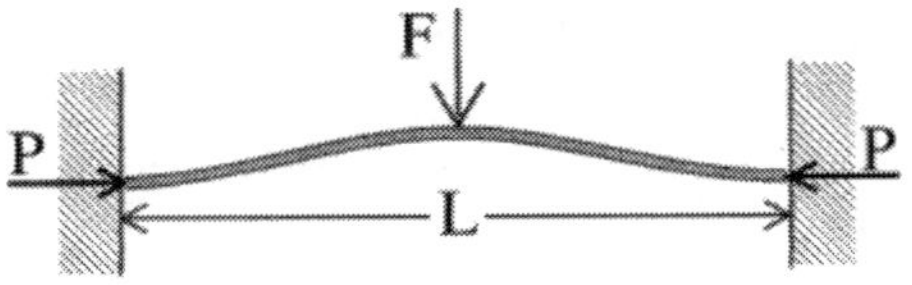


**Figure 3.** Simple buckling beam.

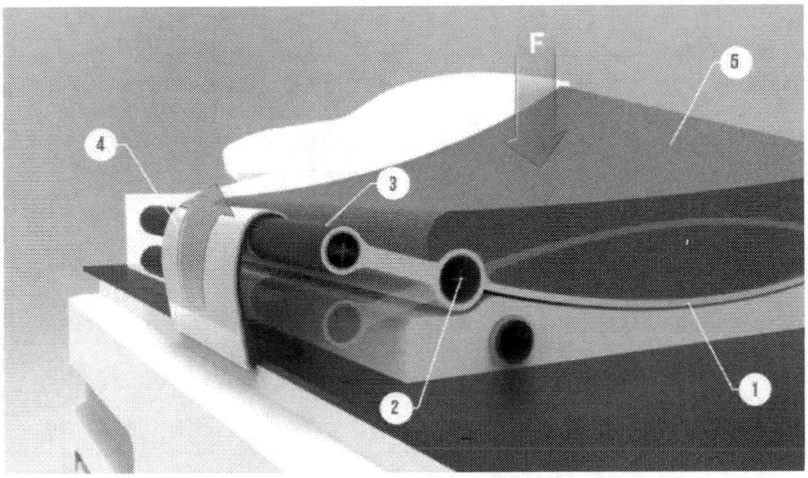


**Figure 4.** Proposed design for the breast board table connection.

Figure 4 gives a design solution that fits the requirements. The weight of the patient's waist (F) on the cushion (5) snaps through the bistable composite plate (1). This movement around the rotational axes (2) clamps the sides (3) against a profile (4) that is connected to the table. The connection can easily be opened by pressing on the handles (3).

Table 1 shows the parameters that can be designed in a bistable composite laminate. The second column indicates the critical parameters of the proposed design, based on the load analysis of the breast-board, material restrictions and system requirements.

**Table 1.** The parameters of the proposed design.

| **Design parameters** | **Critical value** | **Unit** |
|---|---|---|
| Curvature | <60 | mm |
| Snap-through moment | >16 | N.m |
| Buckling force | / | N |
| Internal stress | <35 | MPa |
| Sound intensity | Minimal | dB |
| Snap-through velocity | Maximal | mm/s |
| Service time | 7 | Years |
| Snap through vibration | Minimal | Hz |
| Bending characteristic | / | / |

## 2.3 Materialisation process

Carbon fibre is often used in current applications of bistable laminates, because of its appropriate thermal properties for asymmetric [0°/90°] bistable laminates[10]. The application of the breast board requires an insulating composite, which excludes carbon fibre. Fiberglass is the most appropriate for the application because of the following reasons:

- The radiolucent characteristic ensures minimum distortion.
- The high Young's modulus makes it possible to experiment with the flexibility.
- The integration possibility in the breast-board is better because the breast board is also made from fiberglass.

## 2.4 Construction

When designing a bistable composite laminate, the challenge is to balance the stiffness in order to produce a laminate that is load carrying, shape adaptable and light (Figure 5).

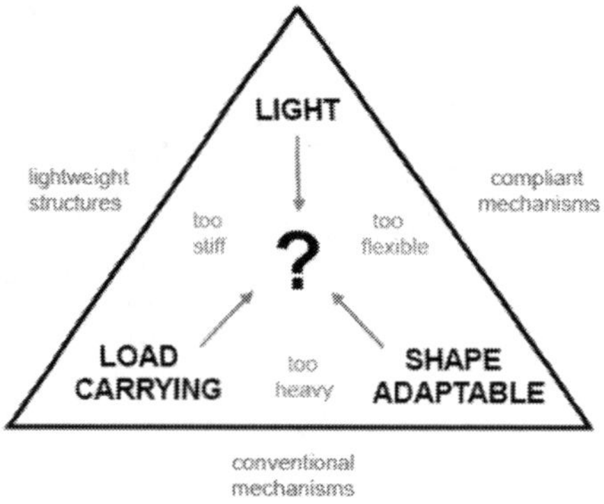


**Figure 5.** The design challenges of a morphing laminate [2].

In the philosophy of this design challenge, we need to consider the following issues for the construction of the laminate:

- The snap-through moment of the laminate must be able to take up the moment created by the weight of the breast board Mbb (Figure 6).
- The laminate has to be stiff enough (at the sides) to avoid that area A bends because of the moment Mbb. This would cause the sides to move downwards and the breast board to tilt over (Figure 6).
- The laminate needs to meet the critical values of the proposed design (Table 1).

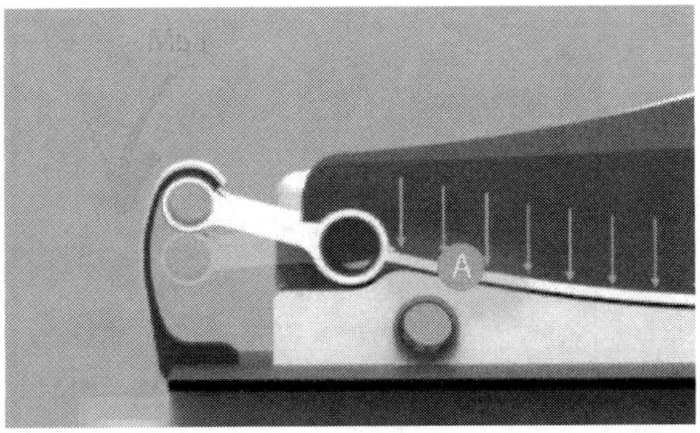


**Figure 6.** Forces on the bistable laminate.

### *2.4.1 Lay-up experiments*

When applying a core in the laminate, the flexible area is reduced and will locally bend with a bigger curvature (Figure 7). Different test samples were produced to analyse the influence of the curvature change on the bistable behaviour (Figure 8)

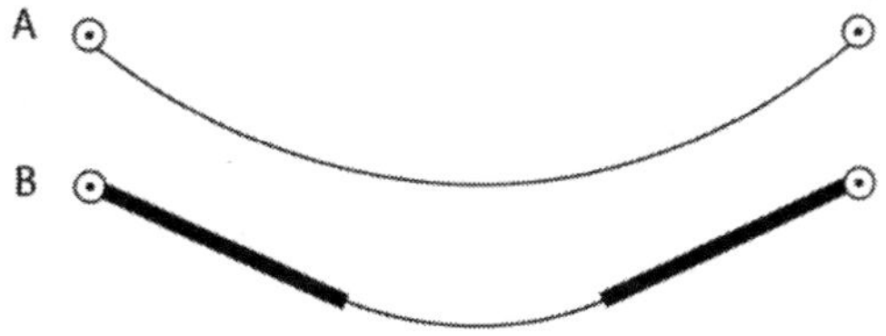


**Figure 7.** Layup without (A) and with (B) core.

**Figure 8.** Laminate constructions with core variations (0°/90°/PVC/0°/90°).

The test results showed that the snap-through moment of the laminate increased with a greater curvature because of the increasing internal stress. Secondly, every laminate with a core showed signs of shear failure at the disrupting transition from fibre to core, which causes local stress peaks.

These results point out that we need to construct the laminate as uniform as possible to avoid stress peaks and to enhance the lifetime of the laminate. To influence the snap-through load, we can variate the curvature or the stiffness of the laminate.

### *2.4.2 Lay-up proposal*

Figure 9 shows the proposed laminate [0°/45°/0°/90°/0°/-45°/0°]. The 0°-fibers will take up the moment of the breast-board and the weight of the patient, the 45°-fibers will take up the torsional forces, and the 90°-fibers ensures stiffness in every direction. Six extra 0°-layers are placed in area 3 to soften the discontinuities and to avoid this area to bend. The fiberglass rotation axes (2) are integrated in the layup and the PVC core at the sides (1) provides extra stiffness.

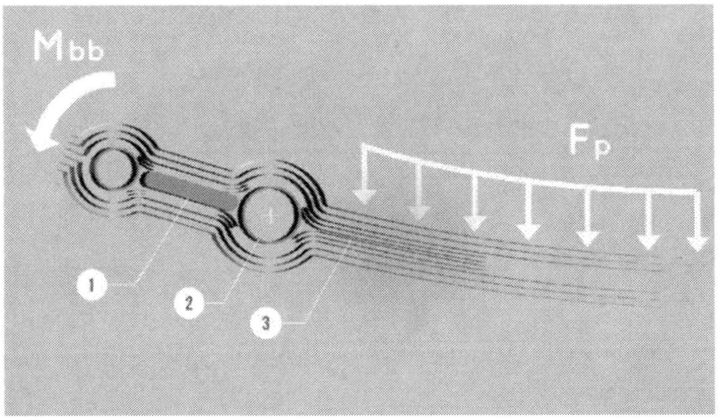


**Figure 9.** The layup proposal.

A test set-up was constructed (figure 10) to simulate the forces of the breast board on the bistable laminate with a curvature of 3cm. The distance of the weight of 30 kg was adjusted to gradually increase the moment M on the connection (figure 11). Two specific findings were made. Firstly, the tests results showed that the bistable laminate could easily resist a moment of 225N.m without snap-through. This is 65 N.m more than required.

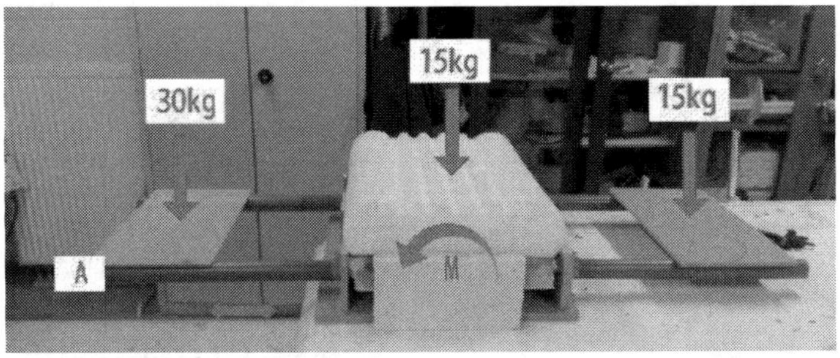


**Figure 10.** Test setup for load simulation on the bistable laminate with a 3cm curvature.

Secondly, the laminate showed deformation because of the bending moment M. This caused the breast board to tilt over, which is very undesirable because of calibration reasons and user comfort (Figure 11). The laminate layup needs to be constructed stiffer to minimalize this effect in future iterations.

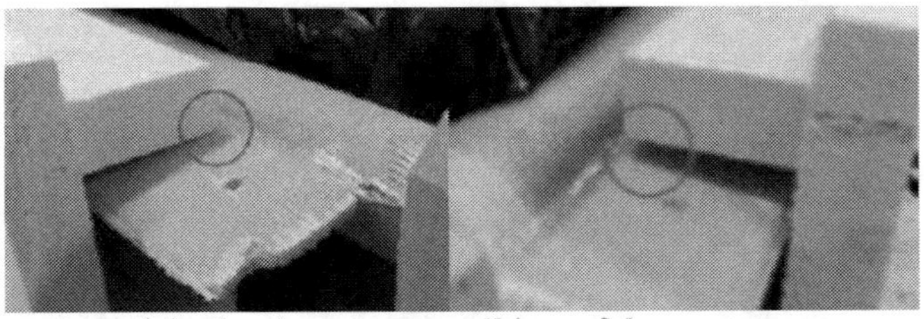

**Figure 11.** The deformation of the laminate causes tilting of the prototype.

It is important to predict the behaviour of the bistable laminate over time. Specific FEM-analyses[11] and test-setups are needed to predict creep and fatigue of the laminate to propose a certain life expectancy. The following tips can be considered to minimize load capacity reduction over time:

- Avoid discontinuities in the laminate
- Design the laminate with a high safety factor (up to 10)
- Using resins with a HTD that is 20° higher than the service time.

#### *2.4.3 Conclusion*

The construction tests of the laminate confirmed the design challenge of bistable composites: the stiffness and internal stress need to be minimized to promote the life-expectancy, but we need to create enough stiffness and internal stress to meet the required snap-through moment and to minimalize the edge flexibility that causes the breast board to tilt over.

### 2.5 Production

#### *2.5.1 Mould*

The mould for the prototype was made in wood to reduce the costs (Figure 12). The production mould could be produced in fiberglass or aluminium, dependent from the batch size.

Fiberglass moulds are perfectly suited for low volume processes that produce less than 1000 parts a year:

- Fiberglass moulds are about 10 times cheaper than aluminum, which makes it cheaper to make adjustments
- Fiberglas moulds can last for up to 1000 parts

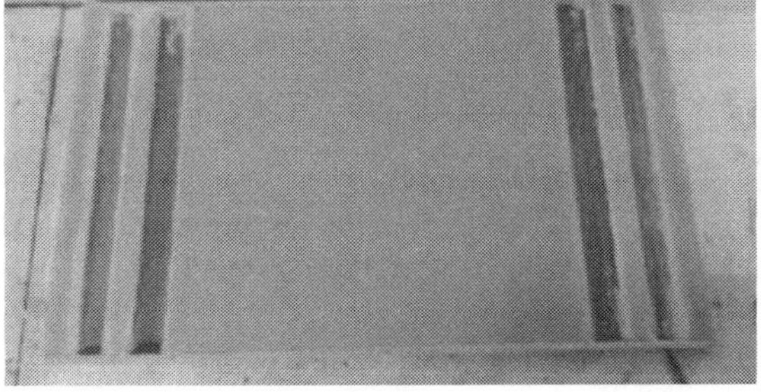

**Figure 12.** Mould for the prototype.

We need to be able to produce connections for different widths of radiotherapy tables. To achieve this, we can choose between 2 options (Figure 13). The first option is to make the clamp profile adjustable to the width of the table. The second option is adjusting the mould dimensions. A split mould has the advantage that we only need to change a small part of the mould.

 DOI: 10.1051/matecconf/201710403001

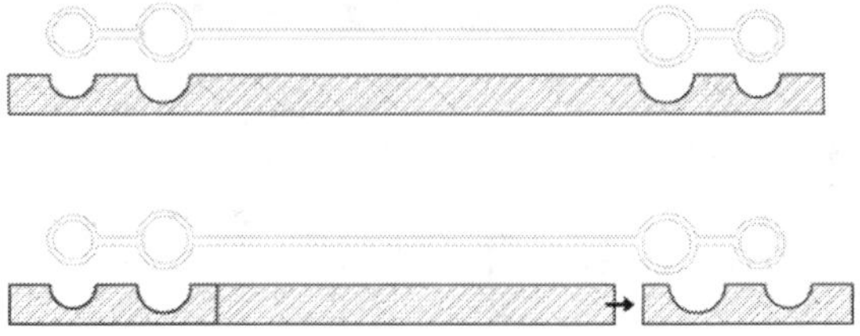

**Figure 13.** One-piece mould versus split mould.

### *2.5.2 Processing*

The production of the laminate allows us to variate with different parameters.

- Fiber/matrix-ratio
- Curing
- Laminate thickness
- Air bubbles
- Surface finish

The production of a bistable laminate for professional applications requires a production process that equally distributes the matrix in the fiber. A good laminate quality is essential for the bistable behaviour over time. The test samples are produced with a resin infusion process to ensure the laminate quality (Figure 14). For professional applications, Prepreg fibers are recommended to obtain an even better resin/matrix ratio.

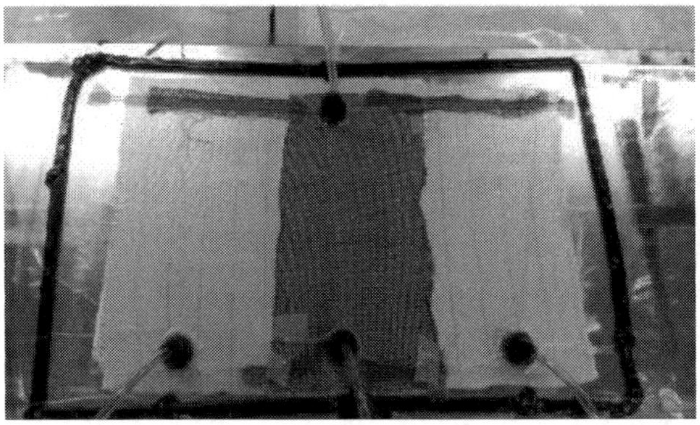

**Figure 14.** Infusion tests.

## 2.6 User tests

A prototype was built to study the person interaction with the bistable connection (Figure 15). The weight of the patient waist causes the laminate (4) to snap-through over the rotation axes (1) and bend upwards against the clamp-profile (3). The connection can be opened by pushing on the handles (2).

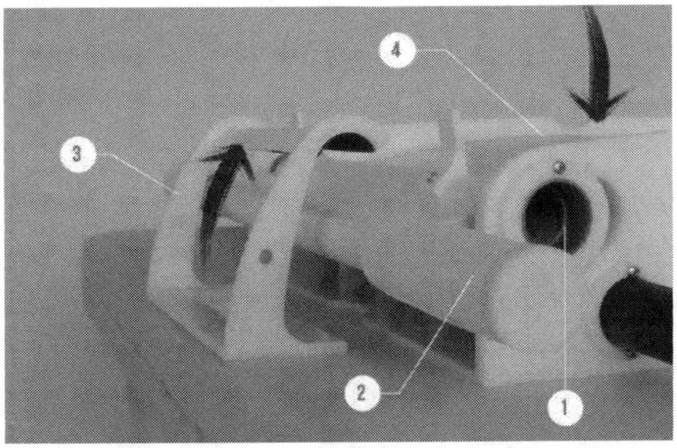


**Figure 15.** User test prototype.

### *2.6.1 Positioning*

The positioning of the prototype took an average of 5.8 seconds more than the current connection. This is caused by the fact that the sides need to be positioned under the clamp profile (Figure 15). This is an issue to take into account for future development.

### *2.6.2 Closure*

After positioning, the nurses will close the connection by pressing the cushion above the bistable plate (Figure 16). If the nurse forgets to close the connection, the system automatically closes in 0.3 seconds when a weight of 5 kg is applied above the connection. This safety factor is an important advantage in hospital environments.

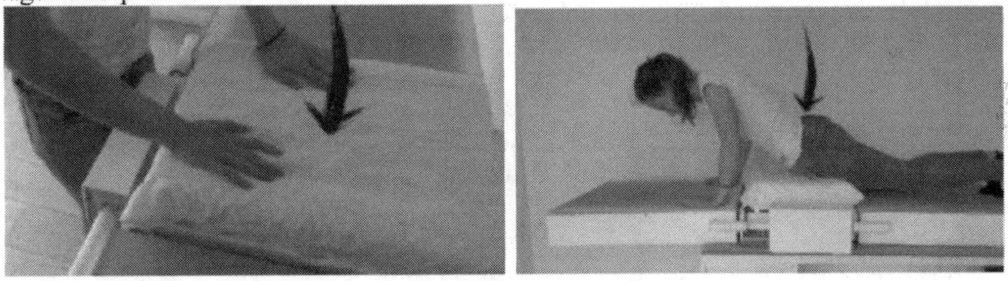

**Figure 16.** Closure by pushing (left) or patient's weight (right).

### *2.6.3 Open*

The opening of the connection (figure 17.) is more intuitive: 5 out of 5 test subjects succeeds to open the connection within 5 seconds by pushing on the handles with a force of approximately 57N.

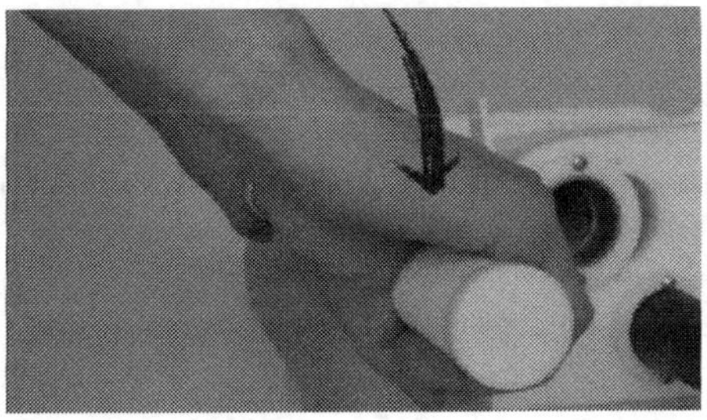

**Figure 17.** Opening the connection at the handles.

DOI: 10.1051/matecconf/201710403001

### 2.7 Integration in breast board

Figure 18 gives a CAD drawing of the overall beast board design integration.

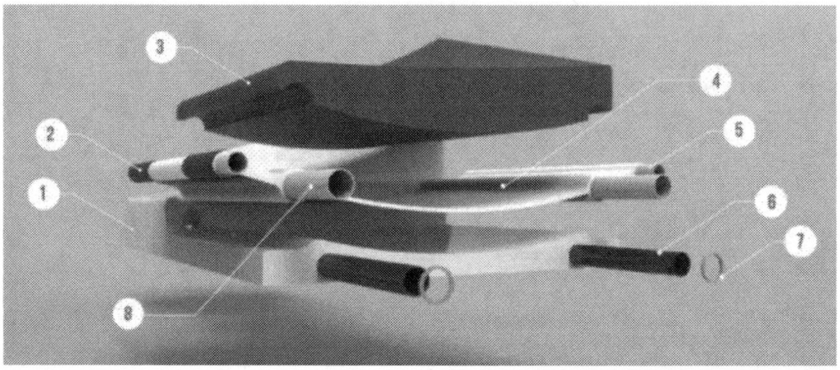


**Figure 18.** Assembly parts of the connection in the breast board.

The bistable composite shell (4) can be built in the breastboard body (1) with the rotation axes (8). Because the connection will have a considerable lower service time than the breast board itself, the integration has to be modular to be able to change the bistable laminate.

## 2 Conclusion and future vision

The bistable behaviour of the laminate shows potential towards feedback and safety in the application of the breast board. The usage of fiber reinforcement makes it easy to tailor the laminate towards the desired design parameters and takes up the load of the patient weight.

The prototype still shows some issues: the positioning of the sides under the clamp profile is too difficult and the undesired flexibility of the laminate when closed can cause the breast board to tilt over. Further study can minimize this effect by lay-up variations. These issues confirm the challenge of designing bistable laminates: they have to be flexible, but load carrying.

For the application of the breast board, extensive failure analysis and user tests are necessary to optimize the bistable laminate and to fundamentally prove that the connection meets the strict safety standards of hospital equipment.

Buckling is an accessible method of obtaining bi-stability in composite laminates, but this technique does not make use of the anisotropic properties of the laminate. Moreover, the buckling stress in the laminate causes creep and is difficult to predict.

Other techniques to obtain bistability in the composite laminate can be reconsidered for more predictability in high-tech applications. Bistable prestressed symmetric laminates [4] [12] are a good alternative because of the similar cylindrical bending characteristic. This technique shows different advantages over buckling:

- Usage of anisotropic properties to create bistability
- Better lifespan
- Better integration possibilities because no rotational axes are needed
- Minimal variability with changes in environmental conditions due to hygrothermal effects.
- Ability to prescribe laminate curvatures by control of the magnitude of applied prestress.

The technique of prestressed laminates can be considered in future work because of the numerous advantages over buckling laminates[13].

In general, bistable composite clamps can be very suitable in applications where a light, quick and heavy loaded connection is required.

## References

1. D. Betts, "Modelling and Optimisation of Bistable Composite Laminates," University of Bath, (2012).
2. P. Weaver, "Challenges of Modelling Multistable Composite Structures," (2012).
3. E. Kebadze, S. D. Guest, and S. Pellegrino, "Bistable prestressed shell structures," *Int. J. Solids Struct.*, **vol. 41**, no. 11–12, pp. 2801–2820, (2004).
4. B. Wang and K. S. Fancey, "SHAPE-CHANGING ( BISTABLE ) COMPOSITES BASED ON," in *10th International Conference on Composite Science and Technology*, (2015), pp. 1–9.
5. H. Li, F. Dai, P. M. Weaver, and S. Du, "Bistable hybrid symmetric laminates," *Compos. Struct.*, **vol. 116**, no. 1, pp. 782–792, (Sep. 2014).
6. D. V Murray and O. J. Myers, "Modeling Bistable Composite Laminates for Piezoelectric Morphing Structures," *ISRN Mater. Sci.*, **vol. 2013**, pp. 1–12, (2013).
7. X. Lachenal, S. Daynes, and P. M. Weaver, "Review of morphing concepts and materials for wind turbine blade applications," *Wind Energy*, **vol. 16**, no. 2. pp. 283–307, (2013).
8. B. Boute, "Optimalisatie van de buikligpositie tijdens borst radiotherapie. Bert Boute," Ghent University, (2014).
9. J. Detand, "Ontwerpen met composieten," *Ghent Univ. syllabus*, 2015.
10. M. Gude and W. Hufenbach, "Design of novel morphing structures based on bistable composites with piezoceramic actuators," *Mech. Compos. Mater.*, vol. 42, no. 4, pp. 339–346, Jul. 2006.
11. D. Fuhong and L. Hao, "Finite element analysis of multi stable structures," *Cent. Compos. Mater. Struct.*, vol. 1, no. 3, pp. 207–220, Jan. 2012.
12. S. Daynes, C. G. Diaconu, K. D. Potter, and P. M. Weaver, "Bistable Prestressed Symmetric Laminates," *J. Compos. Mater.*, vol. 44, 2010.
13. S. Daynes, K. D. Potter, and P. M. Weaver, "Bistable prestressed buckled laminates," *Compos. Sci. Technol.*, vol. 68, no. 15–16, pp. 3431–3437, Dec. 2008.

# An Aesthetic Design Method of Landscape Visualization Restoration for a Rural Highway Around the Nanwan Lake: A Case Study in the Shihe District, Xinyang City, Henan Province

Bojing Liao [1,a] and Feng Wang [1]

[1]*School of Architecture and Urban Planning, Huazhong University of Science and Technology, Hongshan district, Luoyu road, No. 1047, Wuhan, China*

**Abstract:** In this paper, we based our field study to investigate an urban-rural highway around Nanwan Lake, China, and the statistical research results are presented with an ANOVA analysis. For this case study, we collected 80 samples, which we categorized into three groups. Villages and features classify these samples. An ANOVA analysis showed that there is a significant difference between the facility samples. The data analysis of our results showed that the problems of the urban-rural highway landscape are constituted by multiple factors in this area; therefore, we built two visual models to illustrate the problem, and we eventually constructed a model of a methodology to attempt to solve these problems.

## 1 Introduction

The economic growth rapidly of China promoted the integration of urban-rural development since 1990, and constructing the urban-rural highway is necessary as an essential part in this processing. Urban-rural highways across the city and region`s ecological system as the bridge connecting cities and villages, and meet the peripheral natural environment to construct a metabolism cycle [1]. In these areas, which often include valuable protected biotopes, forested hills, preserved woodlands, prime agricultural lands and important wetlands, often provide essential ecosystem services for urban residents [2]. Moreover, land-use patterns, human activities, and their combination strongly influence urban-rural road patterns, that is, the placement of highways in a landscape [3]. Therefore, the rural-highway landscape represent highly complex territorial spaces from economic, environmental and social viewpoints [4].

The quality of the visual landscape is important not only to individual citizens and their health and well-being [5] but also to the rural highway around the lake in the peri-urban. However, the large-scale rural highway development that is a part of urban-rural integration in China often leads to changes in the visual landscape, especially in the lake perimeter of Nanwan Lake. There have been many studies on the environment, geology, atmosphere, noise, community and ecology of this area,

[a] Corresponding author: 471584495@qq.com

but there have been no aesthetic and visual studies in connection with the plans for the urban-rural highway around the lake [6]. Furthermore, extensive excavation has brought about severe damage to the natural landscape, such as cutting hillsides, excavating hill-hillocks, filling up ravines, cutting down trees, draining wetlands, segmenting grasslands and so on, which has caused a lack of integration and coordination between the highway and the peripheral environment, and it has also lead to the separation of the urban-rural highway with the natural landscapes along the route [6]. These issues separated relevance between topography and natural landscape space and visual landscape space, and it also caused various scars in the natural.

There have three main reasons for these questions: (1) urban-rural highway in China are still at the stage of supplying and demanding, people are not so urgent to the beautiful environment and high visual quality compared with the requirements of commutes for the highway [7]; （2） single goals make urban-rural highway to become a standardized product because it lack of considerations to coordinate the highway and natural environment; (3) Landscape architecture was no opportunity to study the perspective of relationship between the highways and natural as well as ecological at the earlier stage of urban-rural highway [6].

In China, however, the development of urban-rural highway landscapes has lagged significantly in the area of highway construction, and it has not received the full attention of society during the rapid construction of cities [7]. Moreover, the study of urban-rural highway landscapes has lacked theoretical investigation in its main research areas and on the research level. Applying urban-rural highway landscape theories from the developed counties has also been restricted in the area of landscape evaluation, and there has been a lack of experience with such theories [1]. The discrepancy between theoretical study and the actual construction of highways formed because these scientific theories applied to highway planning design, and, for several years, construction was rare. This caused great damage to the local landscape and lead to thousands of highways that were constructed in the same way, repeating the same landscape [8].

This paper focuses on changes in the urban-rural highway landscape around Nanwan Lake and the effects of urbanization on urban-rural highways in the area. The research takes the urban-rural highway around Nanwan Lake as an example to provide a method for the restoration of the urban-rural highway landscape visualization system that would result in the construction of highways combined with landscape environment aesthetics and ecological and mass behavior-psychology [9]. In summary, the aims of this paper are:

- To summarize a landscape visual theoretical framework to repair urban-rural highway;
- Devising a method to repair the urban-rural highway landscape based on the theoretical framework to investigate, analysis and characterize the area of Nanwan Lake of China;
- To offer guidelines for future to repair urban-rural highway landscape to some extent.

## 2 Research methods

In this study, we used quantitative holistic technique [10,11] to assess the current situation of urban-rural highway landscape. Through practical survey to analysis the scene, from the view of visual aesthetic, we found that the problems focus on three points, namely the labeling system chaos, slope of desertification as well as the living garbage pollution. Based on the theoretical framework of LES system, we has carried on the on-site investigation from aspects of the Landscape Environment Aesthetics, Ecological Restoration and Mass Behavior-psychology, and tries to analyze the following questions:

(1) In the urban-rural highway landscape, what is the relationship between visual aesthetics, ecological restoration and social activities?

(2) What are the social variables that affect the landscape of urban-rural highway?

Therefore, the study methods followed in this paper can be divided into five distinct parts. First, using geographic information systems [12] to analyze the location, altitude and slope of area of study. Second, we took photos [12] that were intended to cover the situation of slope and facilities within each unit. Third, we categorize these photos and statistic to analyze the characters of each category.

Four, we will create a methodology model according to the analysis results. Finally, we would discuss the above two questions in the model.

### 2.1 Samples

Due to the relatively complete industrial chain in the study area and the above theoretical framework, the collection of samples was selected mainly from the three aspects of on-site investigation processing: (1) the waste recycling facility (landscape environment aesthetics): this part was divided into trash cans, the dump and garbage trucks to segment the statistics; (2) slope desertification (ecological restoration): the urban-rural highway of the lake-ring needs repair because the slopes are concentrated mainly in two sections, which, respectively, are Shui-XiangQiao village to Dong-JiaHe town and Shi-HeGang town to Ma-JiaFan village; and (3) signage system (mass behavior-psychology): the main categories of signs in this area are relatively self-contained, and they include four categories: a) brand of place-name, b) signboard of tea-factory and tea-base, c) signboard of rural pleasures, and d) publicity brand.

### 2.2 Survey methods

To study the current situation of the urban-rural highway landscape in this area, we used the on-board camera to collect sample pictures for four towns (Dongjia-He town, Shihe-Gang town, Tanjia-He town and Shisanli-Qiao town) and seven villages (Shuixian-Qiao village, Huangwan village, Linggang village, Tumen village, Longtan village, Haojia-Chong village, Taoyuan village) along the route. Altogether we collected 194 photos in seven days, of which 111 pictures of the first type sample (53 pictures effectively, the effective rate was 47%), 32 pictures of the second type of sample (23 pictures effectively, the effective rate was 71%), 51 pictures of the third type of sample (51 pictures effectively, the effective rate was 66%).

### 2.3 Statistical analysis

Statistical analysis and comparisons were done by both parametric and non-parametric methods [11].First of all, we consolidated the effectively photos, and a resulting total of 80 samples were surveyed, of which 19 in the first kind of sample, 16 in the second kind of sample, 32 in the third kind of sample.

After that, we integrated the first, second and third class samples to analyze the specific reasons in these samples from the level of sociological research. We categorize all of the samples via statistics, and using ANOVA to analyze whether there have significant difference between the sample groups.

## 3 Results

### 3.1 The statistical results of samples

We conducted a statistical analysis for the distribution of the facilities of each village, the results are as follows: 27.2% of the village facilities distribution is complete, 27.2% of the village facilities distribution is more, and 45.6% of the village facilities is less.

Meanwhile, according to the nature of the facilities, the classification statistics were divided into color tendencies, materials and recognition degree, and the classification statistics of the slope were also divided into unilateral slope, bilateral slope and slope causes. The results of these sample feature classification statistics are as follows:

The first type of samples: color tendency, 52.6 % green, 21.2 % gray, 15.7 % blue and 10.5 % orange; material, 52.6 % iron, 21 % plastic and 26.4% stone; and recognition degree, 78.9 % high and

21.1 % middle.

The second type of samples: desertification tendency, 62.5% bilateral desertification and 37.5 % unilateral desertification; and causation, 62.5 % to the quarry, 25 % to the quarry and 12.5 % to the sandpit.

The third type of samples: color tendency, 34.3 % green and red, 28.2 % blue and 3.2 % orange; material, 50 % plastic, 46.8 % iron and 3.2 % stone; and recognition degree, 75 % high, 18.7 % middle and 6.3% low.

### 3.2 Similarity and difference between results of samples

Through the on-site investigation, the comparison of the samples and the analysis of the data we classified to summarize the similarity and difference between the samples:

(1) The first type of sample (the waste recycling facility): (a) The solid waste could receive better treatment in this area, and garbage on both sides of the urban-rural highway could use a good cleaning; most families state that trash is a problem and that there is nobody to dump it on the side of the road or the river. (b) The colour and shape of the trashcans as a mobile facility are unified along the route, but their appearance was too simple to differentiate distinctive features.

(2) The second type of sample (slope desertification): the slope needs repair in two main sections, namely Shuixiang-Qiao village to Dongjia-He town and Shihe-Gang town to Majia-Fan village. To build the urban-rural highway around the lake, they excavate and remove soil, which renders the slopes to approximately 35 degrees to 40 degrees high, from 15 m to 105 m, with the maximum slope at approximately 160 m.

(3) The third type of sample (signage system): (a) The place nameplates are basically complete, and the main villages and resorts are equipped with basic nameplates; however, several villages do not have them. (b) The signboards of the tea-factory and the tea-base have their own characteristics, due to a lack of unified planning.

By summarizing the similarities and differences among these three groups of samples, we found that the villages` daily activities and behaviors are the main cause of these results as well as their own interests and qualities. We gathered the first group samples and the third group samples for a one-way ANOVA analysis because they belong to the facilities. From our calculations, the results of the ANOVA (Figure 1) showed that there was a significant difference (P-value=0.036198<0.05 and F>F crit) between two group of samples. Meanwhile, the color, materials and recognition degree have significant differences and are the summary statistical comparisons of interblock.

## 4 Implication for practice and advancement of research

This research was conducted to study restoration problems for urban-rural highway landscapes from the perspective of visual aesthetics. First, we interpreted three core elements of the trilism of landscape, which was provided by Prof. Liu [9], and we summarized a conceptual framework to guide our field investigation in the study area. Second, we spent 15 days taking photographs of the three types of samples that we collected, and we counted and analyzed the samples. Subsequently, we classified our summarized results, and we built two visual models to explain our findings. Finally, we established a methodological model according to the investigation results to illustrate the key aspects that are needed to integrate conformance and to facilitate ecological repair. The results of the study show that the restoration of urban-rural highway landscapes requires the repair of seven key points, so that these factors are both in conformance and ecologically sound, namely: color consistency, material consistency, recognition degree of distinction, the embodiment of cultural characteristics, facility appearance consistency, plant shape design and slope greening recovery (Figure 2).

For urban-rural highway landscapes, the aesthetics and artistic charm is reflected mainly in the structures (facilities) along the route. Its appearance (including color, material, shape, and its cultural characteristics) are directly related to the quality of its visual aesthetics. Urban-rural highway landscapes consistently coordinate their style, but different sections of a structure can be changed under the premise of maintaining the landscape style. This enables better integration of the environmental features of the various sections, enriching the scenic views of the highway, and reflecting the harmony between highway landscape and environment. This is our purpose in building this research method.

| | Green | Blue | Orange | Gray | Red | Iron | Plastic | Stone | High | Middle | Low |
|---|---|---|---|---|---|---|---|---|---|---|---|
| Total 1 | 10 | 3 | 2 | 4 | 0 | 10 | 4 | 5 | 15 | 4 | 0 |
| Total 2 | 11 | 9 | 1 | 0 | 11 | 15 | 16 | 1 | 24 | 6 | 2 |

**Analysis of variance: single factor analysis of variance**

SUMMARY

| Groups | Quantity | Summation | Average | Variance |
|---|---|---|---|---|
| Green | 2 | 21 | 10.5 | 0.5 |
| Blue | 2 | 12 | 6 | 18 |
| Orange | 2 | 3 | 1.5 | 0.5 |
| Gray | 2 | 4 | 2 | 8 |
| Red | 2 | 11 | 5.5 | 60.5 |
| Iron | 2 | 25 | 12.5 | 12.5 |
| Plastic | 2 | 20 | 10 | 72 |
| Stone | 2 | 6 | 3 | 8 |
| High | 2 | 39 | 19.5 | 40.5 |
| Middle | 2 | 10 | 5 | 2 |
| Low | 2 | 2 | 1 | 2 |

ANOVA

| Variance | SS | df | MS | F | P-value | F crit |
|---|---|---|---|---|---|---|
| Interblock | 644.4545 | 10 | 64.44545 | 3.157684 | 0.036198 | 2.853625 |
| Intraclass | 224.5 | 11 | 20.40909 | | | |
| Total | 868.9545 | 21 | | | | |

**Figure 1.** The one-way ANOVA results.

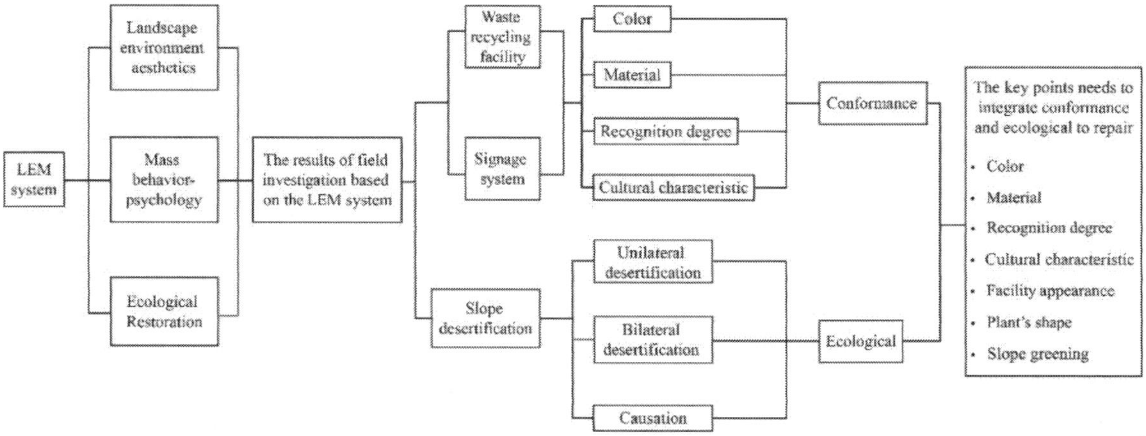


**Figure 2.** A methodological model from the perspective of visual aesthetics to repair the urban-rural highway landscape systematically

# 5 Conclusion

Urban-rural highway construction poses particular challenges to agriculture, forest conservation [13] and local industry. This is a double-edged sword because it can create conveniences for local populations and it can also adversely affect local ecology and culture. Maintaining cultural characteristics and protecting the ecological environment have become particularly important. The results of this study indicate that the landscape is mainly affected by government willpower, villager's preferences and developers' demands. In the approximately 87 km of urban-rural highway around the lake, we selected the three most representative samples to count and analyze. The results of our analysis indicate that they have separate problems that are not related to each other. It is also significant that the highway landscape in this area lacks a systematic design. According to the results, we built a visual model to explain this point, and we also built a methodology model to attempt to solve the problem. From the perspective of visual aesthetics, this study provides strong evidence that the area of urban-rural highway landscapes in China can benefit from the perfect combination of theory and practice.

## References

1. W Deng, L Ning (2010) Highway traffic science and technology (05): 141-143.
2. I Douglas (2006) Peri-urban interface: Approaches to sustainable natural and human resource use: 18-29.
3. R Forman, D Sperling, J Bissonette, A Clevenger (2003) Ecology.
4. Y Lee, J Ahern, C Yeh (2015) Landscape Urban Plan. 139: 137-148.
5. SMEV Dillen, SD Vries, PP Groenewegen, P Spreeuwenberg (2012) J Epidemiol Community Health 66(6): e8.
6. Z Tang. (2008). Research on Harmony between Expressway and Landscape along the Route. Ph.D, Beijing Forestry University.
7. W Cui. (2003). Study of Expressway landscape. Nanjing Forestry University.
8. D Wang. (2012). Landscape restoration of Scenic Area highway landscape——Take Landscape restoration of the Junfu Road for example . Tsinghua University.
9. B Liu (2001) New Architecture (05): 1-3.
10. TC Daniel, J Vining. (1983). Methodological issues in the assessment of landscape quality Behavior and the natural environment (39-84): Springer. (Reprinted).
11. B Chen, OA Adimo, Z Bao (2009) Landscape Urban Plan. 93(1): 76-82.
12. M Arriaza, JF Canas-Ortega, JA Canas-Madueno, P Ruiz-Aviles (2004) Landscape Urban Plan. 69(1): 115-125.
13. C Wu, Y Lin, L Chiang, T Huang (2014) Landscape Urban Plan. 128: 60-71.

# Redesign Spring Bed Based on the Needs of Consumers

Ikhsan Siregar[1,a], Rosnani Ginting[1] and Fitri Agnesia[1]

[1]*Department of Industrial Engineering, Faculty of Engineering, University of Sumatera Utara,Jalan Almamater., University of Sumatera Utara, Medan 20155*

**Abstract.** Redesign is needed to evaluate the results of the design before. Completion of the design can be applied to the existing products. Spring bed became one of the human's needs that helps us to sleep well. This research was conducted at a spring bed company based in Medan, Indonesia. Based on preliminary observations, it is known that there are several consumer complaints about spring bed design that exist today. Therefore, we conducted a research to discover the desire of consumers to spring bed design by using Kansei Engineering Method. Questionnaires were distributed to 97 spring bed product consumer, and the results of questionnaires showed that the fifth design was the best design chosen from among 10 existing designs.

## 1 Introduction

Design is the act of realizing an idea or concept into the real information. The intense competition in this era of globalization requires companies to design products that match and suitable with consumer expectations.

People in the West region had not paid much attention for their sleeping, people slept on a mattress that placed on top of the spring. Sturdy bed was introduced in the late of 1950s, for the prevention of spinal cord disease, since then, many people began to put wooden boards under their mattress, that became a common solution, but not good enough to fix the property of the bed and may cause ventilation problems.

The desire of the consumers is very important in the development of a product. The products used by a consumer must be related closely to them, the consumer nowadays is getting smarter in consuming a product. Thus the desire of consumers should be the basic of a product improvement design. Studies looking at consumer needs very much to do, including the Bali aromatherapy product development towards personal type users [1]. As well as on research in the automotive audio interface that is designed based on user satisfaction [2]. Then there are also some people saying that the design of the product based on the evaluation forms and perceptions of customers [3]. Thus it is clear that the consumer is playing an important role for the development of the product.

Redesign the product is also very necessary to do, to increase the purchasing ability of consumers. Product development can recycle these products and being survived in the market [4]. Design the product also can be making a new product in development, because of the detailed design and specification detail, thereby generating a new product [5]. Design can also be done on the process and

[a] Corresponding author : ikhsan.siregar@usu.ac.id

the equipment, not only limited to products itself [6]. Redesign product also can raise prices, due to factors that are met by consumer satisfaction [7].

This study uses Kansei Engineering Method; this method is also widely used by previous researcher, such as to select the usage of sunglasses that conducted Malaysia [8]. Not only products, Kansei Engineering has also been done on the study, which assesses the satisfaction of service to Indonesian tourists [9]. Then another study presents a method to cope the emotional impact of product attributes in new product development, is also using Kansei Engineering Method [10]. There are other studies that redesign bottle shape, the design of the old bottle had been evaluated and developed through this method [11]. Other studies exist that use Kansei Engineering Method for audio on the eyeglasses [12]. There is even a chocolate snack examines [13]. There are also applied in ceramic souvenirs product manufacture [14].

However, because of there is none spring bed product study before, this study becomes important because Spring bed on of the solutions for a good sleeping equipment. The quality of a human's fitness is strongly influenced by the way the person is rested. If someone resting comfortably and wake up in a fresh condition, then the daily activities could be good and positive.

Based on these problems, companies need to surf the consumers' desire to be done with Kansei Engineering. Kansei Engineering leads us to what he thought about the statement, in which knowledge, emotions, and harmonized desires, people with Kansei is a person full of emotion and passion, and is being able to react quickly and sensitively to something. Kansei Engineering focuses on the perception of an object and thus produce a response. This method combines the feelings and emotions with product development [15].

First type of Kansei Engineering is a fundamental technique of kansei Engineering Methods that use regular process. Everyone can follow the regular process in order to get a conclusion.

## 2 Research method

The experiment was conducted at the spring bed company in Medan, Indonesia. Spring bed as the research object because there are many spring bed companies in Medan and spring bed is a product that fluently purchased by consumers, especially in the Medan City with a population of about 2 million people.

The type of research that being used is survey research to obtain the actual facts regarding customer satisfaction with spring bed products through a questionnaire then get acquired attribute about what point needs to be improved to increase the quality of the product, carried out repair solutions at the company as a form of a repair of the original design.

Spring Bed product attributes that will be examined are:

1. Spring bed elasticity
2. The thickness of the spring bed mattress
3. Spring bed mattress material foam
4. Spring bed backrest material
5. Spring bed foam upholstery material
6. Spring bed color

Closed semantic differential questionnaire distributed to consumers of spring bed products. Engineering samples are used convenience sampling. Large population sizes and numbers are not known with certainty to make the determination of sample size using a technique sample size for proportions. The formula for a large population sample with an unknown amount according to

$$n = \frac{Z^2{}_{\alpha/2}pq}{(e)^2} = \frac{1{,}96^2 x\ 0{,}5\ x\ 0{,}5}{(0{,}1)^2} = 96{,}04\ \approx 97\ \textit{Respondents} \quad (1)$$

Description:
n : Number of Sample
z :z value with 95% confidence level, the value of z = 1.96% (normal distribution table)

p : the proportion will be assessed (by 0.5 if the proportion is unknown)
q : 1 - p
e : error or the maximum error 10%

Steps of First Type of Kansei Engineering are as follows.

1. The first step is the strategy of the company; the company must have a defined concept or strategy for new products. Kansei engineers must utilize this strategy to be applied to new fields.
2. The second step is to collect Kansei words related to new product concept (around 20-30 words Kansei).
3. The third step is the collected Kansei words are going to be arranged at the point Semantic Differential scale 5- or 7-point.
4. The fourth step is to collect samples of the product as a comparison between similar products from different companies and producers (about 10-20 samples).
5. The fifth step is a list of items and categories, items and categories implies the design specifications of product samples collected. All the properties of the products described, for example, an item consists of color, shape, size, brand logos, and others. Categories for example item category colors have yellow, red, green and others.
6. The sixth step is the trial evaluation noted that respondents were asked their feelings with words Kansei for each sample on a sheet of Semantic Differential Scale.
7. The seventh step is analysis, statistically evaluated and analyzed with statistical methods, especially with multivariate statistical analysis.
8. The eight step is the interpretation of the data that being analyzed, that is, all the data is being analyzed to be interpreted from the Kansei Engineering viewpoint. The goal is to find the relationship between human Kansei and product properties. Data that were analyzed every Kansei association was found with the design specifications.
9. The ninth step is Explanation of data, data interpretation must explain to the company's designers to create new designs with the help of a designer.
10. The tenth step of the collaboration of engineers with designers, Kansei motivate the company to create a new product design. This process, engineers design Kansei should support the creation of new products based on the Kansei Engineering data.

## 3 Result and analysis

Stages of data processing with Kansei Engineering begin with determining the decision of strategy is designing products that fit to consumers psychological. Then the collection of Kansei Words derived from the magazine, literature (text book or journal), expert opinions, interviews, namely:

1. Hard-Soft
2. Not Comfortable - Comfortable
3. Not Elegant - Elegant
4. Not Ergonomics - Ergonomics
5. Perishable - Durable
6. Boring - Interesting
7. Not Elastic - Elastic

Validity and reliability tests performed on questionnaires that have been distributed. Testing the validity of the results is shown in Table 1.

**Table 1.** Summary of each design validity.

| Design | Exp-r | Critique-r | Conclusion |
|---|---|---|---|
| I | 0.299 | 0,200 | Valid |

| | | | |
|---|---|---|---|
| II | 0.322 | 0,200 | Valid |
| III | 0.442 | 0,200 | Valid |
| IV | 0.271 | 0,200 | Valid |
| V | 0.375 | 0,200 | Valid |
| VI | 0.404 | 0,200 | Valid |
| VII | 0.278 | 0,200 | Valid |
| VIII | 0.344 | 0,200 | Valid |
| IX | 0.362 | 0,200 | Valid |
| X | 0.409 | 0,200 | Valid |
| XI | 0.299 | 0,200 | Valid |

Reliability test was conducted by using Cronbach's Alpha, the reliability coefficient values were calculated using Cronbach alpha formula is:

$$r = \left(\frac{k}{k-1}\right)\left(1-\frac{\sum \sigma_b^{\,2}}{\sigma_t^{\,2}}\right) \quad (2)$$

$$= \left(\frac{10}{10-1}\right)\left(1-\frac{92,701}{120,330}\right)$$

$$= 0,209$$

R table value with n = 97 and 2 degrees of freedom for a significant level of 5% is 0.200, so the questionnaire otherwise reliable. The next step is statistical analysis using conjoint analysis. With the following steps:

a. Determining the constant value

Constant values obtained using the formula with n = 5 because of the scale used in the questionnaire SD is a scale of 5.

$$c = \left[\frac{1+2+3+\ldots+n}{n}\right] \quad (3)$$

$$c = \left[\frac{1+2+3+4+5}{5}\right] = 3$$

b. Determine the average value

The average value obtained using the following formula:

$$\overline{x} = \left[\frac{x_1 + x_2 + x_3 + \ldots + x_n}{n}\right] \quad (4)$$

Sample calculations for the design of the First Kansei Word, hard – soft

$$\text{Average} = \frac{3+3+4+2+\ldots..+2+5}{97} \quad (5)$$

$$= 3,278$$

Statistical analysis recapitulation using conjoint analysis as follows.

**Table 2.** Statistical Analysis with Conjoint Analysis

| Design | Hard - Soft | Not Comfortable - Comfortable | Not Elegant - Elegant | Not Ergonomics - Ergonomics | Perishable - Durable | Boring - Interesting | Not Elastic - Elastic |
|---|---|---|---|---|---|---|---|
| I | 0.278 | 0.515 | -0.732 | 0.299 | 0.247 | 0.206 | 0.309 |
| II | -0.557 | -0.443 | -0.361 | -0.557 | -0.227 | -0.340 | -0.216 |
| III | -0.021 | 0.165 | 0.021 | -0.134 | 0.052 | -0.010 | -0.113 |
| IV | 0.072 | 0.062 | -0.072 | 0.021 | 0.505 | -0.196 | -0.124 |
| V | 0.247 | 0.588 | 0.485 | 0.320 | 0.134 | 0.536 | 0.588 |
| VI | 0.206 | -0.021 | -0.113 | -0.082 | -0.052 | 0.588 | 0.124 |
| VII | 0.041 | -0.134 | -0.031 | -0.320 | 0.052 | 0.052 | -0.186 |
| VIII | -0.052 | 0.021 | -0.196 | -0.010 | 0.031 | -0.052 | 0.052 |
| IX | 0.175 | -0.113 | 0.216 | 0.082 | 0.247 | 0.216 | 0.278 |
| X | -0.041 | 0.278 | -0.155 | -0.175 | -0.041 | -0.031 | 0.093 |

Table 2 shown that respondent chosen design V from the questionnaire that given to them. Based on the recapitulation of the results of statistical tests, the selected design is V Design with the following attributes.

**Table 3.** Selected design category

| No | Attribute | Information |
|---|---|---|
| 1 | Elasticity | Elastic |
| 2 | Foam Thickness | Thick |
| 3 | Foam Materials | *Polyurethane* |
| 4 | Backrest Material | Seat Foam |
| 5 | Coating Foam Material | Cotton |
| 6 | Color | Brown |

Table 3 shown respondents chosen selected categories that want to improve based on the existing design. Respondent want more elasticity, more thickness, the material foam is polyurethane, need backrest material foam, the material for coating foam is cotton, and last the color chosen for the new design is brown.

## 4 Conclusion

Based on the Semantic Differential questionnaires that have been distributed, it is known that more consumers want the V product design, which has attributes: elastic, thick foam, polyurethane foam materials, materials backrest upholstery foam, cotton foam sheathing, and brown. Selected spring bed product design attributes becomes of draft product design improvements that exist today. The result of this design is entirely based on the largest order of choice of consumers obtained through the method of Kansei Engineering. While the conclusions obtained through this research, now the existing entries have been awarded to the company and will be implemented, this is done to see the extent how far the consumer feedback on products that will be designed according to the tastes of the market.

## References

1. Djatna, Taufik., Balinese aromatherapy product development based on Kansei Engineering and customer personality type, Procedia Manufacturing **4** ( 2015 ) 176 – 183.
2. Gaspar, Jose., User satisfaction modeling framework for automotive audio interfaces, International Journal of Industrial Ergonomics **44** (2014) 662-674.
3. Tang, C.Y., Product form design using customer perception evaluation by a combined superellipse fitting and ANN approach, Advanced Engineering Informatics **27** (2013) 386–394.
4. Aguiar., Jessica de., A design tool to diagnose product recyclability during product design phase, Journal of Cleaner Production **141** (2017) 219-229.
5. Du., Yanbin., Reuse-oriented redesign method of used products based on axiomatic design theory and QFD, Journal of Cleaner Production **39** (2013) 79-86.
6. Fegade, Vishal., *Design for Remanufacturing: Methods and their Approaches*, Materials Today: Proceedings 2 ( 2015 ) 1849 – 1858.
7. Jiang., Yuanchun, Redesigning promotion strategy fore-commerce competitiveness through pricing and recommendation, Int. J.ProductionEconomics **167**(2015)257–270.
8. Chuan, Ngip Khean., *Kansei engineering for e-commerce sunglasses selection in Malaysia*, Procedia - Social and Behavioral Sciences 97 ( 2013 ) 707 – 714.
9. Hartono, Markus., Incorporating service quality tools into Kansei Engineering in services: A case study of Indonesian tourists, Procedia Economics and Finance **4** ( 2012 ) 201 – 212.
10. Huang, Yuexiang., A product configuration analysis method for emotional design using a personal construct theory, International Journal of Industrial Ergonomics **44** (2014) 120-130.
11. Luo, Shi-Jian, A preliminary study of perceptual matching for the evaluation of beverage bottle design, International Journal of Industrial Ergonomics **42** (2012) 219-232.
12. Lu, Weihua., Affective design of products using an audio-based protocol: Application to eyeglass frame, International Journal of Industrial Ergonomics **44** (2014) 383-394.
13. Schütte, Simon. Evaluation of the affective coherence of the exterior and interior of chocolate snacks, Food Quality and Preference **29** (2013) 16–24.
14. Tama, Ishardita Pambudi., Development of customer oriented product design using Kansei engineering and Kano model: Case study of ceramic souvenir, Procedia Manufacturing **4** ( 2015 ) 328 – 335.
15. Falk Steinberg, dkk. 2015. *The Improvement Of Kansei Engineering By Using The Repertory GridTechnique*, Management and Production Engineering Review.

MATEC Web of Conferences **104**, 03004 (2017) DOI: 10.1051/matecconf/201710403004
*IC4M & ICDES 2017*

# Thinking in Terms of Flow in Design of Software Systems

Sabah Al-Fedaghi[1,a]

[1]*Computer Engineering Department, Kuwait University, Kuwait*

**Abstract.** This paper examines conceptual models and their application to the task of software system design. With the software development life cycle in mind, the concern here is the initial phases of the design process, when customer needs are investigated, conceptualized, and included in specifications. More specifically, the paper concentrates on the "patterning aspect of cognition" (i.e., object-oriented) where pattern refers to recurring templates used by designers in thinking. It is proposed that representation of thinking activity be based on flows of things using flow machines formed by stages (states) occurring sequentially in a flow. According to such an approach, a designer's "thought machine" forms a train of thought that excludes other modes such as procedural and object-oriented modes of thinking. The new idea presented here is that flow-based modelling is used not only as an external representation of the design's thinking but also as the style of thinking. A thinking style involves how one organizes thoughts and is a "conscious system of design." The method is illustrated by remodelling examples from the object-oriented paradigm. It seems to have merits that deserve further development.

## 1 Introduction

Among the many functions of the human brain is the ability to think [1]. The cognitive faculty of thinking [2] involves the processes by which we reason and solve problems. In the study of cognition, one aspect is consideration of conceptual models of how we view the world [3]. This paper examines conceptual models and their application to software system design. With the software development life cycle in mind, the concern here is the initial phases of the design process, when customer needs are investigated, conceptualized, and included in specifications. More specifically, the paper concentrates on the "patterning aspect of cognition" [4] (e.g., object-oriented) where pattern refers to recurring templates used by designers in the thinking process.

In general, cognitive models are characterized by great variation depending on factors that include society, environment, and culture (e.g., science, religion) that reflect diverse perceptions of the way the world works. Persons may hold onto more than one of these models and, depending on context, *adopt* a particular view of the world as a framework for thinking. Such selected model plays an important role in thinking by "automatically" constraining meanings (again, e.g., object-oriented). It is like wearing glasses that affect what we see and cannot see. The models adopt a certain way of generating their elements (e.g., space in architectural models), and they direct attention in interactions with the world and affect how we link concepts to objects and vice versa.

[a] sabah.alfedaghi@ku.edu.kw

A model of such type could be founded *predominantly* upon certain types of mental components such as feelings, concepts (ideas), propositions, and pictures. A *conceptual* model is a mental model formed mainly upon concepts as components of thinking. A model can provide a framework for thinking that structures concepts into patterns according to categories. A model can also provide a basis to represent internal thinking in an external form.

Diagrams, as one form of representation, convey a simple and easily communicable method of thinking. "Diagrams are essential representations for thinking, problem solving, and communication in the design disciplines" [5]. Because "Diagrams are effective instruments of thought" [6], this paper suggests imposing a specific style of thought such that (internal) thinking and (external) representation use the same model, thus synchronizing thinking with the method of diagramming. According to such an approach, a designer's "thought machine" forms a train of thought that excludes other modes such as procedural and object-oriented modes of thinking.

The next section introduces a flow-based diagrammatic language to be used both as a thinking style and as a vehicle for drawing the outcome of a design. Section 3 applies such an idea to samples taken from the object-oriented literature.

## 2 Diagrammatic language for thinking and drawing

This section includes two subsections. The first will review a flow-based model (this example is a new contribution) and the second will use such a model to explain the aim of the research in this paper.

### 2.1 Diagrammatic Language to be used in modelling

We propose that thinking and drawing activities be based on *flows of things* grounded in what we call the Flowthing model (FM). FM has been used in a variety of applications (e.g., [7-12]). FM depicts processes using *flow machines* formed by five stages: creation, processing, release, transfer, and receive, as shown in Fig. 1. Hereafter, flow machines will be referred to as machines.

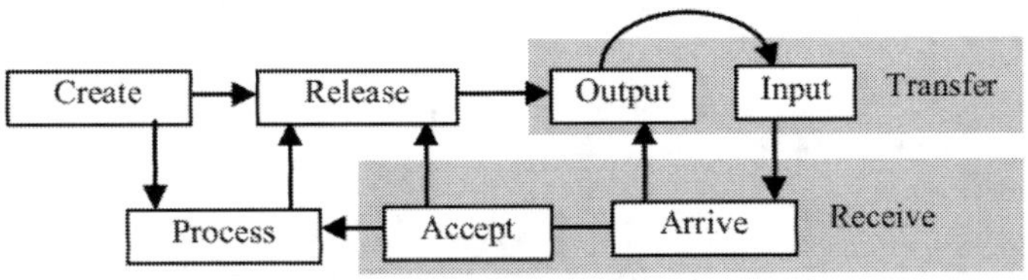


**Figure 1**. Flow machine

FM is a method for modeling things that "flow"; i.e., things that are *created*, *processed*, *released*, *transferred*, and *received* while retaining their individuality throughout the flow. They flow within *spheres*, i.e., the relevant environment that encompasses the flow. Things that flow are referred to as *flowthings*, or simply as *things*. *Flow* in such a model conveys (conceptual) movement in a system. The life cycle of a thing is defined in terms of the five mutually exclusive *stages* of creation, release, transfer, arrival, acceptance, and process (in this last stage the *form of a thing may be changed*, but no new thing is generated). For example, a *flow thing* that is *received* is not permitted to move (flow) to the stage in which it is *created* (i.e., a received thing implies an already existing thing, hence it is not possible to re-create it). A mechanism that can change this succession of flow is called *triggering*. In a propositional sense, flows of things are formulated in terms of the propositions *Create(thing)*, *Process(thing)*, *Receive(thing)*, *Release(thing)*, and *Transfer(thing)*. The reflexive arrow in Fig. 1 indicates flow to the *Transfer* stage of another machine. For simplicity, the stages Arrive and Accept can be combined and termed *Receive*.

FM is a generalization of the model input → Process → output. According to Deepak [13],

> If a system in *any field* is analyzed. It will be observed that it has three basic parts, which are organized in an orderly manner. These three parts can be represented in a model as [input → Process → output]… When a system is designed to achieve certain objectives, it automatically sets the boundaries for itself. The understanding of boundaries of the system is essential to bring clarity in explaining the system components and their arrangements.

**Example**: Peters and Langemann [14] developed a general supply chain of the human brain, illustrated by a type of bubble diagram (not shown), as follows:

> Energy from the remote environment is brought to the immediate environment, it is then taken up by the body, and from there a large part of it enters the brain. In a general supply chain, the flux of energy can principally be determined by the supplier (previous step) or by the receiver (proximate step). The share of the flux which is determined by the supplier is called the "push component" …, the share which is determined by the receiver is called the "pull component" ....

Fig. 2 shows the FM representation of a partial picture of their diagram. The arrow represents the flow of energy. For the sake of simplicity, the energy machine is not enclosed in its own box as a subsphere.

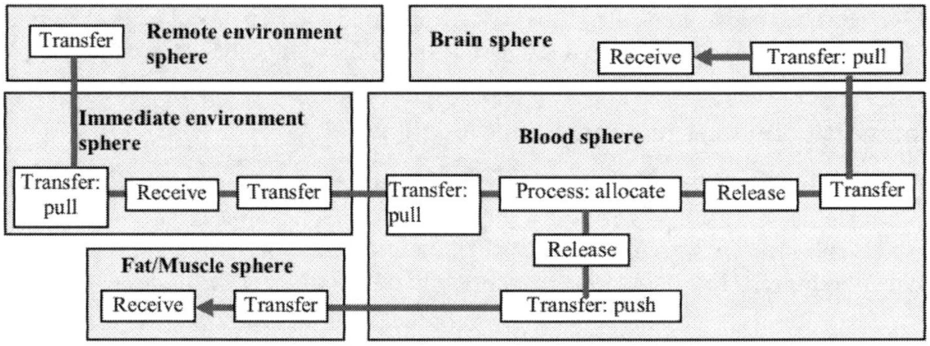


**Figure 2**. FM representation of a general supply chain of the human brain

## 2.2 Aim of This Research

In design, "Drawing is intimately bound with thinking" [5]. We propose imposing flow-based thinking on the conceptual design process instead of other types of thinking such as categories and types that trigger other types (e.g., object-oriented). In forcing this FM-based thinking, we isolate and cut nature up into *things that flow in their spheres* (environment).

Accordingly, we utilize FM in this paper as a machine that unifies the design process to achieve the following dual purposes:

1. A designer's "thought machine" that excludes other modes of thinking such as procedural and object-oriented modes of thinking (see Fig. 3). "Every person is conscious of a train of thought being immediately awakened in his imagination analogous to character or expression of the original object" (Gandy [15], quoted in Crook [16]).

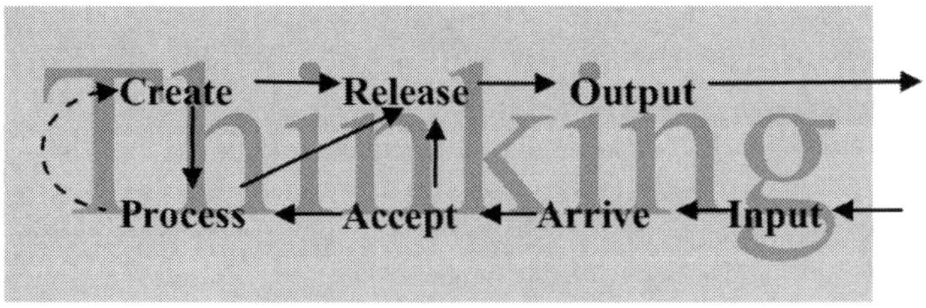


**Figure 3**. Thought machine

DOI: 10.1051/matecconf/201710403004

2. A "design outcome" in the form of a representative diagram that models a portion of reality, as shown in Fig. 4. Note that the figure is a version of the famous semiotic triangle of object, thought, and sign (representation) [17].

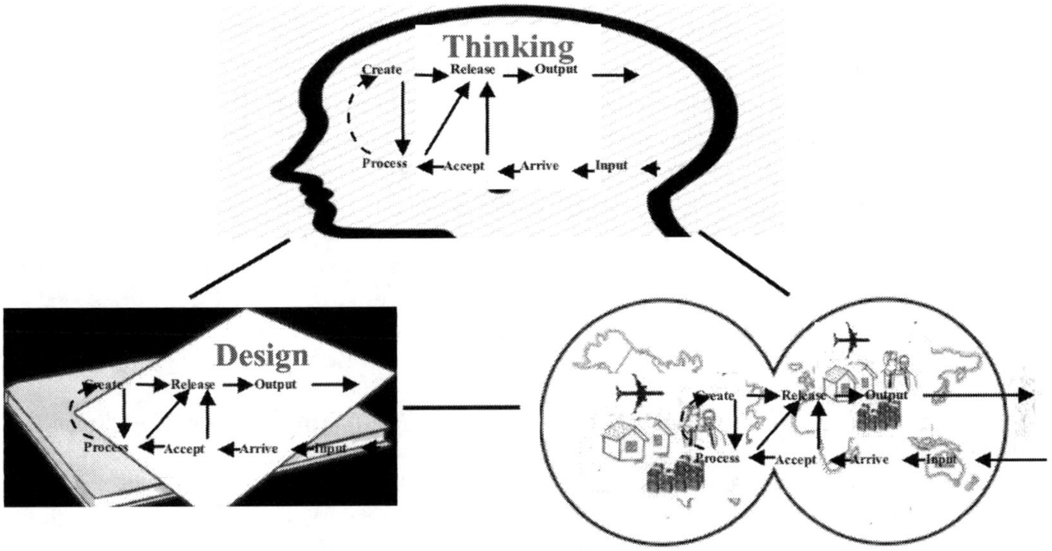


**Figure 4**. FM-based world perception, thinking and design diagramming

## 3 Thinking style

The new idea in this paper is that FM can be employed *explicitly* not only for the drawing of an external representation but also as the *style of thinking*. A thinking style involves how one organizes thoughts. In architecture, according to Crook [16], a style is a "conscious system of design." This scheme would have wide application in many areas besides software design, as follows:

- In mathematics, a *set* can be conceptualized as a flow machine that handles flowthings called elements that flow through and into and out of the system. Handling here refers to transferring, receiving, processing, creating, and releasing elements of the set. A function such as f: X → Y can be represented in terms of a map between sets, conceptualized not in terms of space and movement, but in terms of flows in which the values of input flowthings are changes in the output flowthings. Such a view may have some application in teaching of mathematics [7].
- Physical security can be modeled as a flow machine of interruptions or breaks with an abstract apparatus that represents the flow of things in the physical infrastructure of the organization [9].
- According to Gandhi, "Man is the most wonderful machine in creation." Accordingly, such notions as privacy can be conceptualized as an inner flow machine of a person. It is a natural machine whose function is to control the transferring of (personal) things to the outside [8].

Other applications of flow-based thinking include the areas of aesthetics [10], philosophy [11], and computer programming [12].

According to Penrose [18], "There seem to be many different ways in which different people think – and even in which different mathematicians think about their mathematics." Likewise, in design, a variety of methods of *thinking* apply to design. "You cannot hold a design in your hand. It is not a thing. It is a process" [19].

This paper proposes that the designer think in terms of FM when perceiving reality, in addition to drawing FM diagrams. We speculate that flow-based thinking is a thinking style. Even if this style does not seem "natural," the claim is that thinking and drawing this way lead to a viable design methodology, especially in the area of software development.

One "thought thing" follows another, connected by flows (including triggering), and the connection is driven by such principles as contiguity in time or place, and cause and effect (these principles were laid out by Hume in *An Inquiry Concerning Human Understanding*, 1748).

Accordingly, guidelines for devising an FM design for a particular problem are as follows:

- "Pick up" a thing
- Follow its flow through spheres until
- It triggers another thing that leads to a new stream of flow to follow, etc.

For example, the flow of an *order* leads to the flow of an *invoice*, which leads to a flow of *payment*, which leads to the flow of the *ordered product*.

"Design without drawing seems inconceivable… Sketches can reveal thought, then, ... [and] they can reveal the elements or segments of construction or of thought in a particular domain" [20]. Early in the design process, designers draw diagrams as a way to map their thinking and explore concepts (mental things) and their flows in an iterative thinking process (thinking$_1$ → FM$_1$, thinking$_2$ → FM$_2$, …; see Fig. 4). This process involves capturing ideas and their flow in a succession of conceptual phenomena. Drawing is a vehicle for seeing ideas and their flows and triggering, hence for understanding and organizing their system.

In the context of FM, we say that FM can provide a basin for the flow of thought when the designer thinks in terms of FM, and that "elements and segmentation construction of thought" appear as elements of flow: the conceptual movement of things along the stages of the flow systems. The first phase of design includes identifying end-user motivations and needs, which form the basis of thinking that generates concepts to meet these needs. Arbitrary types of "thinking happenings" drive the appearance of concepts in disconnected thoughts with lack of clear connections between thoughts, just as butterflies, weighing no more than feathers, are blown about in the wind. This paper formulates this phenomenon ("thinking happenings") in a restrained way such that the thinking gives birth and directs *ideas (concepts)* like a controlled wind flow that blows butterflies into a particular air tunnel.

The main notion is that adopting an FM-based *conceptualization* (cognition) precedes its instantiation as a diagram, just as object-oriented thinking leads to object-oriented languages. This idea is analogous to the mapping between abstract classes in the mental domain and objects and their classes as data structures in object-oriented models. The mapping between FM-based thought and an FM-based diagram ought to be possible when the same conceptual pattern underlies each.

With FM, a small number of notions (flow, stages, spheres) with very little verbal description are used to construct a large conceptual picture and physical FM-based depiction. At one level, things can be viewed as concepts, and flows as the conceptual movement of concepts in their spheres. On the other hand, a diagram can be viewed as a composition of arrows and boxes that reflects the conceptual picture through motor elements involved in the act of drawing, i.e., physiological signals and physical actions.

In contrast, the object-oriented (OO) approach focuses on objects that have attributes and do processes; e.g., in GUIs, windows and icons have attributes (locations on the screen), and they can be clicked or moved [13]. In this case, particular "classes" of objects are defined. The class definition is the master description of what data the instances will have and what can be done with that data. We make new instances based on classes [13]. Deepak [13] provides a set of guidelines for formulating an OO design for a particular problem:

- Describe the problem in plain English. Describe what the types of entities in the problem are, what attributes they possess, what actions they can take, and what those actions need to work.
- Write a class for each type of entity, where the attributes are instance variables in the class, and the actions are its methods, and what the actions need to work are passed as parameters to those methods.
- Instantiate instances of the objects you need, and call their methods [13].

## 4 Sample styles of thinking about problems

This section provides some evidence for the viability of FM as a foundation for thinking and drawing through modelling examples from the object-oriented paradigm.

### 4.1 Example 1

Manik [21] gives an example of an object-oriented model, shown in Fig. 5. The *lives in* and *drives* associations, in the OO approach, indicate instances of classes associated with each other. We can add attributes to the association with an association class (connected to an association by a dotted line, not shown). It seems that the model was developed using Deepak's [13] guidelines for OO design reviewed above.

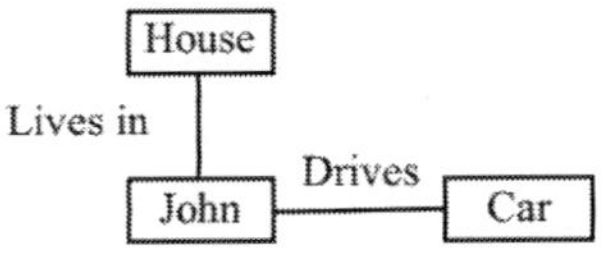


**Figure 5**. An object-oriented model (partial, re-oriented from [21])

According to FM and within the context of the problem under consideration, we first "pick up" a flowthing, say, *John* (Fig. 6a) to " jumpstart" our train of thought. This beginning is complemented by recognizing John's spheres (Fig, 6b and c), then identifying the flow of John; consequently, this step leads us to model the car as shown in Fig 6d. The OO modeling involves conceptual discontinuity that jumps from John to house and to car, resulting in disconnected entities. The designer must examine these disparate concepts pair-wise to jump between John/house and John/car and their relationship. An alternative scenario involves somehow recognizing (e.g., by reading the whole English text – see guidelines in [13]), that *John* is the central entity and can be used as a reference point, then looking around (e.g., reading the text again) to discover the *house,* connect it to John, then discover the *car* and make the second connection. While such an approach is accepted as standard by many designers, this description emphasizes the difference from the FM thinking approach.

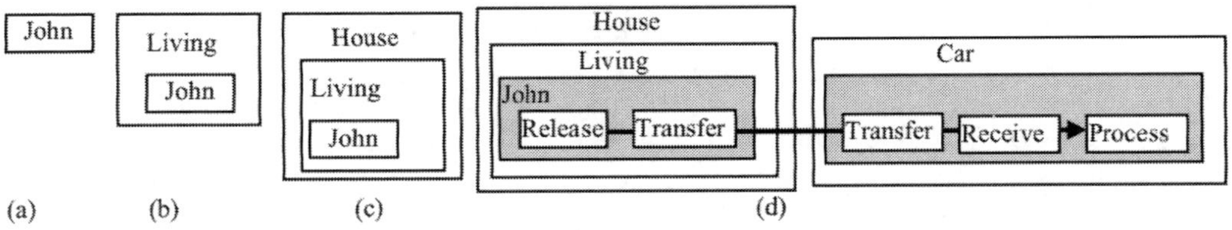


**Figure 6**. Stages in flowthing modelling.

FM modelling is a smooth maneuver along a flow, analogous to holding onto a rope (the flow) and walking along it to reach the next thing to be modeled; e.g., John leaves the *house*, reaches the *car*, starts it up, and drives it away.

Note that Fig. 6c is the "natural" conceptualization of *John lives in a house*. It brings to mind a person *inside* his house, as shown in Fig. 7 (left), with stick figure inside an FM sphere of House. Even searching Google for "lives in + house" results in a similar illustration of a man sitting in a house (Fig. 7 (right)). However, OO modelling reflects a very abstracted portrait of reality developed for the sake of data structuring. If this is the aim of modelling, then to reach such modelling based on data structure, we can:

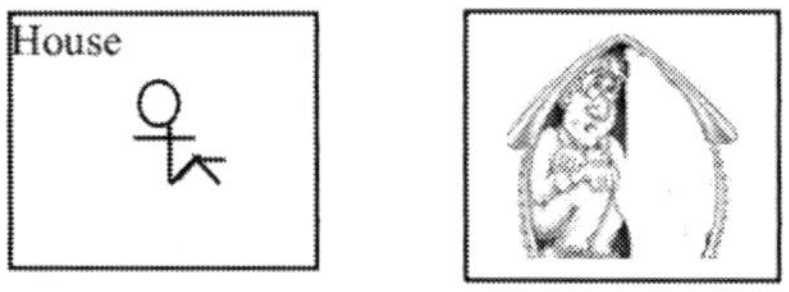


**Figure 7**. FM conceptual representation of "John lives in a house" and image shown in a Google hit when the term ""lives in + house" is searched

- Remove the flow and separate each sphere as in Fig. 6d, resulting in Fig. 8, and
- Replace living and process by lines in that figure, so we get the original OO representation of Fig. 5.

Accordingly, FM representation can be used as a semantic base for OO modelling.

We speculate that FM is closer to ordinary thinking. In fact, the FM representation "pulls" us to follow our thinking to its end when conceptualizing *John lives in a house and drives a car*, as shown in Fig. 9. The figure expresses a portrait of the world of John: John lives in a house, leaves the house to drive a car, drives the car back to the house, gets out of the car, and enters the house.

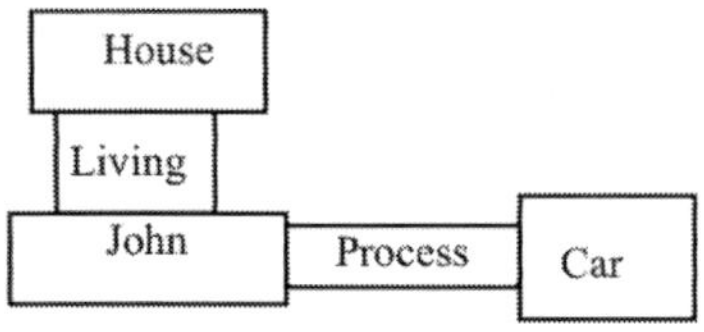


**Figure 8**. Abstracting the FM representation in Fig. 6d.

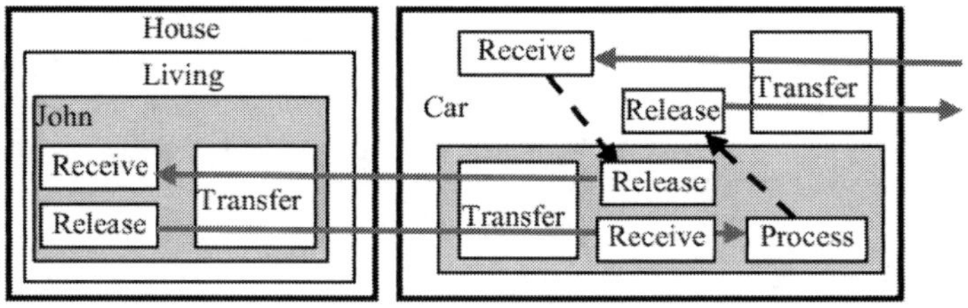


**Figure 9**. Complete description of Jon lives in a house and drives a car.

The result is a complete representation; e.g., when designing a house on this structure, its bounds and a door (representing transfer) immediately come to mind. Similarly, *John lives in a house and John drives a car* brings to mind that John is the common component in this limited world encompassing a house and a car; hence John moves (flows) from the house to the car and vice versa. This is represented in the OO model by having the two lines connected to John.

Additionally, the OO model embeds ambiguity by using lines that represent associations. *Drive a car* has two types of "relationships" with *lives in a house*, in addition to being related to John:

- The car and the house form a "house transportation system" of going away and coming back to the house by use of the car.
- *Lives in a house* and *Drives a car* are separate spheres of *house living* and *driving a car* (to somewhere else).

For example, in the relational database model, the first relationship is represented by a three-tuple real table while the second one is represented by two tables. In the last case, building a universal table may violate the multi-value dependence principle if John lives in more than one house and drives more than one car.

The given FM representation in Fig. 9 assumes the first case, and Fig. 10 shows the second case. Note that *Create* in the sphere of John denotes "come into existence" and is omitted in the spheres of House and Car for simplicity, since specifying the flow indicates their existence. The OO model could be interpreted to be either of these two cases.

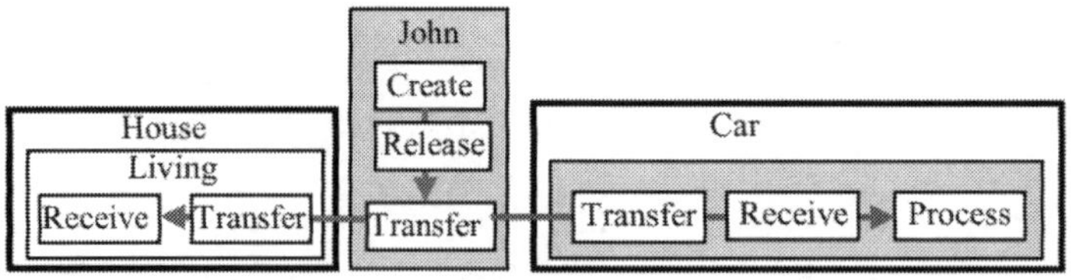


**Figure 10**. *Lives in a house* and *Drive a car* are separate spheres

DOI: 10.1051/matecconf/201710403004

## 4.2 Example 2

In his book *Introduction to Object Oriented Programming*, Budd [22] illustrates the concepts of OO with an example of *sending flowers to a friend in a different city*, as follows.

- Chris is sending flowers to Robin.
- Chris can't deliver them directly. So Chris uses the services of the local Florist.
- Chris tells the Florist (named Fred) the address for Robin, how much to spend, and the type of flowers to send.
- Fred contacts a florist in Robin's city, who arranges the flowers, then contacts a driver, who delivers the flowers (description taken from [23])

The OO approach handles this "real-world situation" by representing it as shown in Fig. 11. "The Object Oriented approach is to view software the same way we view the world – easier for us to understand… The chain reaction that ultimately resulted in the solution to Chris's problem began with a request given to the florist Fred. This request led to other requests, which led to still more requests, until the flowers ultimately reached Chris's friend Robin." [23].

Ferens [24] developed a UML instance diagram for the same problem, shown in Fig. 12. Note the abstracted representation that ignores the differences among *things* in the figure where Chris sends the *Order* and the Driver carries *physical flowers*. Also, the structure is ignored, e.g., the driver is within the sphere of Robin's florist, not any driver.

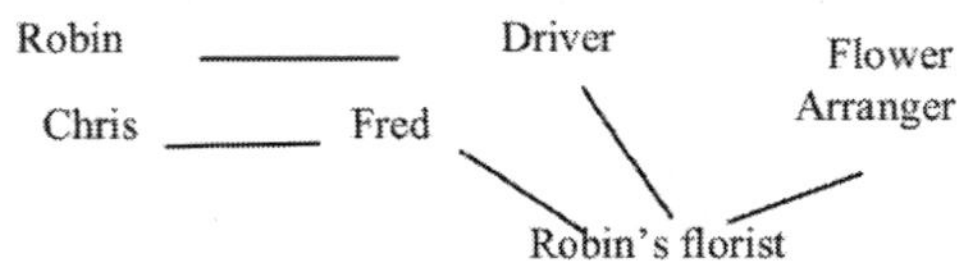


**Figure 11**. Representation of sending flowers to a friend in OO modeling (partial, redrawn from [23])

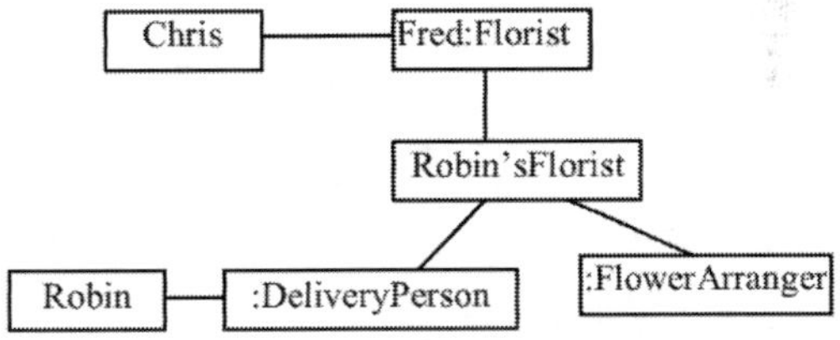


**Figure12**. UML Instance Diagram (partial, redrawn from [24])

According to Weiss [23], the chain reaction that results in the solution to the problem of *sending flowers to a friend* starts with a request to florist Fred that leads to other requests until the flowers reach Robin. This chain is based on the transmission of a message to an agent/object. A message is either a request for information or a request to perform an action.

Fig. 13 shows the FM representation of the sequence of activities involved in sending flowers to a friend.

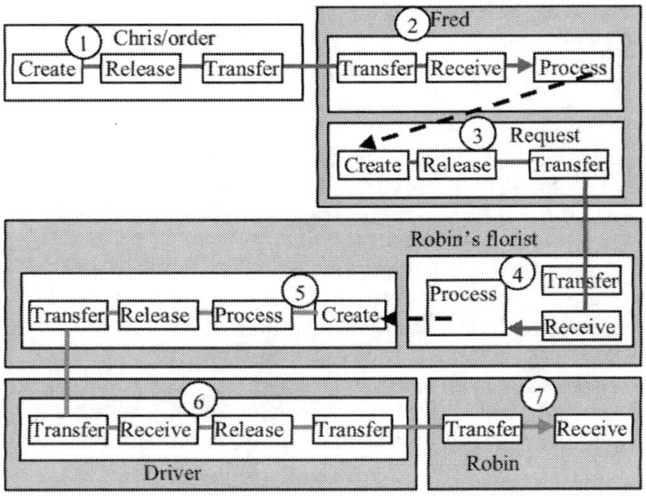


**Figure 13**. FM representation of *sending flowers to a friend*

We assume that the modeler is *thinking of a* starting *flowthing* to follow the *thread of flow* and "pull up" the rest of the flows. Accordingly, the first thing that comes to mind is the first item, *Chris*, and the flow is *order*ing of flowers.

- The *order* (circle 1) flows to Fred (2), who *processes* it. Note that, for simplicity, we depict Chris and order in one box.
- This processing triggers creation of a request (3) that flows to Robin's florist (4).
- Receipt of the request triggers creation of the flower basket per instructions in the order (5).
- The basket flows to the driver (6).
- The basket is delivered to Robin (7).

## 5 Conclusions

This paper has proposed basing the thinking behind system design on flows of things passing through flow machines. According to such an approach, a designer's "thought machine" forms a train of thought that excludes other modes such as procedural and object-oriented modes of thinking. The paper emphasizes this thinking style as a unifying method that could have diverse applications. The re-modelling of examples from the object-oriented paradigm seems to point to merits that deserve further development.

## References

1. T. Parry and G. Gregory, (1998). Designing Brain Compatible Learning. Victoria: Hawker Brownlow Education, pp.168-206 (1998).
2. R. Langacker, Foundations of Cognitive Grammar, vol. 1: Theoretical Prerequisites. Palo Alto: Stanford University Press (1987).
3. M. Wagman, Cognitive Science and Concepts of the Mind: Towards a General Theory of Human & Artificial Intelligence. New York: Praeger (1991).
4. R. C. Anderson, "The notion of schemata and educational enterprise: general discussion of the conference," in R. C. Anderson, R. J. Spiro, and W. E. Montague, Eds., Schooling and the Acquisition of Knowledge. Hillsdale, N.J.: Erlbaum (1977).
5. E. Y.-L. Do and M. D. Gross, "Thinking with diagrams in architectural design," Artif. Intell. Rev., vol. 15, pp. 135-149 (2001).
6. N. Crilly, A. F. Blackwell, and P. J. Clarkson, "Graphic elicitation: using research diagrams as interview stimuli," Qual. Res., vol. 6 (2006).
7. S. Al-Fedaghi, "Thing-oriented learning: application to mathematical objects," 19th IEEE International Conference on Computational Science and Engineering (CSE 2016), Paris (2016).
8. S. Al-Fedaghi, "Privacy as a machine," International Conference on Information System and Artificial Intelligence (ISAI2016), Hong Kong (2016).
9. S. Al-Fedaghi, "Security as a flow machine," IEEE World Congress on Industrial Control Systems Security (WCICSS-2015), London (2015).
10. S. Al-Fedaghi, "Schematization for machine-oriented aesthetics," 15th International Conference on Artificial Intellegence, Las Vegas, USA (2016).
11. S. Al-Fedaghi, "An engineering reading of Deleuze and Guattari's Anti-Oedipus," 2016 International Symposium on Economics and Social Science (ISESS 2016), Bangkok, 2016.
12. S. Al-Fedaghi, "Conceptual understanding of computer program execution: application to C++", Int. J. Comp. Sci. Iss. arXiv, vol. 10, no. 2, March (2013).
13. K. Deepak, Management Information System [blog posting], June 9 (2011). http://deepread.blogspot.com/2011/06/concept-of-system-analysis-design-sad.html
14. A. Peters and D. Langemann, "Build-ups in the supply chain of the brain: on the neuroenergetic cause of obesity and type 2 diabetes mellitus," Front. Neuroenergetics, vol. 1, no. 2, p. 2, February (2009).
15. J. Gandy, "Architecture, its natural model,".Exhibited at the R.A., Soane's Museum, SM XP12(1838) .

16. J. M. Crook, The Dilemma of Style: Architectural. Ideas from the Picturesque to the Post-Modern. Chicago: University of Chicago Press (1987).
17. C. K. Ogden and I. A. Richards, The Meaning of Meaning: A Study of the Influence of Language Upon Thought and of the Science of Symbolism. London: Routledge & Kegan Paul (1923).
18. R. Penrose, The Emperor's New Mind. Vintage, London (1990).
19. B. Gill, Graphic Design as a Second Language. Images Publishing (2004).
20. B. Tversky, "What does drawing reveal about thinking?" in J. S. Gero & B. Tversky, Eds., Visual and Spatial Reasoning in Design. Sydney, Australia: Key Centre of Design Computing and Cognition., pp. 93-101 (1999).
21. S. M. Manik, Object Oriented Concept [slides], Feb 29, 2012. http://www.slideshare.net/smj/object-oriented-concept
22. T. Budd, An Introduction to Object-Oriented Programming. Addison Wesley Longman, (2002). ISBN 0-201-76031-2
23. G. Weiss, "Slides based on Timothy Budds's slides" [course materials], (No date). https://www.google.com.kw/url?sa=t&rct=j&q=&esrc=s&source=web&cd=2&ved=0ahUKEwjb-a37hL3PAhXJAsAKHYbDCUYQFgggMAE&url=https%3A%2F%2Fwww.cs.bgu.ac.il%2F~oosd112%2Fwiki.files%2Fl1.pdf&usg=AFQjCNEms3HJzDe6wOK8mUgUrWrw9v5x5A&bvm=bv.134495766,d.bGg
24. K. Ferens, "Fundamental concepts (principles) of object oriented programming," ECE 3740 lecture notes, Fall (2016).

# Designing Shopping Area Atmospheric Using Choice-based Conjoint Analysis: Case Study at Soekarno-Hatta International Airport

Amalia Suzianti[1] and Hutami Nadya Larasati [2,a]

[1,2]*Industrial Engineering Department, Faculty of Engineering, Universitas Indonesia, Depok, Indonesia*

**Abstract.** As a response to overcapacity issue at Soekarno-Hatta International Airport, the government is recently planning on building larger terminal buildings to transport passengers. However, in designing an excellent terminal building, one should not only consider how to process passengers in the most efficient way, but also how facility-based environmental cues, or atmospheric cues, affect consumers' emotional state and shopping behavior. This research aims at examining passenger preferences towards atmospheric cues in airport shopping area and proposing a shopping area design based on consumer preferences. Choice-based conjoint was conducted towards 500 passengers at Soekarno-Hatta International Airport. Research findings suggested that there is a significant correlation between atmospheric cues and consumers' purchase intention in airport shopping area. Passengers prefer a shopping area with high in-store visibility, cool colors for window display, floors, wall, and ceiling, and bright lighting for airport hallway. This research also concludes that Socio-economic Status (SES) affects passengers' preferences.

## 1 Introduction

Among 189 airports operating in Indonesia, Soekarno-Hatta International Airport (SHIA) recorded the highest airport passenger traffic in Indonesia. SHIA occupies the $8^{th}$ position in world's busiest airports acording to Skytrax. However, SHIA was still ranked the $57^{th}$ for world's best airport according to Airports Council International. This indicates that passenger traffic in SHIA does not offset the service quality. In fact, according to DKMA, airport service quality may affect the airport's total commercial revenues. Airport commercial revenues are considered critical because they contribute to approximately 50% of total revenues, dominated by retail activities which account for 22% of total commercial revenues [1]. However, as the busiest airport in Indonesia, SHIA has not been able to achieve its commercial revenues target yet.

To overcome this issue, the government is planning to enlarge airport capacity both in terms of airfield and terminal building. However, to increase airport commercial revenues, passenger pleasure and arousal should be enhanced [2]. Empirical studies also suggest that atmosphere influence positive emotion then influence satisfaction and intention to purchase [3].

In Indonesia, airport terminal buildings were designed only to facilitate passengers flow as efficient as possible without considering other aspects that actually affect passengers' emotional state,

[a] Corresponding author : hutami.nadya@ui.ac.id

such as atmospheric cues. Therefore, redesigning airport atmosphere based on consumer preferences becomes the key alternative to increase airport commercial revenues.

## 2 Literature review

### 2.1 Atmospheric cues

Kotler defined atmospheric as component of store image, including brightness and crowding [4]. Since then, research on consumer behavior was not only limited on how consumers respond to a product or service, but also on how consumers respond to environment. In the most recent model, atmospheric variables are classified into five categories: exterior variable, general interior variable, layout and design variables, point-of-purchase and decoration variables, and human variables [5].

External variables consist of a number of elements, such as storefront, marquee, window display, and parking area. General interior variables include flooring, wall textures, lighting, scent, music, temperature, cleanliness, and color [5]. Researches in psychophysiology showed that high wavelength or warm colors such as reds, oranges, and yellows, may have physiological effects opposite those of low wavelength or cool colors, such as blues, greens and whites [6]. Respondents perceived the 5000°K lighting as cool and the 3000°K lighting as warm [7]. Layout variables consist of traffic flow, allocation of product and the size of the store.

Product display can be defined as grouping of products, shelf space, allocation of floor space, department allocation, and wall decorations. Excellent product display should have, 1) attractive, creative, and systematic product display, 2) clear information on the products displayed, and 3) store design that allows consumers to see inside the store clearly from the outside.

### 2.2 Choice-based conjoint analysis

Choice-Based Conjoint (CBC) is an application of discrete choice modeling used to analyze consumer decision-making [8]. Discrete choice modeling was firstly introduced as a mathematical formula. Later, it was developed as a mathematical formula of discrete choice modeling with conjoint analysis.

CBC analysis also involves experimental design. Design of experiment is done to obtain profiles or combination of the attributes as well as choice sets that will be examined. The experimental design includes two assumptions that must be fulfilled. The first assumption is that the experimental design must be orthogonal. The second assumption is about the maximum number of tasks that should be tested. To maintain data reliability and prevent respondent fatigue of filling out the questionnaire, a CBC questionnaire should consist of no more than 20 sets of questions [9].

## 3 Research methodology

### 3.1 Pilot study

Pilot study was done by interviewing fifteen experts in order to determine atmospheric variables that may influence consumers purchase intention at an airport, including three professors of architecture and interior design, two practitioners in visual arts and interior design, and ten frequent fliers at SHIA.

The interview consisted of two parts. In the first part, interviewees were asked to categorize each atmospheric variable in the model into two categories. For each variable considered influential on consumer purchase intention at the airport, interviewees were asked to classify the variable as "important" and elaborate their answers, vice versa. After classifying each variable, interviewees proceeded to the second part of the interview, where they were asked to give rankings on atmospheric variables based on the level of importance.

From the interview conducted to fifteen interviewees, twenty different variables were considered influential on purchase intention at an airport. However, only eight variables intersected to one another. They are window display, store signage, cleanliness, product display, store entrance, product signage, color scheme, and lighting.

To clarify the level of agreement among experts, Kendall's W Test was carried out using SPSS. Result showed that level of agreement was acceptable (Kendall's W = 0.51). Therefore, data obtained from previous pilot study could be used for the next step. The next step was to calculate the mean rank for each attribute using Friedman test. This test was performed by SPSS, aim to determine the attributes that will be examined, since CBC analysis should only examine not more than six attributes. Based on Table 1, it can be concluded that the six top-ranked attributes were window display, store signage, product display, store entrance, product signage and color schemes. However, after re-checking with the SHIA managers, store signage, cleanliness, and product signage were eleminated because they did not happen to be airport authority. Therefore, store signage and product signage were eliminated and replaced by lighting variable.

**Table 1.** Friedman Test Result

| Friedman Test | Mean Rank |
|---|---|
| window_display | 2,00 |
| store_signage | 2,13 |
| store_entrance | 5,37 |
| color_scheme | 5,50 |
| product_display | 3,40 |
| lighting | 5,53 |
| cleanliness | 6,67 |
| product_signage | 5,40 |

Consequently, the result of the pilot study was five attributes as follows, window display, store entrance, color scheme, lighting, and product display. To prevent too many profiles and choice sets tested in the questionnaire, the number of levels was limited to two levels for each variable as shown in Table 2.

**Table 2.** Attributes and levels as applied in the questionnaire

| Attribute | Level 1 | Level 2 | Source |
|---|---|---|---|
| Window Display | Cool colors | Warm colors | Somoon & Sahachaisaree, 2012 |
| Store Entrance | Using glass | Using no glass | William, 2002 |
| Color Scheme | Warm colors | Cool colors | Chebat&Morrin, 2007<br>Ballantine & Jack, 2010 |
| Product Display | Products are visible from outside the store | Products are not visible from outside the store | Hussain & Ali, 2015 |
| Lighting | Bright | Dim | Stemer, 2015 |

## 3.2 Main study

The main study of this research includes analysis of consumer preferences using CBC method tested through questionnaire. In previous pilot study, the attributes and levels were all set. The next step was to do experimental design to obtain profile and choice sets that will be examined.

The researchers used a ¼ fractional factorial design, so the number of profiles obtained was 8 profiles with 16 choice sets, where each choice set consists of 2 pieces of alternative options. This experimental design was performed using XLSTAT software.

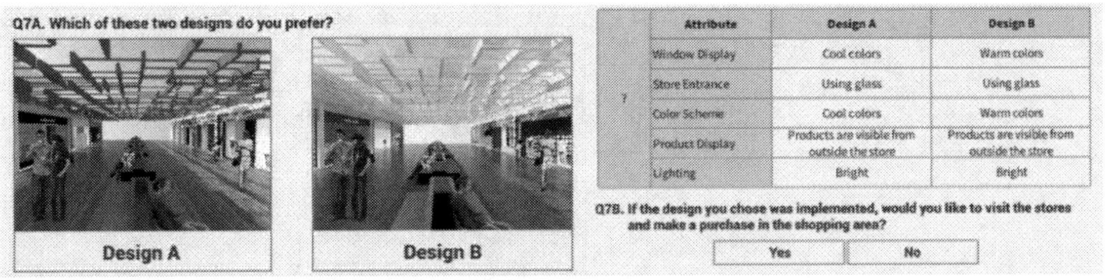


**Figure 1.** Choice-Based Conjoint Questionnaire Design

Figure 1 shows one of choice sets as applied in the questionnaire. Each choice set consists of two types of questions, namely Question A and Question B. Question A examines consumer preferences towards atmospheric design in airport shopping area. In Question A, respondents were asked to choose one out of two alternatives given in the pictures, in which each picture has different atmospheric characteristics. Meanwhile, Question B was intended to analyze the effect of atmospheric design on consumer purchase intention in airport shopping area. In this study, the population size refers to the number of passengers at SHIA. Using margin of errors 0.05, it was calculated that the minimum sample size for this study was 399.96 $\approx$ 400 respondents.

# 4 Results

## 4.1. Consumer preferences towards atmospheric design in airport terminal building

The questionnaires were distributed to 795 respondents with 59% response rate, or in other words there were 500 respondents who completed the questionnaire, consisting of 262 women and 238 men. Data obtained from questionnaires were processed using XLSTAT software. The first analysis was utility value analysis that describes respondent preferences towards the attributes. Higher utility values imply respondent preferences. Table 3 shows the utility values for each attribute and level.

From data processing, it was concluded that respondents prefer window display design with cool colors, such as bluish-white. As for store entrance, respondents prefer a store entrance that uses glass compared to those that does not use glass although both are transparent. For color scheme variable, respondents prefer airport corridor design with cool colors, such as bluish-white, compared to warm colors, such as brown. The next attribute is product display or the level of convenience for consumers when viewing the products in the store from outside the store.

**Table 3.** Utility and importance values of each attributes

| Source | Utilities | Standard deviation | Importance |
|---|---|---|---|
| Window Display-Cool colors | 0,028 | 0,021 | 2,259 |
| Window Display-Warm colors | -0,028 | 0,021 | |
| Store Entrance-Using glass | 0,012 | 0,017 | 0,984 |
| Store Entrance-Using no glass | -0,012 | 0,017 | |
| Color Scheme-Cool colors | 0,094 | 0,019 | 7,691 |
| Color Scheme-Warm colors | -0,094 | 0,019 | |
| Product Display-Products are visible from outside the store | 0,364 | 0,017 | 29,742 |
| Product Display-Products are not visible from outside the store | -0,364 | 0,017 | |
| Lighting-Dim | -0,727 | 0,019 | 59,324 |
| Lighting-Bright | 0,727 | 0,019 | |

DOI: 10.1051/matecconf/201710403005

The data obtained shows that the respondents prefer shopping area with products that are clearly visible from outside the store. The last attribute is the lighting, in which the respondents prefer the design of the airport shopping area with high intensity light. Among these five attributes, lighting variable has the highest importance value. In other words, lighting has the most influence on consumer preferences. Figure 2 shows the clear picture of the most preffered atmospheric characteristics of airport shopping area based on utility value analysis.

**Figure 2.** Consumer preferences towards atmospheric design in airport shopping area

### 4.2. Correlation analysis between atmospheric design and purchase intention

This analysis needs to be done to prove whether shopping area design can affect purchase intention. The analysis was performed with SPSS software using binomial logistic regression.The result of correlation analysis are presented in Table 4. Since p value $< 0.05$, it can be concluded that the atmospheric design has a significant influence on consumer purchase intention at the airport.

**Table 4.** Correlation analysis between atmospheric design and purchase intention

| | | B | S.E. | Wald | df | Sig. | Exp(B) |
|---|---|---|---|---|---|---|---|
| Step 1(a) | Profile | -,112 | ,037 | 8,947 | 1 | ,003 | ,894 |
| | Constant | 1,839 | ,168 | 120,433 | 1 | ,000 | 6,291 |

### 4.3. Effects of Socio-economic Status towards consumer preferences

According to consumer behavior theory, socio-economic class can affect consumer shopping behavior. Therefore, this analysis was conducted to determine whether there are differences on preferences between socioeconomic classes. The analysis was done using XLSTAT software to obtain the utility value of each attribute and level. The analysis was conducted for each socio-economic class divided by the Social Economy Status (SES), SES D, SES C, B SES, SES A, and SES A+.

As shown in Table 5, the findings indicate that each SES has different preferences on atmospheric design. Therefore, airport terminal building should be designed based on SES. By doing this, passengers will feel comfortable inside the terminal building and at the same time driving the passengers to shop at the airport.

## 5 Conclusion

Research was carried out by choice-based conjoint analysis. Survey was distributed to 500 passengers at SHIA, examining five variables including window display, store entrance, color scheme, product display, lighting.

**Table 5.** Consumer preferences based on SES

| SES | Monthly Expenditure | Preference | | |
|---|---|---|---|---|
| | | **Attribute** | **Level** | **Picture** |
| A+ | *Above IDR5,000,000* | *Window Display* | Warm colors |  |
| | | *Store Entrance* | Using glass | |
| | | *Color Scheme* | Warm colors | |
| | | *Product Display* | Products are visible from the outside | |
| | | *Lighting* | Bright | |
| A, B, C | *SES A* (IDR 3 Mio–IDR 5 Mio)<br>*SES B* (IDR 2 Mio–IDR 3 Mio)<br>*SES C* (IDR 1 Mio–IDR 2 Mio) | *Window Display* | Cool colors |  |
| | | *Store Entrance* | Using glass | |
| | | *Color Scheme* | Cool colors | |
| | | *Product Display* | Products are visible from the outside | |
| | | *Lighting* | Bright | |
| D | *Below IDR 1 Mio* | *Window Display* | Warm colors |  |
| | | *Store Entrance* | No glass | |
| | | *Color Scheme* | Cool colors | |
| | | *Product Display* | Products are visible from the outside | |
| | | *Lighting* | Bright | |

Result showed that passangers prefer shopping area with cool colors, high level of store visibility, and high-intensity lighting (200 lx). Binomial logistic regression analysis showed that atmospheric variables significantly influence purchase intention at the airport.

This research also suggests that socioeconomic class influences consumer preferences. Results show that preferences between socioeconomic classes are slightly different to one another. In consequence, airport terminal building should be distinguished based on socioeconomic class.

## References

1. P. Adey, "Airports, mobility and the calculative architecture of affective control,"Geoforum **39**,pp. 438–451 (2008)
2. C. J. Oel and F. Berkhof, "Consumer Preferences in The Design of Airport Passenger Areas,"J. Env. Psy. **36**, pp. 280–290 (2013)
3. S. Gunawan and V. H. Supit, "The Influence of Store Atmosphere on Emotional Responses and Re-Purchase Intention," Bus. Manag. Strat.**Vol.5 No.2** (2014)
4. P. Kotler, "Atmospherics as a Marketing Tool,"J. Retailing **49**, pp. 48–64 (1973)
5. L. W. Turley and R. E.Milliman, "Atmospheric Effects on Shopping Behavior: A Review of The Experimental Evidence," J. Bus. Res. **49**, pp. 193–211, 2000.
6. J. C. Chebat and M. Morrin, "Colors and Cultures: Exploring The Effects of Mall Decor on Consumer Perceptions," J. Bus. Res. **60**, pp.189–196 (2007)
7. N. K. Park and C. A. Farr, "The Effects of Lighting on Consumers' Emotions and Behavioral Intentions in a Retail Environment: A Cross-Cultural Comparison,"J. Interior Design 33, pp. 17–32, 2007.
8. R. Grover and M. Vriens, "The Handbook of Marketing Research: Uses, Misuses, and Future Advances," Thousand Oaks: Sage Publications, Inc. (2006)
9. R. M. Johnson and B. Orme, "Sawtooth Software: How Many Questions Should You Ask in Choice-Based Conjoint Studies," http://www.sawtoothsoftware.com (1996)

# Sensory Integration Training Tool Design for Children with Autism Spectrum Disorder

Lijun Jiang[1a], Maoteng Ye[1b] and Zhelin Li[1]

[1]*South China University of Technology, Higher Education Mega Center, Panyu District, Guangzhou City, Guangdong Province, China*

**Abstract**. This study aims to design a training tool for therapy of children with autism spectrum disorder (ASDs). Typically, ASDs pass through obstacle track several times with sandbags, which should be picked up from starting point and threw into a box at the end during sensory integration therapy. Counting the sandbags can help ASDs to have concept about the progress of mission. We redesign the counting box named "Skybox" which can help counting by playing sound after detect something throw in it. Aims to probe into the sound preference of two main subjects, an experiment with four kinds of sounds is conducted in this paper by using the method of paired comparisons. The result shows they like animals most, followed by human voice and nature sounds, and music instrument is the last. The material preference experiment shows two subjects like acrylic most, wood and paper are secondary while furry is the last. Skybox shortens their training time for 23.53%, 29.87% and 37.37% in three different projects. We consider that Skybox attracts ASDs therefore reduces their distraction and improves their performance in the usability test.

## 1 Introduction

Autism spectrum disorders (ASD) are a group of neurodevelopmental disorders characterized by impairments in social interaction and communication, as well as repetitive behaviors and restricted interests.[1, 2] There are about 35 million ASDs in the world, 1.5 million of them are in China.[3] Part of them can blend into society after early and long-term therapy.

Sensory integration therapy is one of the effective therapy.[4] Typically, one sensory integration therapy (one class) includes serval projects. Many projects have a track with obstacles. ASDs need to climb over obstacles, which set by the therapist, again and again. ASDs are required to pass through the same track 4 to 8 rounds (with sandbags or something else to counting) depending on their capability. The therapy seems boring to ASDs.

Therefore, we aim to design fun training tool to increase ASDs' interest of sensory integration training, and then enhance their training effects. Counting box was chose to improve for its universality, ordinary and high relation to the track.

[a] ljjiang@scut.edu.cn
[b] xiaomieye@163.com

## 2 Training tool design

There aren't any autistic toys could be make as reference for autism sensory integration therapy. Whereas, a case in "Fun Theory Award" is similar.[5] When people throwing rubbish into the bin, it plays the sound of something falling down, that motivates them to collect more rubbish. Our design uses it for reference and make a usefully prototype finally.

Boys R (4-year-old) and Z (8-year-old) diagnosed with ASD are our main subjects. Boy Y, who diagnosed with low severity level of ASD, doesn't have sensory integration class. G is a typical developing (TD) girl who has an autistic brother.

### 2.1 Step 1: Product Concept

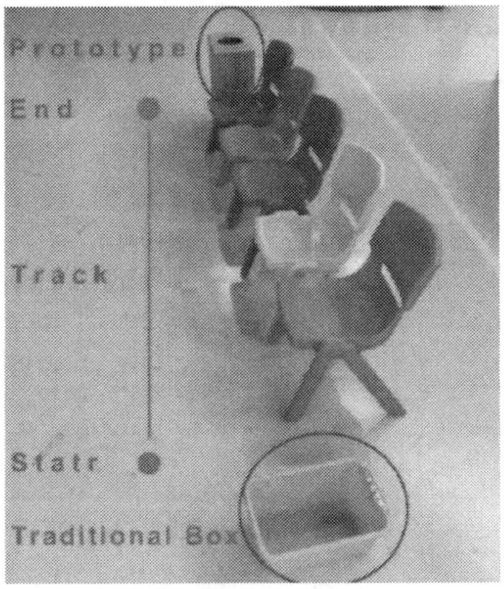


**Figure 1.** Schematic diagram of product concept evaluation

To ensure fun theory [5] works for ASDs, we conduct a product concept evaluation with a paper prototype (Figure 1, end of the track). It consists of paper body and electronic part that can make the sound of falling to the chasm when it detects something passing through the hold. R and Z show interest to the prototype. However, R is afraid of the sound. Y is curious about the box and departs it without talking or paying any attention to experimenters. G ask me how the box works.

Compare with normally developing children, autistic children have different listening preferences in regard to speech.[6]. We haven't find study about ASDs' sound preference, and assume ASDs have different sound preference from TD due to their imbalanced brain functions. Therefore, a sound preference experiment is needed.

### 2.2 Step 2: Sound Preference Experiment

Eight sounds are chose to be the experiment materials. Dog, cat, drop, wave, great, cheer, single-tone of piano, portamento of piano. They can be divided into four kinds of types: animals (dog, cat), natural (drop, wave), human voice (great, cheer) and musical instrument (single-tone of piano, portamento of piano). It also can be divided into two kinds of length: single tone sound (dog, drop, great, and single-tone of piano) and persistent sound (cat, wave, cheer, and portamento of piano). The experiment also sets eight sounds which have the similar volume and various persist time.

To rank the sounds strictly and eliminate the effect of hand-preference (right-hander or left-hander), paired comparison method is used in this experiment. Experiment is divided into two groups, each group has 28 pairs of sounds that can generates 28 choices. Two subjects (Z and R) participate in experiment. The sequence shows in table 1.

**Table 1.** Sounds appear sequence

| Sequence \ Sound | Dog | Cat | Drop | Wave | Great | Cheer | Single-tone | Portamento |
|---|---|---|---|---|---|---|---|---|
| Dog | —— | B28 | B27 | B15 | B14 | B6 | B5 | B1 |
| Cat | A1 | —— | B26 | B25 | B13 | B12 | B4 | B3 |
| Drop | A2 | A3 | —— | B24 | B23 | B11 | B10 | B2 |
| Wave | A14 | A4 | A5 | —— | B22 | B21 | B9 | B8 |
| Great | A15 | A16 | A6 | A7 | —— | B20 | B19 | B7 |
| Cheer | A23 | A17 | A18 | A8 | A9 | —— | B18 | B17 |
| Single-tone of piano | A24 | A25 | A19 | A20 | A10 | A11 | —— | B16 |
| Portamento of piano | A28 | A26 | A27 | A21 | A22 | A12 | A13 | —— |

The experiment conduct in an independent quiet room. One subject participants with one experimenter and one therapist at one time. Two boxes placed on the table. Each box has a button. The box can play a sound once the button is pushed. At the beginning of each test, experimenter pushes two buttons to show which button corresponding to which sound. Then the subject has time to play two boxes by pushing two buttons freely. After subject is familiar with the sounds, therapist asks him which sound he likes. Subject pushes the button to answer. Then experimenter records the data. We adapt this method because ASDs have deficit in speaking. The result shows in table 2.

**Table 2.** Experience result of sound preference

| Item | Animals | | Natural | | Human voice | | Music instrument | |
|---|---|---|---|---|---|---|---|---|
| Sound | Dog | Cat | Drop | Wave | Great | Cheer | Single-tone | Portamento |
| Chose | 14 | 21 | 8 | 19 | 8 | 20 | 8 | 14 |
| Chose | 35 | | 27 | | 28 | | 22 | |

Animals' sounds are the most popular, followed by human voice and natural. Music instrument performs the worst. This outcome meets Anke Prothmann's research in a sense, that their study shows ASDs prefers to interact with dogs rather than people or objects.[7]

Besides, the result shows they prefer persistent sounds (74 times) rather to single tone sounds (39 times). We consider the quantity of stimulus makes this different.

### 2.3 Step 3: Material Preference Experiment

This experiment, by using the method of paired comparisons, aims to probe into the material preference of two subjects. Four common materials are chose in the experiment, include paper, wood, acrylic and plush. To eliminate the color different, we painted the acrylic brown and chose the brown plush which approach to the color of paper and wood looked. The experiment uses four boxes, which have the same size and the same function that play the same sound when detect something drop into the boxes.

The experiment conducts in the real training circumstance. Experimenters place two boxes, which appeared in a random order like step 2, at the end of the track. Subjects throw the sandbags, which they bring from the beginning of the track, into the box they like. After that, experimenters record the data and change the boxes for next throw. The result shows that acrylic is the most popular, followed by wood and paper. Plush is the less.

ASDs show great interest to the interior of the boxes. Acrylic box was sprayed paint, but partly keeps its transmission of light. Paper box and wood box have similar brightness of the interior as well as similar texture because paper is made from wood. Plush box performs worse. ASDs seems scare for a plush box with a dark hole which can make sound after eating a sandbag. It's hard to draw conclusion whether the material or the inner brightness of boxes effect the result. Nonetheless, we choose acrylic for its attractive to ASDs.

## 2.4 Step 4: Prototype

Generally, the autistic children surveyed prefer cool color to warm color; they prefer green, followed by blue, red, purple, orange and yellow.[8] Considering the floor is green, we use sky blue acrylic to compose the body of the prototype. We name it "Skybox" since its color and ASDs also called "children from stars". (Figure 2, 3)

**Figure 2.** Skybox outside

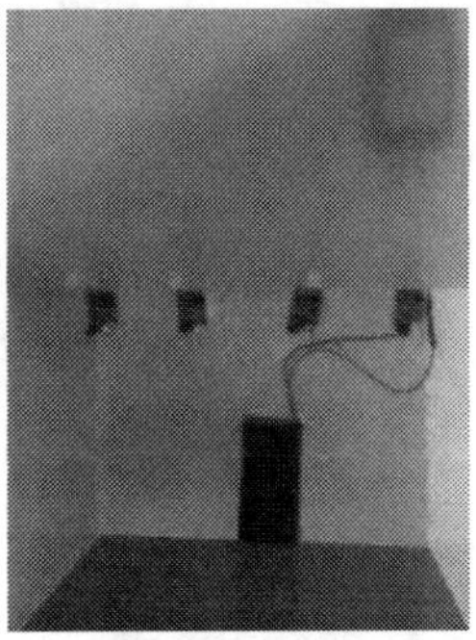

**Figure 3.** Skybox inside

The electronic part of Skybox consists of Arduino, battery pack, transformer, loudspeaker and four IR evading obstacle sensors. All of them are out of children's sight. Battery pack is placed to the bottom of Skybox, so that users can replace batteries conveniently.

Skybox is 30cm height. ASDs can throw the sandbags (or anything else for counting) into Skybox easily whatever they are sitting on the floor, standing on the floor even standing on the chair. The rubber strip which cover the edge of Skybox can make it safe to ASDs. The bottom was remove since we found that it doesn't need to store sandbags but needs to take out them conveniently. ASDs can hear the sound but can't see anything from the opening to the bottom of Skybox. That make it seems dramatic and attractive.

## 2.5 Step 5: Evaluation

**Table 3.** Skybox experiment outcome

| Subject | R | | | | Z | |
|---|---|---|---|---|---|---|
| Mission | Climbing | | Monkey Creep | | Kangaroo Jump | |
| Mode | Tradition | Skybo | Tradition | Skybox | Tradition | After |
| 1st | 1′30″ | 57″ | 49″ | 16″ | 15″ | 14″ |
| 2nd | 1′05″ | 15″ | 15″ | 15″ | 12″ | 15″ |
| 3rd | 1′17″ | 1′11″ | 15″ | 22″ | 17″ | 15″ |
| 4th | 1′28″ | 48″ | 58″ | 34″ | 33″ | 10″ |
| 5th | 52″ | 34″ | 16″ | 30″ | | |
| 6th | 33″ | 27″ | | | | |
| 7th | 27″ | 23″ | | | | |
| 8th | 47″ | 25″ | | | | |
| Summation | 479″ | 300″ | 153″ | 117″ | 77″ | 54″ |
| Average | 59.875″ | 37.5″ | 30.6″ | 23.4″ | 19.25″ | 13.5″ |
| Shorten | 37.37% | | 23.53% | | 29.87% | |

Usability test conducted in the daily training circumstance instead of special circumstance. It is vital to select appropriate training project to evaluate Skybox. "Squat on cushion" is a training project of R, who can't squat for a long time due to the weakness of relevant muscle. R has to finish serval rounds in this project. In each round, therapist count to ten seconds when R keep squatting on cushion. Sometime, when R performs well, therapist may slow down the counting speed stealthily. Therefore, the relationship between consumption of time and the effect of Skybox is blurry in "squat on cushion". We exclude this kind of project and select three projects which have clear relationship between consumption of time and training performance (Table 3).

In the first day, we recorded the consumption of time during each project. The next day, we replace the traditional counting box with Skybox. Each project measure at the same periods of time on two day. The length of the track and the therapist were the same, too. The result showed in table 3.

## 3 Discussion

The result amazes us since we assume about 10% descend before evaluation. This paper think Skybox works by reduce ASDs' distraction. An early feature of autism spectrum disorder is impaired joint attention as manifested by a lack of pointing, showing, or bringing objects to share interest with others, or failure to follow someone's pointing or eye gaze.[2] When ASDs loses the attention to the project, time consumption rises in abundance.

Skybox may provide an interesting point to ASDs and reduces their distraction. That makes the score looks significant. Actually, there is no significant different between two mode when ASDs don't distract. It should be noted that Skybox's evaluation conducted only in two days, the skybox is still fresh to two subjects. The attraction of Skybox to ASDs may drop after a period. Anyway, it improves the training result. Furthermore, reduce of distraction has important meaning for therapy.

## 4 General discussion

The long-term significance of this study aims to help ASDs blend into commend society. ASDs aren't capable enough to use universal smart products like ordinary people does. There are less smart products designed for ASDs therefore they seldom use smart products. Sometime ASDs contact smart produces "passively" such as watching video on TVs or smart phones. In this study, ASDs use smart product "actively", which we consider it significant.

## References

1. Maenner, M.J., et al. JAMA Psychiatry. **71(3)**: p. 292-300. (2014)
2. Association, A.P., *Diagnostic and Statistical Manual of Mental Disorders, Fifth Edition*. (2013)
3. C Shunsen, B Xuejun, Z Risheng. Advances in Psychological Science, **01**: p. 60-72. (2011)
4. Simpson, R.L. Focus on Autism and Other Developmental Disabilities. **20(3)**: p. 140-149. (2005)
5. The World's Deepest Bin. (2009)
6. Klin, A. Journal of Autism and Developmental Disorders. **21(1)**: p. 29-42. (1991)
7. Prothmann, A., C. Ettrich, and S. Prothmann. Anthrozoös. **22(2)**: p. 161-171. (2009)
8. C Shuqin, F Wen, S Aijun. CJSE. **05**: p. 46-52. (2012)

# Integrating Procedural Modelling Process and Immersive VR Environment for Architectural Design Education

Chun-Heng Lin[1,a] and Pei-Hsien Hsu [2]

[1]*NCTU, Architecture Department, 1001 University Road, Hsinchu, Taiwan, ROC*
[2]*NCTU, Architecture Department, 1001 University Road, Hsinchu, Taiwan, ROC*

**Abstract** In the field of architecture, researches are being conducted for the issue of combining VR environment and procedural modelling. Instead of simply using VR as an visualization tool for procedural modelling, as in other researches, we argue that manipulations of 3D elements and adjustments of design parameters should be implemented in VR to provide further supports for a design process. An integrated system is thus proposed based on this augment, with a special focus on architectural design education. This paper presents the design of such a multi-user system, in which 3D modelling, procedural modelling and VR platform are integrated, aiming to support architectural design education.

## 1 Motivation

There has been a long history between architecture and virtual reality. As computer involves into design process, some of researches about digital media propose that virtual reality as medium for architecture design can allow students to experience almost actual perception from the immersive environment[1]. In addition, architecture navigation emerged in many experiments about VR (Virtual Reality)[2-4], they tried to make the scene look real and set up triggers in the scene to improve users' space perception. The device for VR display is changing with time, from computer screen to projection walls[5]. There is a big breakthrough of VR device compare to the old days. Based on the advancement of computing capability, VR equipment became more lightweight and affordable. The new VR device with powerful HMD (Head-mounted display) and natural human motions, brings back designers' imagination in virtual world. Therefore, it is a potential opportunity for architectural design education. The device allow students to access the immersive learning environment.

Procedural modelling is a great leap in CAAD (Computer-Aided Design and Drafting), and it drastically changes the way that designers think. Since 2007 to present, numerous experiments were conducted[6-7]. Procedural modelling has become an important architectural design methodology. With this methodology, designers can set the rules based on design concept to generate diverse form or space. Nevertheless, under the trend of procedural modelling, traditional architectural design education still use sketching and physical modelling as main methods to showcase the ideation. However, unlike experienced designers, students can only imagine or even guess the real scale of projects by current methods. According to the new VR breakthrough, there are some experiments about merging the new VR and procedural modelling process. Only real-time operation to objects, and

[a] Corresponding author : aj71735@arch.nctu.edu.tw

relevant parameters in VR environment are significant enough to provide a comprehensive understanding, in addition to navigation.

It is a new possibility for architecture students nowadays. They can learn in more instinctive and flexible environment with direct interaction to the scene, by the combination of immersive VR environment and procedural modelling process. This study suggests that VR platform is a more efficient tool for students to develop and discuss their design instead of a visualizing tool.

# 2 Context

In the previous paragraph, the relevant resource for our imagination is mentioned roughly, more details will be supplemented in the following contents.

## 2.1 Medium in Architectural Design

Medium contains design concept is employed by architects to share and discuss the ideas. A common way for them to express their design concept is 2D free-hand sketching. This method can be traced back to the time before Renaissance[8]. The ambiguity and amorphousness of sketching help architects to improve ideation process[9]. Sketching is also a simple method for brief discussion and exchange ideas[10]. The small-scale model is another medium. It helps to show design results in more precise way.

After computer got into design process, more and more studies are curious about the effects caused by digital media. Most of the studies think that digital involvement can enhance the efficiency of design process, reducing budget, expanding the complexity of form, and increasing experience[11-12]. As designers become familiar with digital media, they tried to test the tool in diverse directions. At first, some of them applied the traditional design media, sketching and modelling, in digital way with their existed advantages in design[13-14]. On the other hand, some of them developed new design methodology. Create the forms and design results by setting a series of rules, which is procedural modelling. Furthermore, some of them considered about the experience of design process and presentation, so they focused on building an immersive world for design process.

## 2.2 Procedural Modelling in Architecture Design

In traditional CAAD modelling process, designers usually developed the form from concepts in mind first, then, they built it in CAAD tool piece by piece. Despite the concept of procedural modelling has been discussed over thirty years, it is not until about 2007, new digital tools help designers to think outside the box, and they break the linear design process[15]. That is to say, the way that designers develop form can be separated into many steps. Every step is a small function that can be reused in another design process. There are some essential advantages of procedural modelling such as data amplification capabilities, data compression, and unpredictability[16]. Designers can only design their own rules, and connect these rules with real-time data from sensors to design dynamic forms, instead of completing all form in advance. By the flexible procedural modelling tools, designers can also construct their own kit to design, and the kit can be easily share to the public.

## 2.3 Spatial Perception and VR in Architectural Design Education

It is still not common for architectural design education to use VR platform as presentation tool. Currently, students in class propose ideas and discuss by using sketches and models to imagine the project deliverables. In the early stage of architectural design, it depends on brief sketching to develop ideas. As the concept developed, it is time to prototype the abstract idea into space or buildings. In this stage, according to some researches, they think learning in the VR environment can improve novice's spatial cognition [17-18]. There are various studies about learning architectural design in VR[19-21].

VR technique development can be separate into several stages by the constrain of hardware[22]. At the beginning, the idea of immersive environment in real world is practiced by using heavy equipment and shows images on the screen. As projection and 3D image technique mature, CAVE (Cave Automatic Virtual Environment) is available with motion capture to give viewers surrounded by walls full interacting experience[23]. Because of the mobility, well-prepared platform, and the compatibility between display platform and other CADD tools, the new lightweight VR device shows more potential opportunities for architectural design education.

### 2.4 Procedural Modelling in VR

Combining procedural modelling and VR environment in real time is still a new issue. There are a few experiments between Grasshopper, a parametric modelling environment, and VR platform Unity 3D. In Gathmann's demo, he successfully transmits data from Grasshopper to Unity, and the program automatically create the collider of object in the scene[24]. The team, AWARE, tries to return data from Unity to Grasshopper to change the form by capturing the cursor position and object dimensions[25]. Although this case is done by capturing cursor position on 2D screen, the interaction is still a good example for getting operation states in VR platform. The experiment shared by Orange Jellies shows the parameters used in Grasshopper to user interface in Unity. In this case, users in VR platform can revise the value to change the form[26].

## 3 System Overview

As mentioned above, digital media involves in design process more and more often. Designers should also develop their original ideas at the same time, besides using digital media to represent conceptual ideas and to communicate with others. Based on the breakthrough of VR technique, this research expands on existing researches about digital media in design process. Proposing new design system which consists of existing 3D modelling platform and VR platform, and practicing it in architectural design education. It is not sufficient to learn architectural design only by navigation in VR environment, students can gain more feedback by operating and adjusting the form of design objects directly. The system also integrated the parametric operation in VR learning environment to visualize those rules behind forms. The visualization of design rules makes the discussion more understandable in class.

### 3.1 System Workflow

The system proposed by this study is shown as figure 1. 3D modelling platform and parametric modelling environment are setup on the server, and they are linked with VR platform through a specific protocol (TCP/IP), hence making distant connections over the internet feasible. The models in 3D modelling platform will be dynamically read by VR platform in the form of meshes. Some typical operations about objects in VR platform, such as translate, rotate, and scale, are sent back to 3D modelling platform to revise old models, then loading new models into VR platform again. If designers use parametric modelling environment as tool when modelling, they can choose the necessary parameters and show them on VR platform interface. Once the parameter is revised, the new value is sent back to parametric modelling environment, then new models are baked into 3D modelling platform automatically. Finally, those new models will be transmitted to VR platform once again.

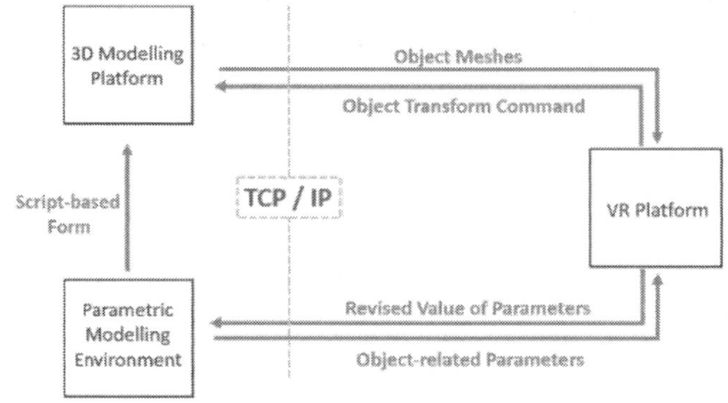


**Figure 1.** Data transmission framework

### 3.2 Role of Designer and Participants in System

There are two major roles in the system: designers and participants. To use the system, students in an architectural design class need to build their own model in 3D modelling platform and parametric modelling environment to present ideas and design concepts. Later upload their files to server. One of the design students in the class displays the model. If the model is built in parametric modelling environment, the student has to choose critical parameters for discussion or concept presentation in VR environment. Teachers and other students in the class, as the participants, can navigate the file with immersive environment. They can manipulate objects of the model at the same time, and the critical parameters chosen by designer are shown in the VR environment for value adjustment to present the concept, which is conducted by procedural modelling. The designers explain their design concepts in the VR environment, and the participants can understand the project not only by the explanation, but also can gain experience from their own perception. Moreover, they can revise the project to leave their advice or to discuss with designer. As a nature of virtual reality, the participants in this framework do not need to be in the same physical space. In addition, multiple version of revised model can be saved during the discussion. Students can also acess different version after the class, and save a new version to leave their revision. As shown in figure 2, with such a system, distance and asynchronous learning can thus be supported.

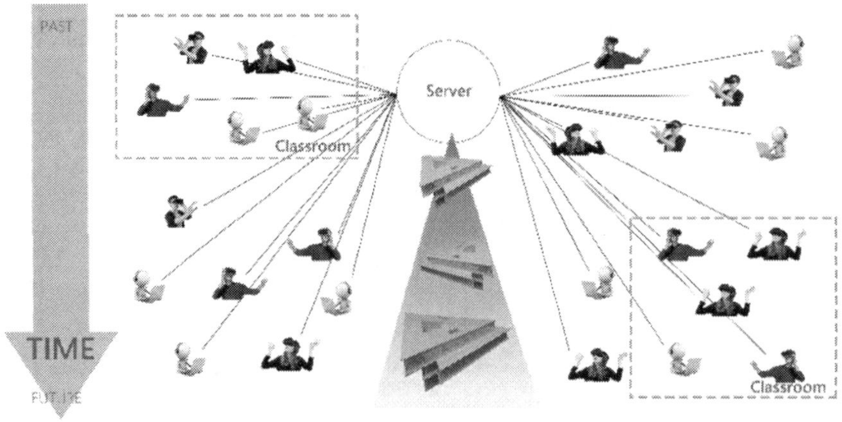


**Figure 2.** Architectural design education architecture

### 3.3 Education-Oriented Operation in VR World

There are some operations need to be designed for the system proposed in this study. It is hard to do precise operations in the VR environment, yet the perception will become instinctive while it is free to control and interact with virtual objects without physical constrain. Therefore, as the scenario shown in figure 3, every part in the scene should be manipulated separately. For example, the walls interior can be moved or extended to test different spatial experience. A window can be put on many vertical surfaces to evaluate the best place for it in the project, and textures should be attached in simple way to get close to real perception of space, as well as the light. Those parameters, chosen by designer, involve in discussion in the VR environment as well. The precise value is not important at all, what is more is the relationship between variables and outcome that make the communication meaningful.

**Figure 3.** Scenario of architectural design education in the VR environment

## 4 Conclusion

Taking procedural modelling and new generation immersive VR as the basis, this paper presents a system framework, which integrates these two components, aiming to support architectural design education. Such a system framework is developed based on a design scenario: students and teachers in a design class simulate and discuss about an architectural design. By interacting with objects, in addition to the design parameters manipulation coming from the linked procedural modelling environment, "inside" an immersive VR environment. This leads to an integrated, multi-user system, which combines the strengths of procedural modelling along with 3D modelling environment and immersive virtual reality. Allowing designers to build 3D models that are automatically linked to an immersive VR environment to be accessed by all design class participants, in order to perform design simulations and discussions, which in turn, through the proposed system, feedback to the designers' 3D models and procedural model. As a work phase, the mechanism of data transmission, for both parametric data and geometric data between modelling and VR platforms, which is the key for the proposed system that have been developed and tested. Solutions to the possible conflicts between design parameters, structured in the procedural modelling platform, and direct object manipulations, supported inside VR environment, need to be developed. In addition to other tasks such as intuitive interfaces design for object manipulation and movement inside the VR environment, as the future steps toward the implementation of whole multi-user system framework for architectural design education.

## References

1. S. Sven, et al. *Educating architecture students to design buildings from the inside out– Experiences from an research-based design studio.* SSS9 Proceedings, Seoul (2013).

2. M. Roupé, P. Bosch-Sijtsema, and M. Johansson. *Interactive navigation interface for virtual reality using the human body.* Comput Environ Urban Syst, **43**, 42-50 (2014)
3. S. F. Kuliga, et al. *Virtual reality as an empirical research tool—Exploring user experience in a real building and a corresponding virtual model.* Comput Environ Urban Syst, **54**, 363-375 (2015)
4. F. Castronovo, et al. *An evaluation of immersive virtual reality systems for design reviews.* Proceedings of the 13th CONVR, 22-29 (2013)
5. S. K. Ong, A. Y. C. Nee. *Virtual and augmented reality applications in manufacturing.* IFAC Proceedings, **46**, 15-26 (2013)
6. S. Dritsas. *Procedural Building Information Modeling for Digital Fabrication.* CAADRIA 2015, 355-364 (2015)
7. D. Davis. *Modelled on software engineering: Flexible parametric models in the practice of architecture.* Diss. RMIT University (2013)
8. J. Whyte. *Towards a new craft of architecture.* BRI, **43**, 2, 263-265 (2015)
9. S. Inoue, et al. *Reductionism as a tool for creative exploration.* Proceedings of 11th EAD, Paris Descartes University, France, (2015).
10. G. Goldschmidt. *Modeling the role of sketching in design idea generation.* An Anthology of Theories and Models of Design. Springer London, 433-450, (2014)
11. D. Wu, et al. *Cloud-based design and manufacturing: A new paradigm in digital manufacturing and design innovation.* Comput Aided Des, **59**, 1-14, (2015)
12. K. Vasylevska, I. Podkosova, H. Kaufmann. *Walking in Virtual Reality: Flexible Spaces and Other Techniques.* The Visual Language of Technique. Springer International Publishing, 81-97, (2015)
13. M. Evans, N. Aldoy. *Digital Design Sketching using the Tablet PC.* The Design Journal, **19**, 5, 763-787 (2016)
14. R. Holden, J. Liversedge. *Landscape architecture: an introduction.* Laurence King Publishing, (2014)
15. P. Hsu *Integration of Procedural Design and 3D Virtual Environment to Support Collaboration for the Design of Built Environment.* (2014)
16. R. M. Smelik, et al, *A survey on procedural modelling for virtual worlds.* CGF, **33**, 6 (2014)
17. A. Angulo. *On the design of architectural spatial experiences using immersive simulation.* EAEA 11 Conference Proceedings, Envisioning Architecture: Design, Evaluation, Communication. Italy: Milan (2013)
18. Y. Chang. *3D-CAD effects on creative design performance of different spatial abilities students.* JCAL, **30**, 5, 397-407 (2014)
19. W. A. Abdelhameed. *Virtual Reality Use in Architectural Design Studios: A case of studying structure and construction.* Procedia Comput Sci, **25**, 220-230 (2013)
20. D. Fonseca, et al. *Visualization methods in architecture education using 3D virtual models and augmented reality in mobile and social networks.* Procedia Soc Behav Sci, **93**, 1337-1343 (2013)
21. A. Brown, T. Green. *Virtual reality: Low-cost tools and resources for the classroom.* TechTrends, **60**, 5, 517-519 (2016)
22. R. A. Earnshaw, M. A. Gigante, H. Jones. *Virtual Reality Systems.* Academic Press, (2014)
23. H. Creagh. *Cave automatic virtual environment.* Electrical Insulation Conference and Electrical Manufacturing & Coil Winding Technology Conference, Proceedings. IEEE, (2003)
24. R. Gathmann. Grasshopper To Unity3D. Retrieved November 25, 2016, from https://vimeo.com/68066366
25. AWARE. Exchanging data between Unity and Grasshopper via UDP. Retrieved November 25, 2016, from http://www.aware-studio.com/exchanging-data-between-unity-and-grasshopper-via-udp/
26. O. Jellies. Send/Receive data between Grasshopper to Unity in realtime. Retrieved November 25, 2016, from https://vimeo.com/153197500

DOI: 10.1051/mateccconf/201710403008

# Intelligent Furniture Design in the Elderly Based on the Cognitive Situation

LU Xinhui[1,2], HE Li[1,2,a]

[1]*Sichuan Agricultural University, 611130 Chengdu, P.R. China*
[2] *Key Laboratory of Wood Industry and Furniture Engineering of Sichuan Provincial Colleges and Universities, 611130 Chengdu, P.R. China*

**Abstract.** This paper analyzes the present situation of Chinese elderly furniture and the elderly has cognitive characteristics that consciousness experiences and recognitions recede, cognitive fuzzy from Information processing. Expounds the elderly intelligent furniture design elements: functional elements required the elderly furniture is easy and simple to handle; Size and shape elements should be biased towards low, light type, reduce multifunction or fold function; colour collocation should use low lightness and low purity natural materials; Emotional elements design should meet the demand of the elderly social emotion. Introduction of intelligent furniture make up the cognitive decline in the elderly, Furniture judge the elderly demand by the inductor, Supplement by hardware control module to solve the special needs of the elderly life. Build design thinking based on the cognitive process and explore the elderly intelligent furniture design. This paper discusses the design process, for example and concludes the design rules: 1.The Operating Experience Pleasure. It is the height matching of user expectation and furniture function. Pleasure in the design of the operating parts mainly embodies in two aspects. Firstly, the Fitts Law; Secondly, it's The Movement Optimization. 2."Unconscious" Design. Intelligent furniture need to delete unnecessary operation module, make it easy to understand, furniture function and cognitive scene match with each other. 3. Modularity Design. Modularization can indirectly regulate the scale and specification of the design. Under the premise of individual character, customization, the compression of the cost, Designer should make the elderly intelligent furniture consistent with the user action.4.Design Consistency. The consistency principle reflected in the appearance, color and operation way consistency.

## 1 Status quo of elderly furniture

An investigation by The China Population and Family Planning Department showed that China social aging degree becomes more obvious in 2015. Currently, China has more than 220 million senior citizens who are above 60, more than 16 percent of the total population. China has entered the phase of aging society. With the growth of economic, the aging people purchasing power are also enhancing. Aged people pay more attention to the quality of life. But the elderly behavior speed generally slows down with age growing[1]. The body organs degradation lead to the decline of information input and output, storage, and reaction rate, the cognitive process is different from young people, so the elderly furniture design shows more particular. The elderly furniture is a serious lack in the market at present. However, Ordinary furniture can't satisfy the specific requirements of the elderly both in physiology

[a]Corresponding author : 490585432@qq.com Project support: 16SA0006

DOI: 10.1051/matecconf/201710403008

and psychology, therefore, the elderly furniture r&d became a problem demanding prompt solution. For providing convenience in using furniture and ease of operation by older people, the addition of intelligent modules will solve such problems effectively.

## 2 Build furniture cognitive situation of the elderly

### 2.1 The cognitive process of the elderly

Cognitive process is a method of study human behavior and provides the basis for furniture design. Cognitive science divides people's cognitive activities into three processes: First, Cognitive mechanism which about information how to input, cognitive mechanism how to output; Second, the mental structure about cognitive how to be stored and in what form; Third, Mental Process about how to find and how to use[2].

Storage and search of memory data are the core of the whole process of information processing (Figure 1).First is the information collection stage, it is a prerequisite for memory formation, the old people's attention as information filter is hard to play normal function; Perceptual experience weakened due to the degeneration of sensory stimulation; The obstruction of information collection result in a loss of recognition.

Followed by the stage of information storage, the form of memory storage leading to confused the feel receiver in logic relationship and cognitive fuzzy between short-term memory and long-term memory due to the decline in visual and auditory sensitivity, Finally is information operation, further cognition process is derived from the sense organs interaction overlap, need a systematic approach to explain how the memory is being collected and stored.

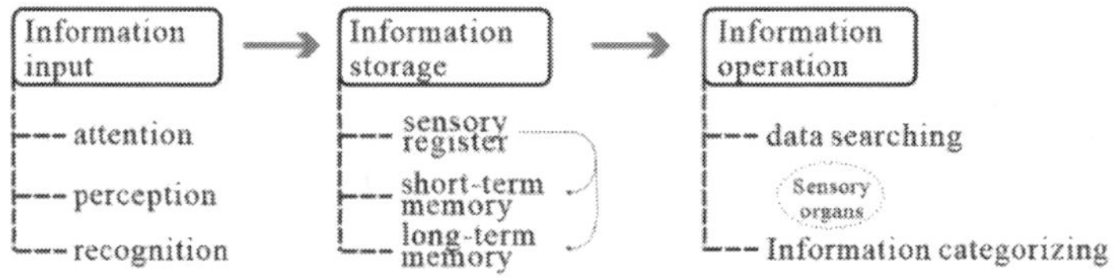


**Figure1.** Information processing

### 2.2 The elderly furniture cognitive situation

Furniture design considers the relationship of user, furniture and environment[3], Cognitive situation builds by interactive scenario based on the elderly psychological characterristics. In the design cognitive situation can be considered as the interactive process which the user is related to the object and the environment, that is to say, the comprehensive perception process of users towards objects and the environment. The ideal cognitive situation is to remove the boundaries between the user information handling and the use of products, which achieve the interaction coordination between the people, the object and the environment. It bring the cognition into correspondence with design elements, Intelligence make the fusion of design elements and cognition (Figure 2). The elderly furniture design elements include the function, form, colour, emotional factors and so on, as the embodiment of the furniture external characteristics. It decided by the particularity of the elderly cognitive process. The process of old people use furniture interacts with various design elements, the premise is to properly handle information. The elderly furniture use process pursues optimal emotional experience, to analyze the elderly cognitive process as a starting point, through the collection of information used to describe the object position to build cognitive situation.

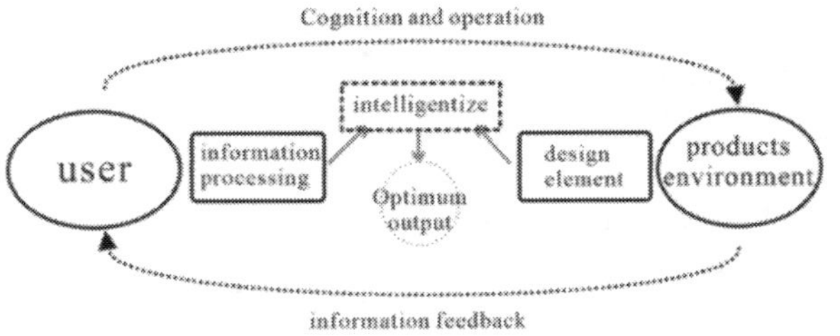


**Figure2.** Intelligent design and cognition process

## 3 Based on cognitive scenarios of the elderly furniture design elements

The elderly cognition is special. Based on the different characteristics of the elderly population, basic life as a starting point; this paper expounded the elderly furniture design elements from the following points:

Functional elements: Behavior is composed of a task, the task consists of an action, the action is composed of operation[4]. Hand, eye, foot as the core of the human body, best action number sequence diagram you can see from figure 3, for example, the hand operating flexibility is in turn of fingers, wrist, forearm, upper arm, shoulder, body, waist. A reasonable operation sequence can simplify the process, improve the efficiency of action. The old behavior has a slower pace; we should consider the ease of operation in the design of furniture, such as arthritis result in fingers activities inflexible. We can increase the support design, convenient old people got up and take things; The furniture should Increase the soft structure and make the corner design smooth to prevent injuring or damaging after being hit.

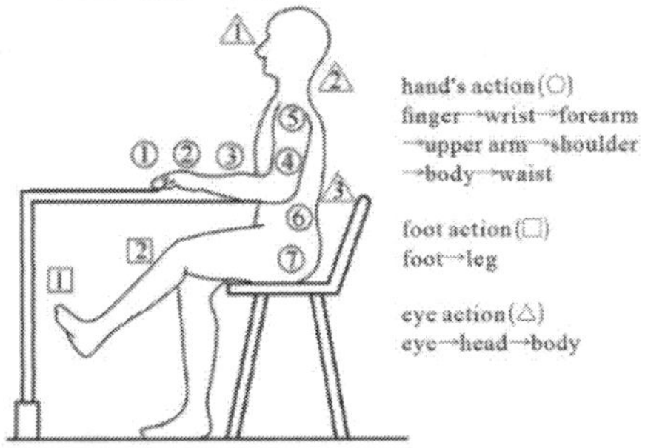


**Figure3.** The best action sequence of human body

Size and shape elements: The elderly furniture should comply with the old standard anthropometric data, because of the body contraction, smaller than middle-young people. Furniture should be biased towards low, light type, reduce multifunction or fold function.

Colour elements: colour collocation should match with the elderly mental feeling, with low lightness and low purity color, reducing the stimulation. Designers can choose natural texture materials to get freedom perception effect.

Emotional factors: Emotion includes three levels. They are instincts, behavior, reflect. Furniture can bring comfort, safety to the elderly, and conform to their aesthetic and meet their daily needs; are interactive on the form and function with the elderly, operate simple and interface easy to use, both entertaining; blend more emotion in the design, assist the elderly social-emotional needs, improve the superiority of life.

## 4 The elderly furniture intelligent design process

The design is a process that finds a problem and solves the problem. The reason why the elderly furniture is combined with intelligent concept is to better solve the various problems of the old life. What the elderly cognitive speed involution can make up for the intelligent design of furniture, and reduce the interaction problem of cognitive process and life scene. Design should be treated more logically and systematically other than traditional methods based on intuition and experience[5].

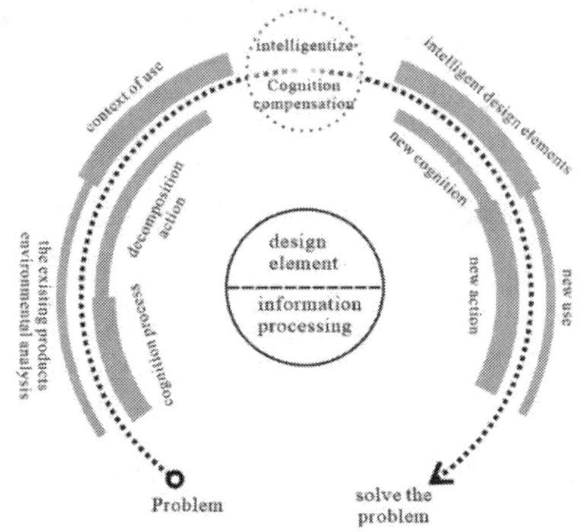


**Figure4.** The design process of elderly intelligent furniture

Intelligent furniture is through the sensor to collect different data information, the controller processes the information according to the preset procedures, so as to realize the automation control process of furniture. It contains two parts: hardware and software, hardware contains furniture itself, data acquisition system, control system, transmission system and other parts, The software includes data acquisition program and control program. Based on the cognitive process of intelligent furniture design, the core is a systemic treatment of product design elements, make design elements and cognitive processes (i.e., information processing) seamless integration (figure 4). The following is the specific design process.

### 4.1 Raise a question

Problem is divided into two levels: In terms of the elderly itself, their cognitive speed, movement speed slow down, the body best action position have changed depending on the specific environment; In terms of the furniture and the environment, the traditional furniture, argued that classification, form, function, size should aim at the older users demand by the functional partition, amounting to little more than a traditional space division, leading to not match the cognitive process and furniture design elements in older people.

### 4.2 Build usage scenarios

Interaction is the center in usage scenarios, classified according to the functional characteristics. Idealized cognitive situation is that cognitive exactly match with furniture operation function in the elderly, It is the unity of the cognitive processes and design elements. Building the old furniture usage scenarios should be accurate to specific groups and space environments, such as the bedside table is a assist function at bedtime which can put on cups, desk lamps, drugs, glasses, etc. The usage scenarios is sitting, lying in bed, when designing in addition to action factors, such as Table layout, air motion, temperature and humidity should be considered. The auxiliary furniture should reduce operating action.

### 4.3 The addition of intelligent elements

Intelligent elements are the key to realizing intelligent furniture. Intelligent furniture realizes by two parts, hardware and software. The feasibility analysis includes the information operation and user demand matching in the software part. In the hardware part it requests that the information collections, transmission parts, furniture are consistent with user operations. How to maximize identify of needs and feedbacks. Taking an example, the elderly watching TV on the sofa, they need a rest after sitting for a long time. Intelligent sofa collects information and need to make a corresponding mechanical adjustment when nodding off. Such as changing to the horizontal state, and regulating temperature for conforming rest. The sofa has low seat height. It can elevate the seat surface when old people get up, etc.

### 4.4 New design whether meets the usage scenarios

Whether the cognitive process matches with design elements, whether the user uses fewer moves to finish the task than before, it is the criterion of intelligent elements implant success and assessment phase of the whole design process.

## 5 The elderly intelligent furniture design principle

Regarding the design of intelligent furniture in elderly, on behalf of the user's demand cognitive situation is implicit. Design elements are dominant and are the function of furniture. Elderly intelligent furniture design on the cognitive scene construction as the breakthrough point, consider the design elements to what form to reflect the user's potential demand. In what ways did the elderly communicate with furniture, is the furniture design principles.

### 5.1 The Operating Experience Pleasure

Pleasure is the height matching of user expectation and furniture function. Whether the furniture function meets the user predict demand can provide the excellent feedback for the furniture. Operation experience makes the product operation interface consistent extensively with the user's cognitive process based on the user's cognitive situation. In the process of the interaction of the interface with intelligent furniture in the elderly, intelligent furniture should bring all the necessary information and tools to users. The furniture multi perceive specific state by sensors and make the user operation of the furniture in the controllable range. The user should be independent in a variety way of perception in interactive. If the intelligent control of the products conflict with the user's demand, should give in to the user.

Pleasure in the design of the operating parts mainly embodies in two aspects. Firstly, the Fitts Law. That is to say when the user is in a particular use state, the operating objects size, distance should be ease to use. Secondly, it's The Movement Optimization. By changing the operation action or improving operations position and the use object height is the action to reduce operation, make more convenient and effective.

### 5.2 "Unconscious" design

Ideally, product is no learning process, make users feel easy to use for the first time. In fact, there is no such thing as absolute without the learning process of operation, so you need to consider the learnability and availability of intelligent furniture. Elderly intelligent furniture must consider using frequency firstly, should make the higher of use, the more simple.

"Unconscious" is the process that objects drive the information storage to achieve operation by a certain characteristic representation. Intelligent furniture on man-machine interface processing should activate the old memories of the experience and associate with the real world in the design. For example in an easier way to arouse the user's cognition, control interface should use common symbols, software algorithm match with the cognitive process, simplify all unnecessary features.

The "unconscious" Operating process is the interactive object to be understood easy and the embodiment of interface humanization. Product interface is the medium of transmission and the exchange of information between people and machine. Is also a media of communication between them[6]. The old memory degradation, intelligent furniture need to delete unnecessary operation module, make it easy to understand, furniture function and cognitive scene match with each other.

### 5.3 Modularity Design

The elderly intelligent furniture can be divided into two categories, software module and hardware modules. Although the functions of furniture make a difference in different usage scenarios, Old people as a kind of user groups, it drives the elderly furniture homogeneity on the intelligent control based on the cognitive process and the usability operating principle.

In the aspect of hardware, its modular mainly reflects on the control plate and the transmission structure. Modular hardware is initial form portion of the intelligent furniture, can adapt to the user's personality needs and customization, the modular control unit has strong learnability, easy to form a good user experience. The transmission structure modularization can indirectly regulate the scale and specification of the design. Under the premise of individual character, customization, the compression of the cost, Designer should make the elderly intelligent furniture consistent with the user action.

### 5.4 Design Consistency

The consistency of the intelligent furniture reflected in how to correctly guide the user operation reliability, and able to comply with the laws of formal beauty. The consistency principle reflected in the appearance, color and operation way consistency. Appearance consistency not only does In line with the aesthetic demand from the perspective of formal beauty but also makes product seriation. Color consistency is mainly aimed to distinguish an operation module with the furniture in the elderly intelligent household and give consideration of aesthetic factors. Operation consistency includes hardware drive and software control. Interact has standard specifications, needs to keep consistent and stable to offset the elderly exclusive, for new things to improve their cognitive acceptance.

## 6 Conclusions

Intelligent furniture is the main way of solving the communication of cognitive process and design elements. The elderly cognitive speed slowed down, need to provide a positive response in the intelligent furniture based on cognitive situation. Furniture judge the elderly demand by the inductor, Supplement by hardware control module to solve the special needs of the elderly life. Build design thinking based on the cognitive process and explore the elderly intelligent furniture design.

## References

1. Hale S, Lima SD, Myerson J, Paychology and aging, **6**,512 ( 1991)
2. Chiu-Shui Chan, Design Cognition——Cognitive Science in Design,15(2008)
3. LU Xin-hui , ICSREI,**5**,(2014)
4. Reimann R, Cronin D, Cooper A, Electronic Industry Press,(2012)
5. Chiu-Shui Chan, Design Cognition——Cognitive Science in Design,29(2008)
6. XU Shi-hu, LU Xin-hui, The packaging engineering,**33**,64(2012)

# Service Redesign Using Factor Analysis, TRIZ, and Service Blueprint (Case study on Garuda Indonesia's sales and service office at Senayan City)

Amalia Suzianti[1], Avicenna[2] and Hutami Larasati[3,a]

[1,2,3] *Industrial Engineering Department, Faculty of Engineering, Universitas Indonesia, Depok, Indonesia*

**Abstract.**Indonesia is a country that has potential to be the one of the biggest airline industry. This potential leads the competition in airline industrymore aggressive. Garuda Indonesia as the leading airline must maintains its position as a leader in Indonesian airline industry. To maintain its position, Garuda Indonesia has repaired all of its aspects, one of them is service. The most important service area that give the most complete facilities to customers and contributions to Garuda Indonesia, is sales and service office (SSO). This research is trying to suggest recommendations of service redesign on Garuda Indonesia's SSO in order to create better customer satisfaction. This research uses factor analysis to analyze the factors that contribute to the customer preferences. The result shows the duration of service delivery, facilities in waiting area and frontliner's competencies are the major factors that contribute to the failure points in sales and service office. In the next stage, the TRIZ contradiction analysis is used resulting twelve inventive principles. Furthermore, new service design of Indonesia airline's sales and service office is proposed following the principles generated in TRIZ through service blueprint.

## 1 Introduction

In last five years, number of passengers at Soekarno-Hatta International Airport increased significantlyboth in domestic and international flight. As a result, Garuda Indonesia as one of the most leading airline in Indonesia, will face more intense competition. In answering this challenge, Garuda Indonesia had been developed the seven drivers strategy. This strategy will maximize its business aspects in order to gain the competitive advanage for Garuda Indonesia. One of those aspects is brand and services.

The brand and service will give five-star service to customers on all of its service areas. The first touchpoint, sales and service office, is the key of serviceas it happens to be the first place customer and frontliner meets[1]. However, based on Garuda Indonesia Airlines, the internal assessment reported that sales and service office scored below average on customer satisfaction. The existing Uplifting Service didn't match the expected service by customer (service failure), while in fact, service has to accomodate the needs of service's stakeholders; customers, frontliners and company, so the service failure can be prevented [2].

[a] hutami.nadya@ui.ac.id

From previous explanation, it can be concluded two problem statements: (1) the need to study the contributed factors of customer preferences in sales and service office as an effort to understand Garuda Indonesia's customers, (2) the need to redesign the service design on sales and service office as an effort to improve the quality of Garuda Indonesia's service. Tthe ultimate goal to be achieved is the creation of a proposed service design of the sales and service office that can meet customer preferences.

## 2 Literature Study

### 2.1 TRIZ

This study uses a TRIZ-based method, using factor analysis as a supporting tool in order to analyze the problem and service blueprint as the service design method for the research output. TRIZ (Theory of Inventive Problem Solving) is a systematization innovation method used for solving business problem with non-experiential domain using the knowledge base [3].

The conceptual framework comprises of five main stages. The initial input is a problem that occurs in service operations. Stage one and two analyze the original problems using various tools, such as cause and effect diagram, in order to provide insightful information for more detailed problem solving. Stage three structures the formulated problems into a typical TRIZ contradiction. Stage four uses TRIZ problem resolution tool to eliminate the contradictions. After the contradiction is successfully eliminated, it collects the innovative solutions to service design. The last stage will evaluate the generated ideas. If solutions are still not found, the problem solving process must be iterated back to the first stage in order to redefine the original problems.

### 2.2 Service Blueprint

Service blueprint is a detailed mapping from all service activities and interactions. Service blueprint is structured into five regions and four boundaries. Five regions includes physical evidence, customer actions, onstage interaction, backstage interaction, and support process. Service blueprint divides service into onstage and backstage activities [4].

## 3 Research methodology

### 3.1 Preliminary problem analysis

Factor analysis method was perform to identify the existing service problems as well as to eliminate the variables that have no effect to customer preferences. The data was collected using questionnaire. The questionnaire was developed through Revised Service Quality (Servqual) and Garuda Indonesia's Uplifting Service concept in five categories, including Product Knowledge, Attitude, Skills, Proess, Facilities, and Premises, with total of 43 variables.

Factor analysis was performed using SPSS 16. The analysis was conducted using principal component analysis with varimax rotation and Kaiser normalisation as an extraction method. The eleven factors identified were chosen since the initial eigenvalues is more than 1.0 [5].The KMO coefficientwas .627, which meant the adequacy for principal component analysis is acceptable. The value of Cronbach's Alphawas .703 which indicated a high degree of reliability.The matrix rotation extracted eleven factors and eliminated nine variables that did not have significant value (loading < 0.55). Table 1 shows eleven factors and the characteristics that will be considered as the customer preferences or preferences in sales and service office.

To solve the service problems, these factors will be analyzed using gap analysis. Factor's gap will be analyzed using pareto analysis in order to obtain the most significant factors that must be improved in sales and service office.

**Table 1.** Factors of Customer Preferences

| No | Variables | | | | | |
|---|---|---|---|---|---|---|
| 1 | Self-service machine | Computer with internet | Number of counters | Signage | | |
| 2 | Customer identification | Friendly | Exciting | | | |
| 3 | Knowledge on promotion | Knowledge on price | Knowledge on route | | | |
| 4 | Advising | Problem analysis | Calling | Confident | Caring | Serving |
| 5 | Aromatherapy | Interior Design | Temperature | Ambiance | | |
| 6 | Printed media | Entertainment | Queuing process | | | |
| 7 | Problem identification | Availability | Processing | Service duration | Seat comfort | |
| 8 | Drink and candy offering | Guidance | Sound | | | |
| 9 | Opening | | | | | |
| 10 | Uniform accessories | | | | | |
| 11 | Knowledge on local event | | | | | |

Based on the pareto chart shown in Figure 1, factor no. 7 and no. 6 must be improved in order to enhance the services in sales and service office for customers. This result will be included as the input on the problem modeling.

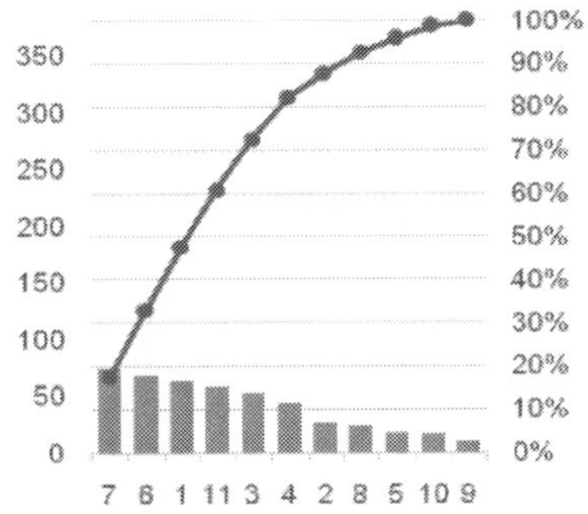


**Figure 1.**Pareto chart

## 3.2 Problem modeling and formulation

Problem modeling was performed using cause and effect diagram. The diagram was built using function-link-function method in order to understand the relationship among the functions in sales and service office. The cause and effect diagram was built through an intensive interview with the experts.

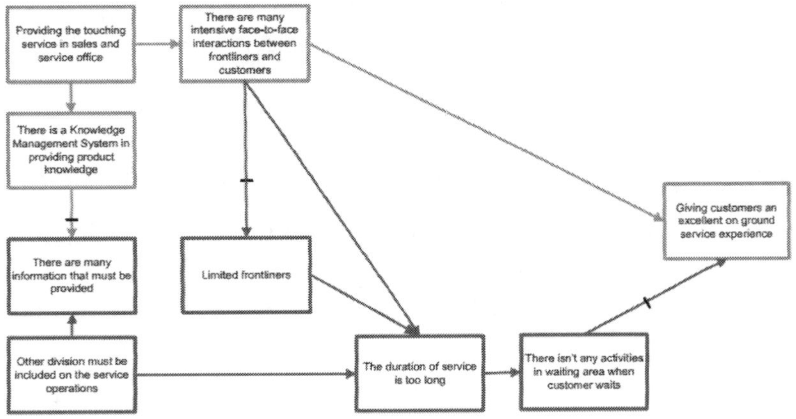


**Figure 2.**Cause and effect diagram

As shown in Figure 2, the cause and effect diagram successfully identified the back-end problems. First, the duration of service is too long which makes the customer wait for a long time.The involvement of other division also affects the duration of service delivery. This means the current knowledge management system still has some limitations that must be improved.There are three failure points that must be improved and will be drawn on the service blueprint. These three failure points are the duration of service, facilities in waiting area, and the knowledge of frontliners.

Based on the cause and effect diagram, problem statements are formulated as shown in Table 2. After problem formulation, this research used TRIZ service parameters on the contradiction analysis.

**Table 2.** Problem Statements

| No | Problem Statement |
|---|---|
| 1 | Find an alternative way to obtain an excellent on-ground service experience by providing many face-to-face interactions and eliminating a long service duration. |
| 2 | Find an alternative way to obtain an excellent on-ground service experience by does not produce a long waiting time. |
| 3 | Find a way to eliminate, reduce, or prevent a long service delivery in order to avoid a long waiting time under the condition of many important information provided by other division. |
| 4 | Find an alternative way to eliminate, reduce, or prevent a long service delivery in order to avoid a long waiting time under the condition of limited frontliners. |
| 5 | Find a way to increase the effectiveness of providing an excellent on-ground service experience to customers. |
| 6 | Find a way to obtain an excellent on-ground service experience by does not require many face-to-face interactions. |
| 7 | Find additional benefits from a long service delivery. |
| 8 | Find additional benefits from a long waiting time |
| 9 | Find additional benefits from the knowledge management system. |
| 10 | Find a way to increase the effectiveness of providing many face-to-face interactions |
| 11 | Try to resolve following contradictions; face-to-face interactions should be in place in order to produce an excellent on-ground service experience and eliminating the limitation of frontliners. |
| 12 | Try to resolve following contradictions; face-to-face interactions should not exist in order to avoid the limitation of frontliner and hindering an excellent on-ground service experience. |
| 13 | Try to resolve following contradictions; many information that is needed in order to produce enhanced-knowledge management system and eliminating the involvement of other division. |

## 3.3 Contradiction analysis

A root cause analysis of the previous problem statements provides some guidance toward following directions. Based on the analysis of previous problem statement, this research concludes that the relevant 11 service parameters for sales and service office are shown in Table 2.

According to the service parameters in Table 2, this research will improve the sales and service office's service parameters. These parameters consist of service duration, waiting area, face-to-face interactions, reliability or availability, and frontliner's competence.

The next step is applying the TRIZ contradiction matrix in order to examine the resolution between the service parameters and the contradictions. TRIZ contradiction matrix of this research as shown in Table 3.

**Table 3.** Service parameter

| Problem-solving directions | | TRIZ-service parameter | |
|---|---|---|---|
| 1 | Improving service duration | 9 | Responsiveness |
| 2 | Enhancing frontliner's knowledge | 14 | Professional competence |
| 3 | Effectiveness of service | 17 | Atmosphere |
| 4 | Quality of waiting area | 18 | Environment quality |
| 5 | Involvement of other division | 21 | Efforts |
| 6 | Improving waiting time | 25 | Waiting time |
| 7 | Availability of frontliner | 27 | Reliability |
| 8 | Frontliner's communication | 28 | Communication |
| 9 | Interaction between customer and frontliner | 30 | Factors |
| 1 | Improving service duration | 34 | Service capabilities |
| 3 | Effectiveness of service | 39 | Performance |

## 3.4 Contradiction elimination

This stage will evaluate all of the possible solutions that should be implemented to the existing condition in sales and service office.The possible solution which is known as TRIZ inventive principle will be chosen as the proposed service resolution. The possible solution that is not chosen will be eliminated. According to the TRIZ contradiction matrix in Table 4, there are 21 underlined TRIZ inventive principles that will be proposed as the service resolution in sales and service office.

**Table 4.** TRIZ Contradiction Matrix

| Improving parameter | Worsening parameter | | | | | | | | |
|---|---|---|---|---|---|---|---|---|---|
| | 9 | 10 | 13 | 19 | 27 | 30 | 34 | 35 | 39 |
| 9 | N/A | 13, 28, 15, 19 | 28, 33, 1, 18 | 8, 15, 35, 38 | 11, 35, 27, 28 | 1, 28, 35, 23 | 34, 2, 28, 27 | 15, 10, 26 | N/A |
| 14 | 8, 13, 26, 14 | 10, 18, 3, 14 | 13, 17, 35 | 19, 35, 10 | 11, 3 | 18, 35, 37, 1 | 27, 11, 3 | 15, 3, 32 | 29, 35, 10, 14 |
| 17 | 2, 28, 36, 30 | 35, 10, 3, 21 | 1, 35, 32 | 19, 15, 3, 17 | 19, 35, 3, 10 | 22, 33, 35, 2 | 4, 10, 16 | 2, 18, 27 | 15, 28, 35 |
| 18 | 10, 13, 19 | 26, 19, 6 | 32, 3, 27 | 32, 1, 19 | N/A | 15, 19 | 15, 17, 13, 16 | 15, 1, 19 | 2, 25, 16 |
| 21 | 15, 35, 2 | 26, 2, 36, 35 | 35, 32, 15, 31 | 16, 6, 19, 37 | 19, 24, 26, 31 | 19, 22, 31, 2 | 35, 2, 10, 34 | 19, 17, 34 | 28, 35, 34 |
| 25 | N/A | 10, 37, 36,5 | 35, 3, 22, 5 | 35, 38, 19, 18 | 10, 30, 4 | 35, 18, 34 | 32, 1, 10 | 35, 28 | N/A |
| 27 | 21, 35, 11, 28 | 8, 28, 10, 3 | N/A | 21, 11, 27, 19 | N/A | 27, 35, 2, 40 | 1, 11 | 13, 35, 8, 24 | 1, 35, 29, 38 |
| 28 | 28, 13, 32, 24 | 32, 2 | 32, 35, 13 | 3, 6, 32 | 5, 11, 1, 23 | 28, 24, 22, 26 | 1, 32, 13, 11 | 13, 35, 2 | 10, 34, 28, 32 |
| 30 | 21, 22, 35, 28 | 13, 35, 39, 18 | 35, 24, 30, 18 | 1, 24, 6, 27 | 27, 24, 2, 40 | N/A | 35, 10, 2 | 35, 11, 22, 31 | 22, 35, 13, 24 |
| 34 | 34, 9 | 1, 11, 10 | 2, 35 | 15, 1, 28, 16 | 11, 10, 1, 16 | 35, 10, 2, 16 | N/A | 7, 1, 4, 16 | 1, 32, 10 |
| 39 | N/A | 28, 15, 10, 36 | 35, 3, 22, 39 | 35, 10, 38, 19 | 1, 35, 10, 38 | 22, 35, 13, 24 | 1, 32, 10, 25 | 1, 35, 28, 37 | N/A |

## 4 Results

From TRIZ method, several things revealed as shown in Table 5. First, they could differentiate their offered service, which will be separated into two main parts, fast-service and full-service counte. Customer can choose the services that suitable for their needs on the queuing machine. In addition, the other supporting part can be served using the self-service machine, for example: check-in kiosk.

Second, they could improve the service experience by installing new facilities in waiting area, such as television and wireless fidelity (wi-fi) that could increase the customer's patience. However, it still needs a limitation, such as fixed channels or low volume, in order to prevent harmful situation.In addition, a facility that could give a new experience like plane's interior showroom or experience room will entertain customers in an innovative way. This facility could also be an alternative way to offer merchandise or partner's product. Combining the cross-selling activity with the experience room would reduce the efforts of frontliners. Moreover, applying a unique decorations when celebrating local event also could give another good experience.

Third, providing feedback channel for customers could increase the customer's trust. Also, they could integrate their feedback channel with knowledge management system. In addition, to shorten the duration of service, a real-time ticket price notification along the chatting feature will be installed on knowledge management system. Frontliners does not need to leave their counter. Last, to initiate the customer researches by internal management, CCTV as a security tool can be used to study their behavior.These proposed principles and the main functions related to them can be found in Table 4. These proposed principles are drawn to the proposed service blueprint.

## 5 Conclusion

There are 11 factors (components) that represent consumer preferences on sales and service office, including sign and additional facilities, main knowledge, problem solver, sensory pleasures, facilities in waiting area, main service aspects, instructions and offerings, opening, accessories, and additional knowledge. Based on the TRIZ analysis, there are three service aspects that must be improved through 12 possible solutions as shown in Table 5.

There are some limitations of this research. This research does not include the solution evaluation. So the impact of this research is still not clear enough. There is not any study that verified the benefit of this service redesign. For further research, the solution evaluation must be included to sharpen the analysis. Furthermore, this research suggests the airline firms that have on-ground physical channels can be generally applied to other airline firms and even other industries providing ticketing services.

**Table 5.**Proposed principle of problem resolution

| Service resolution | Principle | Definition | Main-functions |
|---|---|---|---|
| Differentiating service | 1. Segmentation<br>4. Asymmetry<br>33. Homogeinity<br>35. Transformation of Properties<br>10. Preliminary Action<br>15. Dynamization | Dividing a system into parts<br>Alters a process so that it is different from the others<br>Focuses on the consistency in grouping similar features<br>Utilizes resources for changing the service operations<br>A prior action is performed before the service is launched<br>Makes services more flexible | • Fast-service counter<br>• Full-service counter<br>• Gives frontliner a specific task for each counter<br>• Differentiates customers based on<br>  • Purposes: for work or non-work<br>  • Number of tickets: one ticket or more than one ticket |
| Customer's feedback | 22. Convert Harm into Benefit<br>23. Feedback<br>26. Copying | Using harmful occurences as a way to realize benefits<br>Utilizing market information from market in order to improve a service<br>Imitates another service | • Provides a whistleblower system<br>• Provides a whistleblower signage<br>• Provides internal research using CCTV |
| Self-service machine | 2. Taking off<br>21. Skipping<br>25. Self-service | Eliminate low-value service process from the system<br>Skips unnecessary functions that may cause delay<br>Allows customer to play a role in the delivery of the service | • Installs self-check in kiosk |
| Facilities in waiting area | 32. Color changes<br>17. Another dimension<br>30. Flexible Shells & Thin Films | Focuses on physical characteristics to satisfy customers<br>Encourages thinking about changing the look of a service<br>Isolates harmful issues by using thin barrier | • Installs experience room<br>• Adds electronic entertainment such as TV and wi-fi<br>• Limits the flexibility in using entertainment system |
| Combining service activities | 16. Partial or excessive actions<br>5. Merging<br>31. Porous materials | Doing more or less functions, efforts, or tasks<br>Combine ideas, needs and feedbacks to produce new service<br>Creates holes in a system to improve a service performing | • Adds cross-selling activities<br>• Combines cross-selling with experience room |
| Store's atmosphere | 3. Local quality<br>28. Another Sense | Making changes in the service in different environments<br>Substitutes mechanical means with sensory means | • Decorates sales and service office based on local events |
| Enhancing KMS's features | 19. Periodic Action | Changing a service offers, or delivery from continuous to periodic | • Installs new features in KM such as<br>  • Online courses<br>  • Online tests<br>  • Real-time price's notification<br>  • Chatting to link the front-end to back-end, back-end to other division |

## References

1. Carlzon, J. (1987). Moments of Truth. Massachusets: Ballinger Publishing Company.
2. Teboul, J. (2006). Service is Front Stage - Positioning services for value advantage. New York: Palgrave Macmillan.
3. Savransky, S. (2000). Engineering of creativity: introduction to TRIZ methodology of inventive problem solving. Florida: CRC Press.
4. Shostack, L. (1984). Designing Services that Deliver. Harvard Business Review, 133-139.
5. Hair, J. F., Black, W. C., Babin, J. B., & Anderson, E. R. (2009). Multivariate Data Analysis: 7th Edition. Prentice Hall.

# AUTHOR INDEX